THE VIRTUOUS PHYSICIAN

Philosophy and Medicine

VOLUME 114

For further volumes:
http://www.springer.com/series/6414

THE VIRTUOUS PHYSICIAN

The Role of Virtue in Medicine

by
JAMES A. MARCUM
Baylor University, Waco, TX, USA

Springer

Dr. James A. Marcum
Baylor University
Dept. Philosophy
One Bear Place #97273
Waco Texas 76798
USA
James_Marcum@baylor.edu

ISSN 0376-7418
ISBN 978-94-007-2705-2 e-ISBN 978-94-007-2706-9
DOI 10.1007/978-94-007-2706-9
Springer Dordrecht Heidelberg London New York

Library of Congress Control Number: 2011943759

Printed on acid-free paper

Springer is part of Springer Science+Business Media (www.springer.com)

No one in this world always does right.
Ecclesiastes 7:20 (CEV)

Preface

Fundamentally, medicine is moral (Pellegrino, 2002; Tauber, 1999). It is neither a natural nor a social science, although it often depends upon both for its technical and communal progress. Medicine, especially as a clinical practice, is moral because the defining element of its practice is the patient-physician relationship; and, that relationship is profoundly principled and often based upon ethical rules and duties.[1] The relationship is moral also since the physician's behavior and not just his or her medical knowledge is critical for the patient's wellbeing and possible healing. Finally, it is moral since the ethical mandate of medicine, with respect to the physician's action, is to help—and not to harm—the patient. To harm the patient, either intentionally or unintentionally, is to fail at medicine's primary ethical mandate that dates back to Hippocrates. For both the patient and society call upon the physician to benefit the sick and dying and to assist in the healing process. From this perspective, both the natural and social sciences support the practice of medicine but do not define it. Thus, the physician's behavior, whether good or bad, is not peripheral to the clinical encounter but at the heart of what it means to be a healthcare provider.

Beginning in the late nineteenth century and culminating in the aftermath of Abraham Flexner's 1910 Carnegie Report, the scientific dimension of medicine eclipsed its moral or ethical dimension. This eclipse is the root cause of several crises now facing modern medicine. The first is the quality-of-care crisis. With the advent of a highly technical medicine, often the physician forgets that the patient is a person first. "Our nation's health care system," according to Ralph Snyderman, "has lost its way over the last two decades. It has become so enamored with technology and specialization that it has lost sight of individuals and their needs" (Blumer and Meyer, 2006, p. 5). Patients no longer feel that physicians are concerned about them as persons but only as pathological specimens. A crisis closely associated with the quality-of-care crisis is professionalism in medicine. Many physicians

---

[1] Although moral and ethical are often used interchangeably, an important distinction exists between them. Moral refers to an individual's assessment of personal character or behavior as good or bad, while ethical refers to a social assessment of an action as right or wrong—particularly with respect to codes or rules. I try to keep this distinction in mind; but, I do use ethical quite often as an all-inclusive term.

and even some patients view the medical profession as a collection of technical specialists, and the only skill needed to practice medicine is simply scientific or mechanical knowledge. This technical dimension of medicine can obviously exacerbate the quality-of-care crisis. Many patients often perceive their physicians as cold and uncaring technicians or mechanics, who are only interested in them as diseased body parts and not as individual persons. In an effort to resolve these crises, some physicians attempt to reinstate the humanistic dimension of medical practice (Marcum, 2008). For example, Eric Cassell (2004) champions a notion of patient *qua* person to stem abuses associated with the biomedical model of clinical practice.

To address the quality-of-care and professionalism crises plaguing modern western medicine, I introduce a philosophical notion of virtuous physician. To that end, I discuss in the first chapter the nature of the two crises and contemporary efforts to resolve them, especially with respect to evidence-based and patient-centered medicine. I then briefly introduce the notion of virtuous physician and outline its basic virtues (and corresponding vices of the unvirtuous physician) in traditional terms of metaphysics, ethics, and epistemology. In the next chapter, I introduce and discuss virtue theory, along with virtue ethics and epistemology. I first examine virtue theory, particularly in terms of defining what a virtue is, followed by an analysis of virtue ethics, including its history. Specifically, I discuss and expound upon the notions of ethical or moral action in terms of virtues. At the end of the chapter, I discuss virtue epistemology, with respect to reliabilist and responsibilist intellectual virtues. In a following chapter, I discuss specific virtues, especially as they relate generally to medicine. I first examine the four cardinal virtues of prudence, courage, temperance, and justice. I divide these virtues into two categories, consisting of the epistemic or intellectual virtues (prudence) and the ethical or moral virtues (courage, temperance, and justice). I then examine the theological or transcendental virtues of faith, hope, and love. In addition, I discuss cognate virtues to each of the cardinal and theological virtues.

In the fourth chapter, I explore the ontological priority of caring as the chief metaphysical virtue for grounding a notion of virtuous physician and of uncaring as the main vice of the unvirtuous physician. I then examine two essential ontic virtues of a virtuous physician—care and competence—and the two corresponding vices of an unvirtuous physician—carelessness and incompetence. In order for physicians to be competent in the practice of medicine, they must be genuinely caring (as an ontological attitude or stance). Because by caring physicians care about ($care_1$) patients as persons and strive to be technically or scientifically and ethically or morally competent, which in turn allows physicians to take care of ($care_2$) the individual patient's bodily and existential needs. Although caring (if limited to $care_1$) is inadequate to choose the correct or best course of clinical action, combined with the virtue of technical and ethical competence and with the virtue of $care_2$ it is adequate but still insufficient for practicing right or good clinical medicine. To be sufficient, competence must be transformed into prudence and care into love to form the compound or composite virtue of prudent love. In a following chapter, I examine the transformation of competence into prudent wisdom and care into personal radical love to forge the compound virtue of prudent love, which is sufficient for defining the virtuous physician and the practice of virtuous holistic medicine. In contrast, imprudent lovelessness is the compound vice animating the unvirtuous physician and the practice of unvirtuous fragmented medicine.

In a penultimate chapter, I reconstruct two clinical case stories, both from the medical literature, which illustrate the various virtues and vices associated with medical practice. I utilize these case stories to illustrate the notion of virtuous and unvirtuous physician from an ethical and epistemological perspective. In the final chapter, I discuss how the notion of virtuous physician addresses the quality-of-care and professionalism crises and how the notion of unvirtuous physician exacerbates them. To that end, I utilize the notion of virtuous physician to integrate evidence-based and patient-centered medicine into a virtuous holistic medicine, which I then use in a separate section of the chapter to resolve the two crises. In contrast, the unvirtuous physician practices an unvirtuous fragmented medicine in which evidence-based and patient-centered medicine remain completely divergent. Lastly, I discuss the role of virtues in revising medical education at the undergraduate, graduate, and postgraduate levels and the question of whether the medical faculty can teach and students can learn virtues. My contention is that if the faculty does not teach and students do not learn virtues then the chance for vices to infect medical practice by default remains a viable option. In sum, the prudently loving physician is a genuine medical professional, who practices a holistic medicine that provides the quality of healthcare patients both expect and need.

Finally, I must address my motivation for writing this book and offer a defense for it, especially in terms of a non-clinician advising clinicians how to practice their trade—at least from a philosophical perspective. I have been associated with the healthcare field for almost my entire academic career. I have a doctorate in human physiology from the University of Cincinnati Medical College and conducted basic research on the regulation of hemostasis at Harvard Medical School for well over a decade. For the past decade at Baylor University, I have been participating in a medical humanities program at the undergraduate level. Through these experiences, I have come to appreciate first-hand the quality-of-care and professionalism crises facing modern medicine, especially from a philosophical or theoretical perspective; and, what is needed to address and resolve them. Moreover, I believe that not practicing medicine provides me with the ability to examine and analyze medicine and its practice objectively, without the encumbrance of biases and subjectivity that may attend those who must practice medicine on a daily basis. Lastly, my motivation for writing this book is to help those interested in pursuing or developing a medical career to understand that medicine is a difficult and demanding profession, especially in terms of its moral claim on its practitioners. In this book, I unpack in a philosophical analysis what this claim means in terms of virtuous physician and the practice of virtuous holistic medicine, and contrast it to the failure of the unvirtuous physician to provide quality healthcare in a professional manner.

References

Blumer, R.H., and M. Meyer. 2006. *The new medicine*. New York, NY: Middlemarch Films.
Cassell, E.J. 2004. *The nature of suffering and the goals of medicine*, 2nd edition. New York, NY: Oxford University Press.
Marcum, J.A. 2008. *An introductory philosophy of medicine: Humanizing modern medicine*. New York, NY: Springer.
Marcum, J.A. 2009a. The epistemically virtuous physician. *Theoretical Medicine and Bioethics* 30:249–265.
Marcum, J.A. 2009b. The role of virtues in the practice of clinical medicine. *International Journal of Science in Society* 1:91–97.
Marcum, J.A. 2011a. Care and competence in medical practice: Francis Peabody confronts Jason Posner. *Medicine, Health Care and Philosophy* 14:143–153.
Marcum, J.A. 2011b. “First, do no harm”: Uncaring and incompetent healthcare. *International Journal of Health, Wellness and Society* 1:1–11.
Marcum, J.A. 2011c. The role of prudent love in the practice of clinical medicine. *Journal of Evaluation in Clinical Practice* 17:877–882.
Pellegrino, E.D. 2002. Professionalism, profession and the virtues of a good physician. *The Mount Sinai Journal of Medicine* 69:378–384.
Tauber, A.I. 1999. *Confessions of a medicine man: An essay in popular philosophy*. Cambridge, MA: MIT Press.

Acknowledgments

As true for most if not all authors, I have acquired a great many debts to others for their help. I am indebted especially to Kay Toombs and Bob Kruschwitz, as well as to Mike Attas, Bob Roberts, Bob Rose, Bill Stempsey, and Fred Tauber, for their invaluable discussions and insightful comments on earlier drafts of the book as a whole or in parts. Several chapters contain material reworked from articles appearing in *Theoretical Medicine and Bioethics* (Marcum 2009a), *International Journal of Science in Society* (Marcum, 2009b), *Medicine Health Care and Philosophy* (Marcum, 2011a), *International Journal of Health Wellness and Society* (Marcum, 2011b), and *Journal of Evaluation in Clinical Practice* (Marcum, 2011c). And, I thank Baylor University for supporting my scholarship with a sabbatical award and a URC grant. Next, I thank my daughters, Meg and Meredith, for their loving support, which has enriched both this book and my life. Finally, I dedicate the book to my students, who hope to practice medicine some day sensitive not only to the physical needs of patients but also to their emotional, psychological, and spiritual needs, and to the memory of my sister, Nancy, who first introduced me to the healing arts.

Contents

Chapter 1
Medicine's Crises

Modern western medicine, especially in the United States of America, is facing a number of crises, including economic, malpractice, healthcare policy, quality-of-care, professionalism, public or global health, primary or general care and critical care, and healthcare insurance—to name a few (Cutler, 2004; Daschle, 2008; Relman, 2007a). For example, the cost of medical care in the United States is the highest in the world. The total amount of money Americans spent on healthcare in 2007 was over 2.4 trillion dollars, and economists project the total healthcare costs to exceed well over 4 trillion U.S. dollars by 2018 (Sisko et al., 2009).[1] Related to the crisis of healthcare cost is the crisis of healthcare access. Until the recent passage in March 2010 of the US healthcare reform bill, over 45 million Americans were uninsured medically and often have little, if any, access to healthcare (Anonymous, 2009). Under dire circumstances, many uninsured use hospital emergency facilities; and they are generally unable to pay fully for the healthcare they receive, which exacerbates the cost crisis (Kotlikoff, 2007). Solutions to these crises are not readily apparent and pundits debate about how best to resolve them—although the current US healthcare bill, dubbed "Obamacare" by critics, promises to rectify the problems (Bristol, 2010). Two other important crises include quality-of-care and medical professionalism. Although efforts have been made to resolve them, with little success, I introduce the notion of virtuous physician to address them.

1.1 Medicine's Crises

In this section, I expound upon the quality-of-care and professionalism crises to provide a backdrop for motivating the need for a notion of virtuous physician. Although medical science and technology have produced, during the twentieth century, "miraculous" cures for many diseases such as infectious diseases, patients are often dissatisfied with the quality-of-care received from modern medical professionals. "I have a deep concern," acknowledges Arthur Kleinman, "...that at the same time that we are enabling doctors to become technologically effective we are

[1] On average, the amount of money spent individually on healthcare in 2006 was over $7400.

J.A. Marcum, *The Virtuous Physician*, Philosophy and Medicine 114,
DOI 10.1007/978-94-007-2706-9_1, © Springer Science+Business Media B.V. 2012

disabling them from being humanly compassionate and responsive" (Blumer and Meyer, 2006, p. 8).[2] A related crisis is professionalism, in which a variety of factors deprofessionalizes medical personnel. According to Edmund Pellegrino, "the [medical] profession is losing its commitment to the kind of character traits requisite for protection of the welfare and interests of patients" (2002a, p. 384).

The above crises are products of two clashing cultures—the scientific and the humanistic. Patients seek from professional caregivers not only scientific or technical cures for or management of their physical or organic ailments but also humane care for the psychological, emotional, and existential dimensions of those ailments. Modern medicine, however, often emphasizes technical cure and management of disease over humane care, and a technical professional character over a humane one. This emphasis on the scientific or technical begins with the education and training of physicians, who are obliged to meet scientific requirements as undergraduates for entrance into medical school—generally with no requirements in the humanities. And, once in medical school, the scientific and technical training in medicine of prospective physicians often brackets from the clinical consultation the patient's illness experience and any emotions associated with it. This training generally results in a medical professional who is emotionally detached from what the patient feels or experiences and who thereby appears uncaring to the patient.

1.1.1 Quality-of-Care Crisis

Although physicians always presume the highest quality of healthcare in their practice, quality became a critical issue in the early 1970s with the proposal of the federal HMO Act (Caper, 1988; Gruber et al., 1988). The U.S. Senate Health Subcommittee convened hearings to assess the impact of HMOs on healthcare quality. The concern was how best to measure the quality-of-care patients receive in order to determine whether HMOs would compromise that quality. To assess quality care quantitatively required a precise definition of it. Unfortunately, members of the committee discovered that no such definition, exhibiting community consensus, was available.[3] In response, concerned pundits and professional communities offered a plethora of definitions for quality-of-care throughout the decade. For example, the pediatric community assembled a committee to formulate an operational definition for quality pediatric healthcare. That community defined such quality accordingly: "Quality pediatric medical care embodies a scientific approach to health supervision; the

[2] David Weatherall also acknowledges a care crisis: "the art of medicine, in particular the ability of doctors to *care* for their patients as individuals, has been lost in a morass of expensive high-technology investigation and treatment. . .In short, modern scientific medicine is a failure" (1996, p. 17, emphasis added).

[3] Philip Caper (1974) also acknowledged the "elusive" nature of quality-of-care and proposed objective standards for medical procedures to define it, especially standards established through clinical trials.

establishment of a diagnosis of deviation from optimum health; institution of appropriate therapy; and management designed to satisfy the overall needs of the patient" (Osborne and Thompson, 1975, p. 625).[4] Towards the end of that decade, Avedis Donabedian—who Grant Steffen (1988) called the "Dean of Quality Assessment"—defined quality-of-care as "the application of medical science and technology in a manner that maximizes its benefits to health and minimizes its risks" (1979, p. 278).[5]

In the following decade, the healthcare quality issue took on new life (Caper, 1988). What added to that life were spiraling healthcare costs and efforts to contain them. The fear, not only in the governmental sector but also in the private and public sectors, was that cost containment would inevitably lead to reduced quality care. The need for a precise definition of quality-of-care became urgent so that interested parties could measure such quality accurately and ensure the highest possible quality of medical care at a reasonable cost (Lohr et al., 1988). Again, pundits and professional societies rose to the occasion and proposed a variety of definitions for quality-of-care. For example, the American Medical Association (AMA) formulated a definition of quality healthcare at its 1984 annual meeting. The AMA defined quality care as that care "which consistently contributes to improvement or maintenance of the quality and/or duration of life" (Council on Medical Service, 1986, p. 1032). The AMA also included eight characteristics in their definition of quality healthcare, ranging from adequately documenting the patient's record to optimally improving the patient's condition. Interestingly, the focus on quality-of-care represented the AMA's effort to renew its commitment to excellence in healthcare, particularly through initiatives in medical education, ethical reflection, preventive medicine, quality assurance, among others (Anonymous, 1986).

In an effort to bring some semblance of accord to the debate over quality-of-care, Steffen (1988) criticized select definitions that he claimed are representative of the myriad definitions proposed for quality healthcare. Specifically, he claimed that the AMA's definition is circular in nature in that the AMA used quality as both the definiendum and definiens. In addition, he complained that the AMA's definition fails to connect the eight characteristics of quality healthcare in order to define quality-of-care adequately. Although he largely agreed with Donabedian's definition, he quibbled with whether quality is a property medical care exhibits

[4] David Rutstein and colleagues proposed another widely recognized definition of quality medical care by defining quality as "the effect of care on the health of the individual and of the population," where care pertains to "the application of all relevant medical knowledge" (Rutstein et al., 1976, p. 582).

[5] Donabedian (1990) identified seven characteristics, or what he called pillars, that define the nature of quality-of-care. The first is efficacy, which represents the ability to affect a cure or to improve a patient's wellbeing, while the second is effectiveness and involves the realization of a cure or an improvement of a patient's wellbeing. The next is efficiency, which represents maximal treatment at minimal cost, while the next related characteristic is optimality and pertains to optimal balancing of risks and benefits. Two subsequent characteristics are acceptability, representing patient's approval of medical goals, and legitimacy, involving society's sanction of those goals. The final characteristic is equity or fair distribution of medical care.

in varying degrees to balance risks and benefits. According to Steffen, quality is not a metaphysical property of care but rather a preference or value for choosing a particular type of care. Finally, he criticized David Rutstein's definition for equating quality with the outcome of medical care. For Steffen, quality represents the capacity to achieve such an outcome. He goes on then to define quality medical care as "the capacity to achieve the goals of both the physician and the patient" (Steffen, 1988, p. 59). The goals represent legitimate medical outcomes negotiated by both the patient and physician. Steffen concluded with the hope that the proposed definition provides "a point of departure rather than a final statement in this dialogue on the nature, assessment, and improvement of quality medical care" (1988, p. 61).[6]

During the 1980s, the spiraling costs of the Medicare program prompted the U.S. Congress to examine quality medical care and its assessment. The Congress charged the Institute of Medicine (IOM), a division of the U.S. National Academy of Sciences, with the task. In 1990, the IOM released a two-volume report, *Medicare: a strategy for quality assurance*, detailing its findings and recommendations.[7] Kathleen Lohr was the editor of the report. Lohr and colleagues at the IOM were responsible for the most recognized and often cited definition of quality healthcare (Blumenthal, 1996a; Bowers and Kiefe, 2002; Zipes, 2001). "Quality of care," according to the IOM, "is the degree to which health services for individuals and populations increase the likelihood of desired health outcomes and are consistent with current professional knowledge" (Lohr, 1990, p. 4). The definition represents the consensus of around one hundred definitions in the literature. In addition, the report cites eight characteristics of quality medical care, selected from eighteen. The characteristics range from the scale of quality to technological restraints on care. Although the definition struck a chord with many, it failed in one important aspect—it over emphasized the technical or practical nature of medical care.

In the lead article to a series of six articles on the quality issue in medicine, which appeared in the *New England Journal of Medicine*, David Blumenthal (1996a) reviews the history of efforts to define quality-of-care in medicine.[8] Blumenthal notes that many of the definitions for quality medical care emphasize the technical or practical dimension of such care. This dimension includes two important aspects. The first is the appropriateness of the medical care, whether the care is what the patient needs to improve or restore health. The second aspect of technical care is the physician's performance of that appropriate care. In other words, quality technical care represents the best clinical practice done correctly. However, Blumenthal identifies another aspect of care, especially with respect to the patient-physician

---

[6] In utter frustration with the healthcare community's efforts to define quality-of-care precisely or even adequately, Caper (1988) declared a ban on the word quality from discussions on assessing healthcare. In its place, he proposed a pragmatic approach in which interested parties simply measured the "components" of care, such as efficacy and appropriateness.

[7] For a summary statement of the IOM report, see Lohr and Harris-Wehling (1991).

[8] For a conceptual analysis of quality healthcare definitions, see Harteloh (2003).

relationship. That aspect involves "the quality of their communication, the physician's ability to maintain the patient's trust, and the physician's ability to treat the patient with 'concern, empathy, honesty, tact and sensitivity'" (Blumenthal, 1996a, p. 892). Although he does not name or develop this aspect of quality medical care, others do.

Besides the technical or practical dimension of quality medical care, Donabedian (1979, 1988) acknowledges what he calls its interpersonal dimension. This dimension involves "conformity to legitimate patient expectations and to social and professional norms" (Donabedian, 1979, p. 277). Donabedian identifies the interpersonal dimension with the art of medicine, in contrast to the technical or scientific dimension of quality medical care. Caper (1974, 1988) also acknowledges an interpersonal dimension of quality healthcare.[9] He identifies this dimension with the "process of caring for the patient—the interpersonal, supportive and psychological aspects of the physician-patient relationship" (Caper, 1974, p. 1137). Both Donabedian and Caper claim that medical professionals too often ignore the interpersonal dimension of healthcare but that patients do not. Part of the reason why physicians are likely to ignore this dimension of quality healthcare is that the interpersonal dimension is too subjective, making it almost impossible to measure accurately by standard criteria used to measure the technical dimension of quality medical care.

Although problems certainly exist with defining quality-of-care in terms of assessing and improving it, these problems did not precipitate a crisis for technical quality care but rather only a debate; however, problems with defining technical quality healthcare, in order to measure and improve it, did contribute to the malpractice crisis. Thus, the contemporary crisis in quality-of-care is not a result of its technical dimension per se but of its existential dimension. "Measuring the quality of medical care predominantly by heart beats and body heat," Eric Cassell notes, "is one of the reasons modern medicine got into its current difficulties—focused more on diseased organs and technology than on the goals of sick persons. Patients do not simply want to survive," he goes on to stress, "they want to survive in order to live a life in which they can recognize themselves and in which their values are preserved" (1997, p. 130). The real crisis of quality care pertains then to patients' perceived indifference by physicians to their existential needs. As Brian Berman so succinctly articulates these needs, "Yes, you want your physician to be highly skilled, to be extremely knowledgeable, in medicine. But in addition to that, you want them to know you as a person" (Blumer and Meyer, 2006, p. 6).

In response to the interpersonal dimension of quality healthcare, Steffen proposes the addition of a ninth characteristic of care to the AMA's eight: "care of high quality includes assessment of patient goals and values" (1988, p. 57). Although the

[9] See also Campbell et al. (2000) for further discussion of the interpersonal dimension of quality medical care.

physician and patient share the goal of improving a patient's health or returning a patient to full health, this goal, which is often technical in nature, may not be the only important goal for the patient. Other goals, especially existential ones, may also be significant to the patient vis-à-vis holistic healing. According to Steffen, these goals "refer to the nontechnical or interpersonal aspect of care, the art of medicine; these goals usually are not achieved by tests or therapies but by attention to those patient values that generated them" (1988, p. 59). Although physicians cannot measure these goals accurately, their awareness of them can improve the overall quality of healthcare and lead not only to patient wellness but also to physician satisfaction.

During his career, Donabedian strove to combine the technical and the interpersonal dimensions of quality healthcare to attain a single, unified definition.[10] According to Donabedian, such a unified definition of quality medical care "is expected to maximize an inclusive measure of patient welfare, taking account of the balance of expected gains and losses that attend the process of care in all its parts. This is a process fundamental," emphasizes Donabedian, "to the values, ethics, and traditions of the healthcare professions: at the very least to do no harm; usually to do some good; and ideally, to realize the greatest good [i.e. quality healthcare] that is possible to achieve in any given situation" (1979, p. 278). Of course, he is well aware that realization of such quality-of-care belies the complexity of healthcare itself. Although Donabedian is apprehensive about the possibly of realizing such an ideal quality care, he recognizes that such care is context dependent and subject to the norms and values of the medical profession, patients, and the society in which both reside.

In summary, medical care exhibits two dimensions that are relevant to the current quality-of-care crisis. The first is the technical dimension of quality medical care. Although physicians and patients are concerned about this dimension, physicians are confident that quality healthcare is adequately definable to ensure precise measurement in order to improve healthcare quality—even though they acknowledge that the nature of quality is often a moving target (Blumenthal, 1996b; Zipes, 2001). The second dimension is interpersonal and existential in nature. This dimension represents the patient's emotional or psychological needs and is responsible for the quality-of-care crisis that permeates the healthcare industry—at least from the patient's perspective. For, the interpersonal dimension of quality of medical care concerns patients most. In commentary on the future of quality healthcare, David Blumenthal and Arnold Epstein (1996) claim that the physician, especially in terms of the patient-physician relationship, represents a key component in the resolution of this crisis. My contention is that the notion of a virtuous physician helps to address and resolve it.

---

[10] Caper also envisions a unified notion of quality healthcare: "In medical care, its objective [technical] and subjective [interpersonal] characteristics are woven into a single fabric" (1974, p. 1137).

1.1.2 Professionalism Crisis

A related crisis to the quality-of-care crisis in modern medicine is professionalism (Relman, 2007b; Smith, 2005; Swick et al., 2006).[11] A physician's unprofessional behavior may result in the delivery of poor or less than adequate healthcare. The question that fuels this crisis is what type of physician best addresses the patient's total healthcare needs and fulfills medicine's social contract. The answer to this question involves the physician's professional demeanor or character. Importantly, that demeanor is a function of the social contract between the medical profession and the larger public. In the early 2000s, several medical societies launched the Medical Professionalism Project to update the social contract for twenty-first century medicine (Project of the ABIM Foundation et al., 2002). The professionalism crisis is currently reaching fever pitch, and commentators are spilling much ink over the crisis. For example, special issues of *Academic Medicine* and *Perspectives in Biology and Medicine* recently published articles devoted to the nature of professionalism and especially to whether professionalism can be taught (Humphrey, 2008; Whitcomb, 2007a).[12] In this section, I examine briefly the history of professionalism in medicine, especially in the United States. I then discuss the nature of the current medical professionalism crisis, especially the efforts to define medical professionalism. I also discuss the Medical Professionalism Project's charter and its impact on the practice of medicine.

According to Matthew Wynia (2008), the date of medicine's nascence as a profession is the same as medical professionalism's nascence. Although many look traditionally to the Hippocratic period—especially in terms of the Hippocratic Oath—for the birth of medical professionalism, several problems confront this interpretation of medical professionalism's origin. For example, Hippocratic physicians did not represent the majority of Greek physicians in terms of medical standards of practice, especially in terms of end-of-life issues (Miles, 2004). According to Wynia, neither the Medieval Age nor the Renaissance represents the origin of medical professionalism because no universal standards for medical behavior were operative then. In fact, he even rejects Thomas Percival's *Medical Ethics* as the source of medical professionalism for the same reason. Not until the AMA's 1847 Code of Medical Ethics is medicine's nascence achieved, claims Wynia, although maturity is several decades off. What initially defined this code is a tripartite contract between physicians and patients, physicians and colleagues, and physicians and their professional societies. However, what eventually became problematic for the medical practitioners is how best to define their professionalism. Is professionalism simply reducible to an ethical code or does it require more?

---

[11] Professionalism, along with evidence based medicine and patient safety, is part of a quality-of-care movement in modern medicine (Hafferty and Levinson, 2008).

[12] Holly Humphrey (2008) notes that over 1500 articles appeared in the literature during the six years intervening from the founding of the Medical Professionalism Project to her introductory essay for the *Perspectives in Biology and Medicine* special issue on professionalism.

Although defining medical professionalism is elusive, just as defining quality-of-care is, Herbert Swick (2000) provides a contemporary definition, which has been influential. According to Swick, "medical professionalism consists of those behaviors by which we—as physicians—demonstrate that we are worthy of the trust bestowed upon us by our patients and the public, because we are working for the patients' and public's good" (2000, p. 614). Since he strives for a normative definition of medical professionalism, Swick identifies nine behaviors essential to the definition. The first is the subordination of the physician's interests to those of others, especially patients, for their betterment. The next behavior requires physicians to endorse and adhere to high ethical standards. The third behavior pertains to the social contract between medicine and society. That contract, in terms of its duties and obligations, must guide a physician's behavior in meeting the medical needs of patients. The next behavior involves humanistic values, such caring, compassion, honesty, integrity, among other, which should animate a physician's solicitude for a patient. Physicians should also be accountable to peers and patients for their professional behavior. They should also be committed to both excellence and scholarship with respect to their technical competence. Finally, given the complexity and uncertainty of medical practice, physicians should develop reflective and deductive practices and skills to dispense healthcare in a just and objective manner. As Swick admits, the purpose of his effort to provide a normative definition is to stimulate dialogue in the medical community to strive towards a consensus definition. That effort bore fruit in a number of definitions for medical professionalism, at the level of medical communities (Hafferty and Levinson, 2008).

In 1999, the American Board of Internal Medicine, the American College of Physicians and American Society of Internal Medicine, and the European Federation of Internal Medicine formed the Medical Professionalism Project.[13] Members of the project's committee included Troy Brennan as chair, along with members from each of the societies and several special consultants. The goal of the Project was "to develop a 'charter' to encompass a set of principles to which all medical professionals can and should aspire" (Project of the ABIM Foundation et al., 2002, p. 244). The charter, called a Physician (or Physicians') Charter, was published simultaneously in 2002 February issues of the *Annals of Internal Medicine* and the *Lancet*.[14] In the charter, committee members identify three principles, with nine associated professional responsibilities. The first principle is the primacy of the patient's welfare, in which the physician exhibits an altruistic stance towards patients and their healthcare needs. The next principle is patient autonomy, in which physicians empower patients to make the best decisions as to their healthcare. The

[13] Frederic Hafferty and Dana Levinson (2008) identify such efforts to address professionalism in the practice of medicine by professional communities as the fourth wave of medicine's professionalism movement. The first three waves include the emergence of professionalism as an issue vis-à-vis challenges such as the commercialization of medicine, attempts to define professionalism, and efforts to measure it.

[14] The charter also appeared in the May issue of the *European Journal of Internal Medicine*.

final principle is social justice, in which physicians endeavor to distribute healthcare resources fairly. The nine associated responsibilities include commitments to enhancing professional competence and scientific knowledge, developing honest and appropriate relations with patients and maintaining patient confidentiality, improving quality and access to healthcare, distributing healthcare resources justly, and managing conflicts of interest to maintain patient trust. The charter ends with recognition of medicine's social contract and the maintenance of the contract though rededication to the above principles and responsibilities of professionalism.

In a 2003 May issue, the *Annals of Internal Medicine* carried a 15 month report of the charter's impact upon the medical profession (Blank et al., 2003; Eldar, 2003). Around a dozen professional medical journals published the charter, including *American Journal of Obstetrics and Gynecology*, *American Journal of Surgery*, *Canadian Medical Association Journal*, *Clinical Medicine* (journal for the Royal College of Physicians in the U.K.), *European Journal of Internal Medicine*, *La Reveu de Médecine Interne*, and *Medical Journal of Australia*. The charter was translated into almost a dozen languages, including German, Italian, Japanese, and Turkish. Members of the charter's committee gave more than one hundred professional presentations at national and international medical conferences, grand rounds, seminars, and workshops. Moreover, almost one hundred professional associations and societies, medical schools, and certifying boards adopted or endorsed the charter. Finally, the charter received considerable public attention through newspaper, radio, television, and internet coverage.

Although the charter's overall impact upon the medical community was positive, critics did raise objections and challenges to the charter. For example, the 2003 May issue of *Annals of Internal Medicine* included an editorial by Stanley Reiser and Ronald Banner and over a half-dozen letters to the editor. In their editorial, Reiser and Banner (2003) praise the charter for addressing contemporary issues facing medical professionalism; however, they raise concern over the lack of input from patients and patient advocacy groups in the charter's formulation. The outcome is "rhetoric [that] portrays physicians at a distance from patients" (Reiser and Banner, 2003, p. 845). Their recommendation is to revise the charter to include the patient's voice and to include the patient as partner.[15] Letters to the editor raise a number of objections and concerns with the charter. For example, the first letter, as do others, points out that the Hippocratic Oath is sufficient for defining medical professionalism—no additions or revisions are necessary. Another letter chastises the charter's framers for advocating conflicting principles. Specifically, the principles of patient welfare and of social justice conflict since the former principle presupposes individual rights while the latter group rights and both are exclusive. Other letters are more conciliatory, attempting to extend the charter's role in the discussion on medical professionalism. For example, one correspondent claims

---

[15] In commentary on the charter, Laine Ross (2006) points out that the charter's framers marginalized the patient by shifting from a principle of respect for persons to respect for patient autonomy.

medicine is a lifestyle and that the charter should include the mentoring role of physicians. In response to these criticisms, the charter's framers express delight that the charter opened lines of communication over the professionalism crisis.

The charter continues to influence the medical community's efforts to address the professionalism crisis (Hafferty and Levinson, 2008). Its most significant impact, however, is paving the way for an organizational approach to the medical professionalism crisis.[16] For example, in one of the letters to the editor in the 2003 May issue of *Annals of Internal Medicine*, the correspondent—a healthcare administrator—encourages the charter's framers to revise the charter to include the organizational or administrative level of healthcare. Rather than focusing just on the behavior or responsibilities of individual physicians, medical professionalism must also include its social responsibilities—especially in terms of medicine's social contract (Cohen, 2007). Hafferty and Levinson (2008) provide an important instance of the organizational dimension of medical professionalism with respect to the social contract—conflict of interest (COI). Although COI issues are not new to medicine, they highlight the problems associated with modern medicine as it attempts to negotiate the demands to provide quality healthcare for patients in a profession that includes business and industry interests. Recently, the chair of the Physician Charter, along with several colleagues, issued a policy proposal for medical schools and academic centers to limit the influence of the healthcare industry vis-à-vis COI (Brennan et al., 2006).

Although the organizational dimension does include the social realm, its focus is still on the individual physician and his or her ability to maintain professional integrity in the face of compromise, e.g. COI situations. According to Hafferty and Levinson (2008), the problem is that efforts to frame medical professionalism so far fail to capture the complexity of modern medical practice. To address this complexity, they propose a systems approach to medical practice and professionalism.[17] Hafferty and Levinson "suggest reframing the issue of professionalism. . .from a matter of individual motives, or even as an object of remedial actions at the organizational level, to that of a complex, adaptive system where social actors, organizational settings, and environmental factors interact" (2008, p. 608). To illustrate their proposal, they discuss medical school as a complex system vis-à-vis medical professionalism. Specifically, they locate professionalism within the complexity of formal, informal, and hidden medical school curricula. For the formal curriculum, professors can teach professionalism in the classroom setting. However, to leave the conveyance of professionalism at that level would open the possibility for distorting what it means to practice medicine professionally. Medical students also need to learn professionalism at an informal level in which medical professors mentor students. Finally, professors must model what medical professionalism looks

[16] Hafferty and Levinson (2008) identify such efforts as the fifth wave of medicine's professionalism movement.

[17] Hafferty and Levinson (2008) denote this systems approach as the sixth wave of medicine's professionalism movement.

like on the hospital or clinical wards, at level of the hidden curriculum.[18] Only by approaching medical professionalism from a systems perspective, conclude Hafferty and Levinson, in which actors (professors and students) engage each other at various organizational settings (classroom, hospital, or clinic), given specific environmental factors (COI or other ethical dilemmas), can students adequately learn to practice medicine professionally.

Finally, in an analysis of the charter, Swick and colleagues, review its foundations and offer an alternative foundation (Swick et al., 2006). According to these commentators, the conceptual foundation of the charter is a duty-based ethic. The charter's language is contractual in nature, reflecting a relationship of distrust between the patient and physician. Swick and colleagues endeavor to switch the charter's duty-based ethic to one of virtue, since "medicine is to a large extent a moral enterprise precisely because the physician must merit the patient's trust" (Swick et al., 2006, p. 267). To that end, they enlist William Osler (1849–1919). Although Osler appreciates the new science and its advantages for medical practice, he also argues that the physician's character and the virtues animating that character are crucial for keeping the profession of medicine from becoming simply a trade. For Swick and colleagues, a duty-based ethic is inadequate for a higher form of medical professionalism. That higher form of professionalism, as opposed to a basic form, is covenantal in nature and places the patient's interests first. "The Physician Charter is one important step toward finding a common ground for understanding medical professionalism," conclude Swick and colleagues, "but the profession must move beyond the Charter's somewhat narrow focus on duty and competence to embrace the ideals, the genuine sense of selfless service, and the deep commitment to patients that have for so long epitomized the highest values of medicine" (Swick et al., 2006, p. 273).

In summary, just as pundits divide quality-of-care into two types so they divide professionalism (Swick et al., 2006). The first type of professionalism is practical or technical in nature. Just like technical quality-of-care, practical professionalism pertains to the competent application of current standards or guidelines of medical care in terms of diagnosis and therapy. For the most part, it became the defining feature of professionalism as medicine yoked itself to the natural sciences. Physicians are confident that improving their practical or technical professionalism is simply a matter of teaching medical students, interns, and residents to dispense such healthcare in a professional manner. No real crisis in professionalism exists with this type of professional, except for isolated issues. The second type of professionalism, which is moral in nature, does represent a crisis in medicine. This moral professionalism represents a selfless service to the sick and demands at times an altruistic attitude on the part of medical practitioners, which technical professionalism trumped several

[18] Importantly, Hafferty and Levinson (2008) argue that medical educators cannot successfully change the current hidden curriculum, with its negative impact on professionalism, without changes to both the formal and informal levels of the medical curriculum.

decades ago. Again, as for the quality-of-care crisis the notion of virtuous physician can help to address and resolve the professionalism crisis.

1.2 Resolving Medicine's Crises

In this section, I discuss the medical profession's attempts to resolve the quality-of-care and professionalism crises. To that end, I first explore evidence-based medicine (EBM) to improve healthcare quality and effectiveness. EBM is the consensus means for enhancing the technical dimension of quality medical care and hence of medical professionalism. To address the existential dimension of the quality-of-care crisis and the moral dimension of the professionalism crisis, I next examine patient-centered medicine (PCM). PCM represents the attempt of healthcare professionals to humanize the biomedical model, especially in terms of ethical virtues, thereby enhancing both the quality of medical care and medicine's professionalism. While the technical dimensions of both quality care and professionalism are measurable and thereby quantifiable, the existential or moral dimensions are not easily measurable. Improving the latter represents a challenge. Finally, how best to teach both existential or ethical quality-of-care and moral professionalism also represents a challenge.

1.2.1 Evidence-Based Medicine

Within the past several decades, medicine appears to be undergoing a revolution or paradigm shift—to turn a Kuhnian phrase—with respect to its practice. Rather than medical practice based upon an older paradigm of pathophysiology and clinical experience or expertise, advocates of a new paradigm, EBM, claim that medical practice should incorporate the best contemporary scientific data or evidence. Proponents of EBM rely less on traditional medical authority and more on systematic clinical and laboratory observations and data, especially obtained from randomized clinical trials and interpretation of that evidence though meta-analysis. This revolutionary claim is not without its critics; and the debate, although not as intense as it once was several years ago, still continues in some sectors of medicine over whether EBM is truly revolutionary.[19] For the most part, EBM enjoys a certain level of support and consensus within modern medicine. In this section, I examine what EBM is and how it addresses the quality-of-care and professionalism crises.

What is EBM? David Sackett and colleagues, who formed the Evidence-Based Medicine Working Group chaired by Gordon Guyatt, provide the best or most widely recognized and accepted definition of EBM: "the conscientious, explicit,

---

[19] Commentators also debate the proper philosophical framework for articulating EBM. For example, W.V. Quine and Larry Laudan's philosophy of science represent competing frameworks (Kulkarni, 2005; Sehon and Stanley, 2003).

and judicious use of current best evidence in making decisions about the care of individual patients" (Sackett et al., 1996, p. 71). According to EBM advocates, best evidence represents results from randomized clinical trials and meta-analysis of those trials. For example, the Cochrane Collaboration provides periodic reviews of up-to-date evidence from clinical trials for how best to proceed in terms of clinical practice (Chalmers, 1993). Sackett and colleagues identify five steps for the practice of EBM: (1) articulating clinical question(s) concerning a patient's disease state, (2) discovering the relevant evidence within medical literature databases to answer the question(s), (3) appraising the evidence with respect to its validity or soundness and its clinical usefulness, (4) applying the evidence to the patient's clinical problem, especially in terms of the patient's values, and (5) formally evaluating the four steps to determine the effectiveness of the process (Sackett et al., 1998).[20]

Critics of EBM abound (Cohen and Hersh, 2004; Timmermans and Mauck, 2005). Leonard Gibbs and Eileen Gambrill (2002) divide the criticisms or objections leveled against EBM into six categories: objections from ignorance of EBM's nature, objections from misinterpretation of EBM standards, objections from appeal to traditional medical practice, objections from ad hominem arguments, objections from ethical concerns, and objections from philosophical problems.[21] For example, one of the popular criticisms of EBM, especially given the various steps required to practice it, is that it would result in "cookbook" medicine (Farquhar et al., 2002; Wood, 1999). The basis of this objection, according to Gibbs and Gambrill, is a misconception of EBM's nature. EBM proponents charge that the criticism fails to recognize that EBM's guidelines are not recipes that require strict adherence to ingredients but rather they are roadmaps for negotiating the bumpy terrain involved in patient care (Farquhar, 1997). Another important criticism is that EBM does not meet its own standards for evidence (Goodman, 2002). In other words, evidence from clinical trials is not available to demonstrate the effectiveness of EBM over traditional approaches to medical practice. EBM proponents argue that such criticisms help to define what EBM is not (Sackett et al., 1996; Pronovost et al., 2002).

Although critics of EBM abound, so do its defenders (Gibbs and Gambrill, 2002; Straus and McAlister, 2000).[22] However, the nature of EBM is evolving in light of these criticisms. For example, William Ghali and Peter Sargious (2002) note that a frequent criticism of EBM is that it is often impractical and unrealistic for busy physicians who may not have the time to evaluate critically clinical trials in order to

[20] Porzsolt and colleagues identify an additional step after the first step: attempting to answer the clinical question(s) based on a clinician's current level knowledge or experience (Porzsolt et al., 2003). This additional step helps the clinician to identify how best to incorporate EBM into a patient's care.

[21] For philosophical basis of and issues facing EBM, see Goldenberg (2006), Guyatt and Busse (2006), Howick (2011), and Sehon and Stanley (2003).

[22] Although EBM proponents claim they welcome criticism in order to advance EBM's application to the practice of medicine, opponents claim they are often ignored or marginalized from the discussion (Buetow et al., 2006; Miles and Loughlin, 2006).

apply them to their medical practice. They illustrate EBM's adaptation to this criticism with a "diagnostic critical pathway" employed in the University of Calgary's teaching hospitals to assess a patient presenting with pulmonary embolism symptoms. After an initial assessment of the patient, the critical pathway dictates the procedure for further assessment in terms of diagnostics tests. The pathway's basis is the most recent clinical evidence justified through clinical trials. Interestingly, the physician need not know or critically evaluate the evidentiary basis for the diagnostic procedures. Thus, the physician only needs to appropriate the outcomes of the critical pathway to the patient and not to engage the initial steps for practicing EBM in terms of critically evaluating clinical tests results. As Ghali and Sargious conclude, "the paradigm of EBM is evolving to more realistically accommodate clinicians" (2002, p. 111).

In celebration of its 125 years, the *Journal of American Medical Association* (*JAMA*) republished articles in 2008—appearing in the journal published during its tenure—that were responsible for shaping the practice of medicine. One of those articles was the Evidence-Based Medicine Working Group's contribution in a 1992 issue of *JAMA*. In commentary on that original article, Victor Montori and Gordon Guyatt (2008) recount EBM's rapid development in the medical literature. They also expound on several key areas of EBM's evolution. The first involves access to and dissemination of clinical trial information, especially through the Cochrane Collaboration's publications of systematic reviews. In addition, electronic textbooks are also instrumental in making large segments of clinical findings accessible to busy physicians. Another key area of EBM's evolution is the critical assessment of recommendations for patient treatment based on EBM principles. Proponents call the area, Grades of Recommendation Assessment, Development and Evaluation (GRADE). GRADE is an important branch in EBM's evolution, according to Montori and Guyatt, because it "highlights the importance of clear specification of the [diagnostic] question with the identification of all patient-important outcomes and the necessity for systematic summaries of all the best evidence to guide recommendations" (2008, p. 1815). The authors conclude cautioning that EBM needs to continue to evolve in order to counter abuses of EBM's application to clinical practice.

Most physicians would not question the link between quality-of-care and EBM. In fact, the relationship between them is obvious and often "taken for granted" (Dickenson and Vineis, 2002, p. 244). The quality-of-care referred to here is technical in nature and concerns generally only the proper treatment of patients according to technical guidelines, which represent a consensus among practitioners within a professional community. A deep-seated assumption pervades modern medicine, in which the best evidence is that obtained only from clinical trials (Dickenson and Vineis, 2002). "When based on EBM principles," as Montori and Guyatt acknowledge, "quality improvement science can realize the reliable application of evidence and make health care a high-value proposition" (2008, p. 1816). In other words, quality care requires medical practice based on the best available evidence derived from randomized clinical trials. Invoking the responsibility of commitment to enhancing scientific knowledge promulgated in the charter on medical professionalism, Ghali and Sargious (2002), for example, propose enhancing quality-of-care through developing and implementing professional guidelines, such as the diagnostic critical pathway for assessing pulmonary embolism. Such guidelines are important not only for enhancing healthcare quality but also for ensuring fair or equitable distribution of quality healthcare in all cases and at all times. Thus, EBM resolves the quality-of-care crisis, at least its technical dimension, by ensuring that physicians provide the best possible care based on the best possible clinical evidence available.

Besides improving technical quality-of-care, EBM is also an important means for improving medical professionalism. Again, this professionalism is technical in nature and pertains to the physician's professional competence to deliver quality healthcare based on the best available evidence obtained from randomized clinical trials. In other words, physicians practicing EBM exhibit greater professionalism by making better or more accurate diagnoses and by prescribing better or more efficacious therapeutic modalities. The basis of this professionalism is the acquisition of expert skills (Epstein and Hundert, 2002; Trenti, 2003). Besides basing or measuring medical professionalism on the utilization of EBM, evidence-based practitioners also prescribe measuring professionalism with respect to a physician's involvement in EBM's development. For example, Ghali and Sargious propose measuring medical professionalism based on "the extent to which individuals adopt a commitment to contributing to the knowledge industry—of which EBM is a central movement—and to enhancing quality of care in our clinical settings with the tools of EBM" (2002, p. 111). As for the physician's professional behavior, besides technical skills, the physician must maintain "appropriate professional relations with patients" (Project of the ABIM Foundation, 2002, p. 245). Thus, EBM helps to resolve the medical professionalism crisis, at least its technical dimension, by ensuring that physicians provide the best care possible for their patients in a skillful and appropriate manner.

Teaching medical students, residents, and interns is also an important means for addressing both the quality-of-care and medical professionalism crises. The approach is the development of formal curricula for teaching quality medical care and professionalism. For example, over the past few years several medical schools developed and implemented such curricula to improve the quality of healthcare (Jotkowitz et al., 2004). Formal curricula for teaching medical professionalism are also important, especially in terms of professional codes of conduct such as the physician charter on medical professionalism (Cohen, 2006). In addition, professionalism is a moving target and changes as students progress from medical school to postgraduate studies and finally to full-time practice. Educational curricula must reflect the specific needs and requirements for each of these particular levels of medical professionalism (Woodruff et al., 2008). Besides formal curricula for teaching quality medical care and professionalism, informal curricula are also important, especially mentoring programs in which senior medical practitioners exhibit and model the attributes required for delivering quality healthcare in a professional manner (Hafferty and Levinson, 2008).

In summary, proponents of EBM argue that EBM is the best means for resolving both the quality-of-care and professionalism crises in modern medicine. EBM leads to the development of professional guidelines, rules, and standards, which ensure the best possible quality of medical care consistently dispensed under varying conditions. These guidelines, rules, and standards pertain to the physician's technical skills at practicing medicine, such as the proper skill for conducting a surgical protocol or for utilizing a diagnostic protocol. Besides technical skills, EBM also promotes cognitive skills. Cognitive skills include the acquisition of professional knowledge, the sharpening of critical thinking, and the commitment to lifelong professional learning (Maudsley and Strivens, 2000). Cognitive skills for obtaining clinical knowledge and then critically evaluating it are important for choosing the best evidence available for providing preeminent quality healthcare for the patient in a professional setting. A commitment to lifelong professional learning ensures that physicians are able to provide such care when guidelines, rules, and standards are updated and improved. EBM, then, makes available the foundation for instilling and maintaining professional conduct that ensures quality medical care.

1.2.2 Patient-Centered Medicine

In contrast to EBM, some physicians advocate patient-centered medicine (PCM) to address the quality-of-care and professionalism crises (Maizes et al., 2009; Wilson, 2008). Although PCM's roots extend deep into medicine's history, its contemporary roots are less than a century old (McWhinney, 2003). For example, Michael and Enid Balint and colleagues championed PCM in the mid twentieth century (Brown et al., 2003; Lipsitt, 1999). Early in his career, Michael Balint, a psychiatrist, argued that a crisis in medical practice is looming on the horizon. That crisis is the result of healthcare providers treating patients simply as illnesses. His solution to the crisis is to treat the individual patient "as a whole" (2002, p. 15). The Balints contrast illness-oriented medicine with patient-centered medicine. The practitioner of the former focuses on the illness or disease to obtain a "traditional diagnosis," while the practitioner of the latter strives for an "overall diagnosis" in which the physician knows the patient well enough so that the patient is "understood as a unique human being" (Balint, 1969, p. 269). The Balints' introduction of psychiatry into general practice represents a paradigm shift; and, the "idea that physicians should attend to their own emotional development as well as the emotions of the patient was revolutionary in its day" (McWhinney, 2003, p. 28). The impact of the Balints on the practice of medicine was significant, with the founding in 1975 of an International Balint Federation (Salinsky, 2002).

Students and colleagues of the Balints continue to develop PCM, especially in terms of its theory; however, the limitation of theoretical PCM is its practical application, i.e. "what the clinician must do and how the process is to be validated" (McWhinney, 2003, p. 28). Ian McWhinney and colleagues address the practical dimension of PCM, with development of the patient-centered clinical method (PCCM) at the University of Western Ontario (Stewart et al., 2003). The task of PCCM is to understand both the patient and the illness (Levenstein et al., 1986).

Two agendas define the task: the patient's agenda and physician's agenda. And, two methods are necessary to accomplish those agendas: a patient-centered method and doctor-centered method. The former agenda and method involve understanding the patient's illness experience through expectations, feelings, and fears, while latter revolves around the pathophysiology of the disease through making a differential diagnosis based on the patient's history, physical exam, and lab results. PCCM represents the integration of these two agendas and methods.

The challenges facing PCCM's practical development are whether proponents could define it precisely in operational terms. Certainly, what PCCM is not is clear: it is not "technology centered, doctor centered, hospital centered, disease centered" (Stewart, 2001, p. 444). Moira Stewart proposes a global or international definition of patient-centered care. According to this definition, the patient-centered physician

(a) explores the patients' main reason for the visit, concerns, and need for information;
(b) seeks an integrated understanding of the patients' world—that is, their whole person, emotional needs, and life issues;
(c) finds common ground on what the problem is and mutually agrees on management;
(d) enhances prevention and health promotion; and
(e) enhances the continuing relationship between the patient and the doctor (Stewart, 2001, p. 445).

The underlying assumption of this definition is holism in which the patient and physician enter into a covenantal relationship for the benefit of the patient, as opposed to the assumption of reductionism for disease-centered medicine (Evans, 2003).

From the above definition, Stewart and colleagues define six components for practicing PCCM (Stewart et al., 2003). The first involves the identification of the disease by the physician through the patient's medical history, physical exam, and lab results. However, the physician does not stop simply with the physical dimension but also includes the patient's illness experience through probing the patient's feelings, fears, and expectations concerning the illness. The physician's (and patient's) aim is to understand not just the diseased body part but also the patient as a whole person—the next component of practicing PCCM. Understanding the patient qua whole person also includes knowledge of the patient's life story in various contexts—even the ecosystem. The third component pertains to identification of common ground between the patient and physician, which includes identifying and defining the patient's healthcare issue, an appropriate treatment or management plan, and the appropriate roles of both patient and physician in that plan. The next component involves utilizing each patient-physician consult as an occasion for improving the patient's health or preventing illness. The fifth component stresses the need to deepen the patient-physician relationship, with compassion and trust, to ensure the patient's healing. The final component includes a realistic appraisal of time and resources, by both patient and physician.

Critics raise several problems or concerns with PCM, especially PCCM, including "takes more time; focuses primarily on the patients' psychological issues versus their diseases; requires acquiescing to patients' demands; means seeking out the patient's 'hidden agenda'; expects sharing all information and all decisions with patients" (Brown et al., 2003, p. 13). Proponents of PCM see these criticisms not so much as problems in need of solving but as misconceptions in need of correcting. For example, the criticism that PCM would involve much more time for consultation in contrast to conventional medicine is unfounded. In a case study presented by a general practitioner, a middle-aged female patient presented with nondescript headaches (Balint, 1969). In just two additional consults, with the longest lasting 20 minutes, the physician helped the patient gain access into her troubled past and the relationship of that past to her current medical complaint. The outcome of these consults was a "radiant" person whom the physician did not recognize initially at the final consult, during which the patient recounted for the physician the insights she had into her illness and life story.[23] The advantage of PCM is that time spent in consultation with a patient is generally higher quality and often leads to the root cause of a patient's chief complaint, if no simple organic cause is palpable.

Teaching PCM, especially PCCM, involves more than simply adding courses to an already overcrowded medical curriculum; rather, it involves a transformation not only of the medical curriculum but also of medical students. To that end, the foundation for teaching PCCM relies on a dialogue metaphor in which students actively participate in their medical education and not on a transmission metaphor in which the student passively obtains medical competence (Weston and Brown, 2003a). The protocol for teaching PCCM then mimics its practice. Just as six steps compose PCCM, so six steps compose the learner-centered method: "exploring both learning need and aspirations; understanding the learner as a whole person; finding common ground [between learner and teacher]; building on previous learning; enhancing the learner-teacher relationship; being realistic" (Weston and Brown, 2003b, p. 167).[24] The outcome of PCCM pedagogy is threefold (Weston and Brown, 2003a). First, the medical student gains mastery or competence in terms of both the art and science of medicine. This mastery shapes the next outcome, which is the student's professional identity. That identity is the last outcome, which is that of a healer who is not simply interested in curing a patient's disease but in restoring the patient's wholeness.[25] "Learning to be a patient-centered doctor," conclude Wayne Weston and Judith Brown, "challenges young physicians to develop their skills and, more importantly, themselves" (2003a, p. 166).

---

[23] The general practitioner presented the case study at a Balint group meeting. After he reported the first consultation and the enormity of the patient's problems, the consensus among the physicians was that the patient simply had too much history to resolve the case in a timely fashion.

[24] The challenges facing PCCM pedagogy reflect the criticisms and misconceptions of PCCM (Weston and Brown, 2003c).

[25] Robbie Davis-Floyd and Gloria St. John (1998) also advocate a transformative journey for physicians from technico-scientific doctors to holistic healers.

An important component of PCM curriculum is the incorporation of medical humanities courses, which is part of a larger movement to humanize modern western medicine (Marcum, 2008; Weston and Brown, 2003d). The movement began almost a half-century ago, with the resurgence of Renaissance humanism by moral philosophers and theologians and medical personnel concerned over the technocratization of medicine (Carson, 2007). It represents a shift in the conceptual foundations of medicine (Marcum, 2008). The movement's goal is to humanize medical practice by exposing medical students and residents to the art of medicine. To that end, the metaphysical foundation of medicine is shifting from a reductionistic perspective to a holistic (or minimally to a dualistic) one. In other words, the humanistic physician treats the patient as a whole person rather than simply as a diseased body part (Cassell, 2004). Epistemologically, the shift is from strictly objective medical knowledge, which is justified scientifically, to the inclusion of subjective knowledge, such as the patient's personal narrative of the illness experience. Finally, the ethical shift in the humanistic movement is from the traditional emotionally detached concern of the physician for the patient's disease state to an emotionally engaged empathic care for the patient's wellbeing. The outcome is a patient-physician relationship grounded in mutual respect and commitment to achieve the goal of restoring the patient to wholeness.

To achieve humanization of modern medicine and thereby to address the quality-of-care and professionalism crises, medical schools and even some undergraduate institutions recently instituted medical humanities programs (Carson, 2007; Stempsey, 2007a).[26] Courses that comprise these programs range from the traditional philosophical and historical disciplines to the avant-garde artistic and theatrical disciplines. Such programs are not simply interdisciplinary in which the disciplines intermingle with each other to produce a tertium quid, but rather they are multidisciplinary in which disciplines maintain their distinctive features but focus on exposing medical students to the human dimension of medical practice (Stempsey, 2007b). One of the chief goals or tasks of these courses is to "illuminate the practice of medicine (and, perhaps, medical theory) using ideas and insights distinctively associated with humanities or social science disciplines; especially doing so in a way that is not usually accessible through scientific descriptions and explanations" (Evans, 2007, p. 369). Besides formal coursework, medical humanities courses are also hands-on or practical. For example, engaging in theatrical performances allow medical students to reflect on the profounder personal issues that surface in medical practice (Savitt, 2002). Besides reflection, medical humanities courses provide students with perspective—not only the patient's but also society's (Stern et al., 2008). Finally, bioethics looms large in educating medical students about the many ethical quandaries that often face clinical practitioners.[27]

---

[26] The nature or definition of medical humanities is a rather debatable point in the literature (Campo, 2005).

[27] William Stempsey (1999) cautions that medical humanities course must provide the conditions for the development of medical students' critical skills to assess the values that shape medical

Although the biomedical model, especially its latest manifestation as EBM, is responsible for many of modern medicine's successes and triumphs, that model is also accountable for many of its failures and crises. To resolve the crises facing medicine, especially the quality-of-care and professionalism crises, healthcare providers and humanists, especially philosophers of medicine, attempt to humanize modern medicine and its biomedical model in terms of PCM (Marcum, 2008). For example, according to Steffen, "quality of care will be improved if patient and physician goals are mutually understood and pursued" (1988, p. 61). PCM provides the opportunity for both the patient and physician to negotiate each other's goals to achieve a common goal. The outcome is that "patient-centered medicine is more rewarding for both doctors and patients" (Weston and Brown, 2003c, p. 185). This reward often translates into enhanced quality of healthcare for the patient and greater job satisfaction for the healthcare provider.

Finally, PCM and medical professionalism go hand-in-hand, especially in terms of humanism (Cohen, 2007; Stern et al., 2008). Medicine is not just factual but it is also deeply moral in terms of its professional values that under gird medical and healthcare practice. These values animate the professional demeanor of healthcare providers, who fulfill their "moral imperative to serve" (Swick, 2007, p. 1026). "Without a solid foundation of humanism to animate it," warns Jordan Cohen, "professionalism is overly dependent on good intentions, and it has little chance to prevail under the intense lure of self-interest rife in contemporary medical practice" (2007, p. 1031). Thus, PCM and the associated humanistic movement strive to reinsert the human dimension into medical practice and thereby enhance both medicine's quality-of-care and professionalism.

1.3 Summary

According to some medical pundits, EBM and PCM are conflicting approaches to the practice of medicine. For them, both EBM and PCM are incommensurable positions with little or no common ground between them. For example, EBM utilizes population studies to obtain the best possible evidence for the most efficacious treatment of patients as a large homogeneous population, whereas PCM utilizes the individual patient or, at most, subpopulations of patients (Bensing, 2000). From a study of family practitioners' responses to questions concerning EBM and PCM, Naomi Lacy and Elisabeth Backer (2008) denote the conflict between the two approaches to medical practice as an either/or model in which a healthcare provider uses either EBM or PCM to practice medicine. In other words, EBM and PCM are mutually exclusive. Lacy and Backer depict the model as two non-overlapping circles, as in a Venn diagram. Healthcare practitioners advocate either EBM or PCM for several reasons. Thea Vliet Vlieland (2002), for example, argues that EBM is

practice. In addition, Joanna Rogers (1995) warns that medical humanities should apply not simply to physicians and medical school education but to all healthcare providers and their education.

inadequate for treating many patients with chronic diseases—given the complexity of these diseases and the difficulty in studying them in clinical trials.[28]

For many healthcare practitioners and medical commentators and pundits, EBM and PCM are not conflicting or incommensurable approaches to healthcare. Rather, these approaches do interact or intersect with one another. From their study, Lacy and Backer (2008) identify three models for EBM and PCM interactions. The first is a cyclical model. This model is akin to the either/or model, in that no overlap between the two approaches to healthcare exists. However, the two models are different from each other in that the practitioner of the cyclical model revolves in an iterative fashion between EBM and PCM. In other words, the healthcare provider flips from one approach of medical practice to the other while treating a patient. Thus, the cyclical model is dynamic and represents "a process of active movement between two modes of operating" (Lacy and Backer, 2008, p. 419).

The next two models for the interaction between EBM and PCM, according to Lacy and Backer (2008), are the continuum and the integration models. For the continuum model, EBM and PCM form poles on opposite ends of a spectrum. In between these poles, healthcare practitioners balance the advantages and disadvantages of EBM and PCM for treating patients. As Lacy and Backer quote one of the participants in their study, "Somewhere in the middle is where good care occurs. . .I mean, part of what makes good practice is balancing and weighing the population risks with the individual's value of those risks" (2008, p. 419). Importantly, practicing healthcare at either pole or extreme, EBM or PCM, would represent an unbalanced practice. Either the evidence would trump the patient's preferences or the patient would trump the evidence's validity and potential usefulness in restoring the patient's health. Either extreme would represent a distortion of optimal medical practice in terms of the best available medical evidence vis-à-vis the patient's healthcare needs.

The second—and final—interactive model is integration, which is the most popular model among healthcare practitioners. According to Lacy and Backer (2008), overlapping circles—as in a Venn diagram (which Sackett and colleagues originally introduced)—best represents the integration model. Lacy and Backer call the overlap between EBM and PCM, evidence-based patient-centered care (EBPCC).[29] Although they admit no consensus definition for EBPCC, they acknowledge a possible definition in the future after integrating the philosophical assumptions underlying EBM and PCM. Other practitioners also propose integrating EBM and PCM (Barker, 2000; Engebretson et al., 2008; Hasnain-Wynia, 2006; Moore, 2010; Sidani et al., 2006; Vliet Vlieland, 2002; Wagner et al., 2005). For example, Jozien Bensing (2000) suggests bridging the gap between EBM and PCM by making EBM

---

[28] Recently, however, Vliet Vlieland (2007) also acknowledges EBM is a possible direction for treating patients with chronic diseases. Moreover, Edward Wagner and colleagues claim that a combination of EBM and PCM is the way to proceed in terms of treating chronically ill patients (Wagner et al., 2005).

[29] An intermediate position of the continuum model or the iterative process of the cyclical model also represents EBPCC.

more patient-centered and PCM more evidence-based. To that end, Bensing advocates augmenting EBM by individualizing treatment decision through inclusion of the patient's preferences.[30] He recommends strengthening PCM by providing banks of analyzed evidence accessible to both practitioner and patients. However, the key to integrating EBM and PCM, according to Bensing, lays in communication, especially in terms of communication between healthcare providers and patients and between biomedical science and clinical practice. By improving lines of communication, especially through communication research, practitioners can integrate EBM and PCM based on the best evidence from communication studies.

Besides the challenge of communication research, other challenges and barriers face the integration of EBM and PCM and its possible future for influencing medical practice, especially in terms of medicine's crises. For example, Lacy and Backer (2008) catalogue into nine categories the responses from practitioners, when asked about challenges to the integration of EBM and PCM. The main categories include system-related factors, such as not enough time to accomplish the demands of both EBM and PCM; relational issues, such as new or challenging patients; and problems finding common ground, such as when patient's preferences run counter to what the best evidence dictates. Another challenge not listed by Lacy and Backer involves how best to motivate the integration of EBM and PCM, especially to address the medical crises of quality-of-care and professionalism. The aim of this book is to address that challenge by identifying and expounding upon the virtues of healthcare providers. My contention is that the notion of virtuous physician represents a good means for integrating EBM and PCM in order to improve both healthcare quality and professionalism.

The basis for the notion of virtuous physician resides in the humanistic movement of contemporary medicine, especially attempts to humanize physician's behavior (Marcum, 2008). David Solomon (2004) identifies three ways to modify such behavior: agent's character, type of action, and resulting consequence. The causal relationship among these three components is as follows:

$$\text{Agent} \rightarrow \text{Action} \rightarrow \text{Consequence}$$

In this book, I am not interested in just the type of action or its consequence, which are the result of the agent's character, since my aim is to provide a foundation for motivating an approach to integrate EBM and PCM in an attempt to resolve the quality-of-care and professionalism crises. To that end, two main approaches are available for modulating an agent's behavior. The first is through rules or guidelines, which involve deontological approaches to behavior modification. Unfortunately, rules are historically not very efficacious in producing long-lasting or quality change in behavior. The second is through virtues, which pertain to an agent's character and the development of that character with respect to values a society deems important.

---

[30] Patient's values and preferences are part of EBM's evolution (Montori and Guyatt, 2008, p. 1815).

In some sense, both of these are important but the second is most important for the present purpose of initiating change in a healthcare provider's behavior, in order to address medicine's crises of quality-of-care and professionalism through integration of EBM and PCM. The reason why virtues are more important than rules is that virtues provide an ontological basis for following rules. In other words, virtues make possible rules and not rules virtues.

I propose the notion of virtuous physician to address the medical crises of quality-of-care and professionalism. The notion represents the combination of virtue ethics and virtue epistemology, grounded in a metaphysical analysis of the nature of virtues. To that end, I discuss in the next chapter the metaphysics of virtue theory, followed by virtue ethics and epistemology. In a subsequent chapter, after expounding in a separate chapter upon traditional virtues and vices, I explore the metaphysics of a virtuous physician in terms of caring, which is the ontologically prior virtue of such a physician. I contrast this virtue with the vice of uncaring. I next discuss the two ontic virtues of care and competence (and the two ontic vices of carelessness and incompetence), especially in terms of their relationship to one another. The chief virtue of an epistemically virtuous physician is competence, especially technical competence. Furthermore, the care that justifies the ethically virtuous physician is not simply a general concern about ($care_1$) a patient but rather a deeply felt solicitude to take care of ($care_2$) the individual patient based on technical and ethical or moral competence for meeting both the individual patient's bodily ($care_{2a}$) and existential ($care_{2b}$) needs. Combined with the virtue of competence, the virtue of $care_2$ provides the needed guidance for right or ethical action with respect to taking care of an individual patient. Although care and competence are necessary virtues for medical practice, they are insufficient and must be transformed into the compound virtue, prudent love, which is sufficient for practicing quality healthcare in a professional manner. I discuss this transformation in the fifth chapter, along with the compound vice, imprudent lovelessness, of the unvirtuous physician. In a penultimate chapter, I analyze two medical stories from the literature, utilizing the forgoing discussion of virtuous and unvirtuous physician. In a final chapter, I explore how the notion of virtuous physician facilitates the integration of EBM and PCM into what I call virtuous holistic medicine and the resolution of the quality-of-care and professionalism crises in modern medicine through the notions of virtuous physician and virtuous holistic medicine.

References

Anonymous. 1986. The 'new' AMA: A rededication to excellence. *Journal of American Medical Association* 256:1038–1040.

Anonymous. 2009. The USA's crisis of the uninsured. *Lancet* 373:782.

Balint, E. 1969. The possibilities of patient-centered medicine. *Journal of the Royal College of General Practitioners* 17:269–276.

Balint, M. 2002. The crisis of medical practice. *American Journal of Psychoanalysis* 62:1–15.

Barker, P. 2000. Reflections on caring as a virtue ethic within an evidence-based culture. *International Journal of Nursing Studies* 37:329–336.

Bensing, J. 2000. Bridging the gap: The separate worlds of evidence-based medicine and patient-centered medicine. *Patient Education and Counseling* 39:17–25.
Blank, L., H. Kimball, W. McDonald, and J. Merino. 2003. Medical professionalism in the new millennium: A physician charter 15 months later. *Annals of Internal Medicine* 138:839–841.
Blumenthal, D. 1996a. Quality of care?—What is it? *New England Journal of Medicine* 335:891–894.
Blumenthal, D. 1996b. The origins of the quality-of-care debate. *New England Journal of Medicine* 335:1146–1149.
Blumenthal, D., and A.M. Epstein. 1996. The role of physicians in the future of quality management. *New England Journal of Medicine* 335:1328–1331.
Blumer, R.H., and M. Meyer. 2006. *The new medicine*. New York, NY: Middlemarch Films.
Bowers, M.R., and C.I. Kiefe. 2002. Measuring healthcare quality: Comparing and contrasting the medical and marketing approaches. *American Journal of Medical Quality* 17:136–144.
Brennan, T.A., D.J. Rothman, L. Blank, D. Blumenthal, S.C. Chimonas, J.J. Cohen, J. Goldman, J.P. Kassirer, H. Kimball, J. Naughton, and N. Smelser. 2006. Health industry practices that create conflicts of interest: A policy proposal for academic medical centers. *Journal of American Medical Association* 295:429–433.
Bristol, N. 2010. US passes landmark health-care bill. *Lancet* 375:1149–1150.
Brown, J.B., M. Stewart, W.W. Weston, and T.R. Freeman. 2003. Introduction. In *Patient-centered medicine: Transforming the clinical method*, 2nd edition, eds. M. Stewart, J.B. Brown, W.W. Weston, I.R. McWhinney, C.L. McWilliam, and T.R. Freeman, 3–15. Abington, UK: Radcliffe Medical Press.
Buetow, S., R. Upshur, A. Miles, and M. Loughlin. 2006. Taking stock of evidence-based medicine: Opportunities for its continuing evolution. *Journal of Evaluation in Clinical Practice* 12:399–404.
Campbell, S.M., M.O. Roland, and S.A. Buetow. 2000. Defining quality of care. *Social Science & Medicine* 51:1611–1625.
Campo, R. 2005. The medical humanities, for lack of a better term. *Journal of American Medical Association* 294:1009–1011.
Caper, P. 1974. The meaning of quality in medical care. *New England Journal of Medicine* 291:1136–1137.
Caper, P. 1988. Defining quality in medical care. *Health Affairs* 7:49–61.
Carson, R.A. 2007. Engaged humanities: Moral work in the precincts of medicine. *Perspectives in Biology and Medicine* 50:321–333.
Cassell, E.J. 1997. *Doctoring: The nature of primary care medicine*. New York, NY: Oxford University Press.
Cassell, E.J. 2004. *The nature of suffering and the goals of medicine*, 2nd edition. New York, NY: Oxford University Press.
Chalmers, I. 1993. The Cochrane collaboration: Preparing, maintaining, and disseminating systematic reviews of the effects of health care. *Annals of the New York Academy of Sciences* 703:156–163.
Cohen, J.J. 2006. Professionalism in medical education, an American perspective: From evidence to accountability. *Medical Education* 40:607–617.
Cohen, J.J. 2007. Linking professionalism to humanism: What it means, why it matters. *Academic Medicine* 82:1029–1032.
Cohen, A.M., and W.R. Hersh. 2004. Criticisms of evidence-based medicine. *Evidence-Based Cardiovascular Medicine* 8:197–198.
Council on Medical Service. 1986. Quality of care. *Journal of American Medical Association* 256:1032–1034.
Cutler, D.M. 2004. *Your money or your life: Strong medicine for America's healthcare system*. New York, NY: Oxford University Press.
Daschle, T. 2008. *Critical: What we can do about the health-care crisis*. New York, NY: Thomas Dunne Books.

Davis-Floyd, R., and G. St. John. 1998. *From doctor to healer: The transformative journey*. New Brunswick, NJ: Rutgers University Press.

Dickenson, D., and P. Vineis. 2002. Evidence-based medicine and quality of care. *Health Care Analysis* 10:243–259.

Donabedian, A. 1979. The quality of medical care: A concept in search of a definition. *Journal of Family Practice* 9:277–284.

Donabedian, A. 1988. The quality of care: How can it be assessed? *Journal of American Medical Society* 260:1743–1748.

Donabedian, A. 1990. The seven pillars of quality. *Archives of Pathology and Laboratory Medicine* 14:1115–1118.

Eldar, R. 2003. Charter on medical professionalism. *Croatian Medical Journal* 44:502–503.

Engebretson, J., J. Mahoney, and E.D. Carlson. 2008. Cultural competence in the era of evidence-based practice. *Journal of Professional Nursing* 24:172–178.

Epstein, R.M., and E.M. Hundert. 2002. Defining and assessing professional competence. *Journal of American Medical Association* 287:226–235.

Evans, R.G. 2003. Patient centered medicine: Reason, emotion, and human spirit? Some philosophical reflections on being with patients. *Journal of Medical Ethics: Medical Humanities* 29:8–15.

Evans, H.M. 2007. Medical humanities: Stranger at the gate, or long-lost friend? *Medicine, Health Care and Philosophy* 10:363–372.

Farquhar, D.R.E. 1997. Recipes or roadmaps? *Canadian Medical Association Journal* 157:403–404.

Farquhar, C.M., E.W. Kofa, and J.R. Slutsky. 2002. Clinicians' attitudes to clinical practice guidelines: A systematic review. *Medical Journal of Australia* 177:502–506.

Ghali, W.A., and P.M. Sargious. 2002. The evolving paradigm of evidence-based medicine. *Journal of Evaluation in Clinical Practice* 6:109–112.

Gibbs, L., and E. Gambrill. 2002. Evidence-based practice: Counterarguments to objections. *Research on Social Work Practice* 12:452–476.

Goldenberg, M.J. 2006. On evidence and evidence-based medicine: Lessons from the philosophy of science. *Social Science & Medicine* 62:2621–2632.

Goodman, N.W. 2002. Evidence-based medicine needs proper critical review. *Anesthesia and Analgesia* 95:1817.

Gruber, L.R., M. Shadle, and C.L. Polich. 1988. From movement to industry: The growth of HMOs. *Health Affairs* 7:197–208.

Guyatt, G.H., and J.W. Busse. 2006. The philosophy of evidence-based medicine. In *Evidence-based endrocrinology*, ed. V.M. Montori, 25–33. Totowa, NJ: Humana Press.

Hafferty, F.W., and D. Levinson. 2008. Moving beyond nostalgia and motives: Towards a complexity science view of medical professionalism. *Perspectives in Biology and Medicine* 51:599–615.

Hasnain-Wynia, R. 2006. Is evidence-based medicine patient-centered and is patient-centered medicine evidence-based? *Health Services Research* 41:1–8.

Harteloh, P.P.M. 2003. The meaning of quality in health care: A conceptual analysis. *Health Care Analysis* 11:259–267.

Howick, J.H. 2011. *The philosophy of evidence-based medicine*. New York, NY: Wiley-Blackwell.

Humphrey, H.J. 2008. Medical professionalism. *Perspectives in Biology and Medicine* 51:491–494.

Jotkowitz, A.B., S. Glick, and A. Porath. 2004. A charter on medical professionalism: A challenge for medical education. *European Journal of Internal Medicine* 15:5–9.

Kotlikoff, L.J. 2007. *The healthcare fix: Universal insurance for all Americans*. Cambridge, MA: MIT Press.

Kulkarni, A.V. 2005. The challenges of evidence-based medicine: A philosophical perspective. *Medicine, Health Care and Philosophy* 8:255–260.

Levenstein, J.H., E.C. McCracken, I.R. McWhinney , M.A. Stewart, and J.B. Brown. 1986. The patient-centered clinical method. 1. A model for the doctor-patient interaction in family medicine. *Family Practice* 3:24–30.

Lacy, N.L., and E.L. Backer. 2008. Evidence-based and patient-centered care: Results from an STFM group project. *Family Medicine* 40:417–420.

Lipsitt, D.R. 1999. Michael Balint's group approach: The Boston Balint group. *Group* 23:187–201.

Lohr, K.N., and J. Harris-Wehling. 1991. Medicare: A strategy for quality assurance, I: A recapitulation of the study and a definition of quality of care. *Quality Review Bulletin* 17:6–9.

Lohr, K.N., K.D. Yordy, and S.O. Their. 1988. Current issues in quality of care. *Health Affairs* 7:5–18.

Lohr, K.N. (ed.). 1990. *Medicare: A strategy for quality assurance*, volume 1. Washington, DC: National Academy Press.

Maizes, V., D. Rakel, and C. Niemiec. 2009. Integrative medicine and patient-centered care. *Explore* 5:277–289.

Marcum, J.A. 2008. *An introductory philosophy of medicine: Humanizing modern medicine*. New York, NY: Springer.

Maudsley, G., and J. Strivens. 2000. Promoting professional knowledge, experimental learning and critical thinking for medical students. *Medical Education* 34:535–544.

McWhinney, I.R. 2003. The evolution of clinical method. In *Patient-centered medicine: Transforming the clinical method*, 2nd edition, eds. M. Stewart, J.B. Brown, W.W. Weston, I.R. McWhinney, C.L. McWilliam, and T.R. Freeman, 17–30. Abington, UK: Radcliffe Medical Press.

Miles, S.H. 2004. *The Hippocratic oath and the ethics of medicine*. New York, NY: Oxford University Press.

Miles, A., and M. Loughlin. 2006. Continuing the evidence-based debate in 2006: The progress and price of EBM. *Journal of Evaluation in Clinical Practice* 12:385–398.

Montori, V.M., and G.H. Guyatt. 2008. Progress in evidence-based medicine. *Journal of American Medical Association* 300:1814–1816.

Moore, L.A. 2010. Being empathetic: Benefits and challenges for the clinician and client. *Topics in Stroke Rehabilitation* 17:20–29.

Osborne, C.E., and H.C. Thompson. 1975. Criteria for evaluation of ambulatory child health care by chart audit: Development and testing of a methodology. *Pediatrics* 56:625–692.

Pellegrino, E.D. 2002a. Professionalism, profession and the virtues of a good physician. *The Mount Sinai Journal of Medicine* 69:378–384.

Porzsolt, F., A. Ohletz, A. Thim, D. Gardner, H. Ruatti, H. Meier, N. Schlotz-Gorton, and L. Schrott. 2003. Evidence-based decision making—the six step approach. *Evidence Based Medicine* 8:165–166.

Project of the ABIM Foundation, ACP-ASIM Foundation, and European Federation of Internal Medicine. 2002. Medical professionalism in the new millennium: A physician charter. *Annals of Internal Medicine* 136:243–246.

Pronovost, P., S. Berenholtz, T. Dorman, E. Martinez, and W. Merritt. 2002. Evidence-based medicine needs proper critical review: Response. *Anesthesia and Analgesia* 95:1817.

Reiser, S.J., and R.S. Banner. 2003. The charter on medical professionalism and the limits of medical power. *Annals of Internal Medicine* 138:844–846.

Relman, A.S. 2007a. *A second opinion: Rescuing America's healthcare*. New York, NY: Perseus Books.

Relman, A.S. 2007b. Medical professionalism in a commercialized health care market. *Journal of American Medical Association* 298:2668–2670.

Rogers, J. 1995. Being skeptical about the medical humanities. *Journal of Medical Humanities* 16:265–277.

Ross, L.F. 2006. What is wrong with the physician charter on professionalism. *Hastings Center Report* 36:17–19.

Rutstein, D.D., W. Berenberg, T.C. Chalmers, C.G. Child, 3rd, A.P. Fishman, and E.B. Perrin. 1976. Measuring the quality of medical care: A clinical method. *New England Journal of Medicine* 294:582–588.

Sackett, D.L., W.M.C. Rosenberg, J.A.M. Gray, R.B. Haynes, and W.S. Richardson. 1996. Evidence-based medicine: What it is and what it isn't. *British Medical Journal* 312:71–72.

Sackett, D.L., W.S. Richardson, W. Rosenberg, and R.B. Haynes. 1998. *Evidence-based medicine: How to practice and teach EBM*. London: Churchill Livingstone.

Salinsky, J. 2002. The Balint movement worldwide: Present state and future outlook: A brief history of Balint around the world. *American Journal of Psychoanalysis* 62:327–335.

Savitt, T.L. (ed.). 2002. *Medical readers' theater: A guide and scripts*. Iowa City, IA: University of Iowa Press.

Sehon, S.R., and D.E. Stanley. 2003. A philosophical analysis of the evidence-based medicine debate. *BMC Health Science Research* 3:14.

Sidani, S., D. Epstein, and J. Miranda. 2006. Eliciting patient treatment preferences: A strategy to integrate evidence-based and patient-centered care. *Worldviews on Evidence-Based Nursing* 3:116–123.

Sisko, A., C. Truffer, S. Smith, S. Keehan, J. Cylus, J.A. Poisal, M.K. Clemens, and J. Lizonitz. 2009. Health spending projections through 2018: Recession effects add uncertainty to the outlook. *Health Affairs* 28:w346–w357.

Smith, L.G. 2005. Medical professionalism and the generation gap. *American Journal of Medicine* 118:439–442.

Solomon, W.D. 2004. Ethics. III. Normative ethical theories. In *Encyclopedia of Bioethics*, 3rd edition, volume 2, ed. S.G. Post , 812–824. New York, NY: Thomson Gale.

Steffen, G.E. 1988. Quality medical care: A definition. *Journal of American Medical Association* 260:56–61.

Stempsey, W.E. 1999. The quarantine of philosophy in medical education: Why teaching the humanities may not produce humane physicians. *Medicine, Health Care, and Philosophy* 2:3–9.

Stempsey, W.E. 2007a. Medical humanities: Introduction to the theme. *Medicine, Health Care, and Philosophy* 10:359–361.

Stempsey, W.E. 2007b. Medical humanities and philosophy: Is the universe expanding or contracting? *Medicine, Health Care, and Philosophy* 10:373–383.

Stern, D.T., J.J. Cohen, A. Bruder, B. Packer, and A. Sole. 2008. Teaching humanism. *Perspectives in Biology and Medicine* 51:495–507.

Stewart, M. 2001. Towards a global definition of patient centered care. *British Medical Journal* 322:444–445.

Stewart, M., J.B. Brown, W.W. Weston, I.R. McWhinney, C.L. McWilliam, and T.R. Freeman. 2003. *Patient-centered medicine: Transforming the clinical method*, 2nd edition. Abington, UK: Radcliffe Medical Press.

Straus, S.E., and F.A. McAlister. 2000. Evidence-based medicine: A commentary on common criticisms. *Canadian Medical Association Journal* 163:837–841.

Swick, H.M. 2000. Toward a normative definition of medical professionalism. *Academic Medicine* 75:612–616.

Swick, H.M. 2007. Professionalism and humanism: Beyond the academic healthcare center. *Academic Medicine* 82:1022–1028.

Swick, H.M., C.S. Bryan, and L.D. Longo. 2006. Beyond the physician charter: Reflections on medical professionalism. *Perspectives in Biology and Medicine* 49:263–275.

Timmermans, S., and A. Mauck. 2005. The problems and pitfalls of evidence-based medicine. *Health Affairs* 24:18–28.

Trenti, T. 2003. Evidence-based laboratory medicine as a tool for continuous professional improvement. *Clinica Chimica Acta* 333:155–167.

Vliet Vlieland, T.P.M. 2002. Managing chronic disease: Evidence-based medicine or patient-centered medicine? *Health Care Analysis* 10:289–298.

Vliet Vlieland, T.P.M. 2007. Non-drug care for RA—is the era of evidence-based practice approaching? *Rheumatology* 46:1397–1404.
Wagner, E.H., S.M. Bennett, B.T. Austin, S.M. Greene, J.K. Schaefer, and V. Vonkorff. 2005. Finding common ground: Patient-centeredness and evidence-based chronic illness care. *Journal of Alternative and Complementary Medicine* 11:S7–S15.
Weatherall, D. 1996. *Science and the quiet art: The role of medical research in health care*. New York, NY: Norton.
Weston, W.W., and J.B. Brown. 2003a. Teaching the patient-centered clinical method. In *Patient-centered medicine: Transforming the clinical method*, 2nd edition, eds. M. Stewart, J.B. Brown, W.W. Weston, I.R. McWhinney, C.L. McWilliam, and T.R. Freeman, 153–166. Abington, UK: Radcliffe Medical Press.
Weston, W.W., and J.B. Brown. 2003b. The learner-centered method of medical education. In *Patient-centered medicine: Transforming the clinical method*, 2nd edition, eds. M. Stewart, J.B. Brown, W.W. Weston, I.R. McWhinney, C.L. McWilliam, and T.R. Freeman, 167–183. Abington, UK: Radcliffe Medical Press.
Weston, W.W., and J.B. Brown. 2003c. Challenges in learning and teaching the patient-centered clinical method. In *Patient-centered medicine: Transforming the clinical method*, 2nd edition, eds. M. Stewart, J.B. Brown, W.W. Weston, I.R. McWhinney, C.L. McWilliam, and T.R. Freeman, 185–197. Abington, UK: Radcliffe Medical Press.
Weston, W.W., and J.B. Brown. 2003d. Developing a patient-centered curriculum. In *Patient-centered medicine: Transforming the clinical method*, 2nd edition, eds. M. Stewart, J.B. Brown, W.W. Weston, I.R. McWhinney, C.L. McWilliam, and T.R. Freeman, 229–243. Abington, UK: Radcliffe Medical Press.
Whitcomb, M.E. 2007a. Professionalism in medicine. *Academic Medicine* 82:1009–1010.
Wilson, H. 2008. Becoming 'patient-centered': A review. *NZFP* 35:164–170.
Wood, B.P. 1999. What's the evidence? *Radiology* 213:635–637.
Woodruff, J.N., P. Angelos, and S. Valaitis. 2008. Medical professionalism: One size fits all? *Perspectives in Biology and Medicine* 51:525–534.
Wynia, M.K. 2008. The short history and tenuous future of medical professionalism: The erosion of medicine's social contract. *Perspectives in Biology and Medicine* 51:565–578.
Zipes, D.P. 2001. Quality of care: A moving target worthy of pursuit. *Journal of American College of Cardiology* 38:1766–1769.

Chapter 2
Virtue Theory, Ethics, and Epistemology

What is virtue? The answer to this question is not as simple as it first appears. The initial intuition for virtue pertains to the quality of a person's character or behavior, but what that character or behavior should be is generally elusive under scrutiny. In like manner, the question of what constitutes the nature of vice is not easily answerable and is often unclear. However, many see the two questions on the nature of virtue and vice as intimately and inherently related or connected to each other; and, the answer to one generally influences or depends on the answer to the other. Broadly and loosely construed, virtue is a type of excellence that is required for living a fulfilled and productive life, while vice is a deficiency or an excess that leads to an unfulfilled or a depraved life. But, these broad construals are fraught with problems, which require philosophical scrutiny and analysis. The answers to the questions about the nature of virtue and vice constitute part of contemporary virtue theory, which represents a metaphysical analysis of virtue and of vice.[1]

2.1 Virtue Theory

Again, let us entertain the question: what is virtue? As noted, the answer to the question falls under the aegis of virtue theory (Pence, 1991; Pojman and Fieser, 2008; Richter, 1999). To answer the question also includes an answer to the question of what vice is, which I examine in the next section. In this section, I endeavor to answer the virtue question through historical reconstruction of attempts to explicate the nature of virtue. Beginning with the ancient Greeks, I examine the various definitions of virtue up to the present day, in order to achieve a workable—if not also an inclusive and a comprehensive—definition of virtue. In addition, I analyze examples of the different types of traditional virtues, along with various vices, in the next chapter. This analysis sets the stage for examining, in Chapters 4 and 5, the chief ontological virtue of the virtuous physician, caring, and its two ontic derivatives, care and competence, along with the compound virtue of prudent love. In

[1] Besides the metaphysical dimension of virtue and vice, a psychological dimension also plays a critical role in defining virtue, especially in terms of personal character and personality.

J.A. Marcum, *The Virtuous Physician*, Philosophy and Medicine 114,
DOI 10.1007/978-94-007-2706-9_2,

like manner, the chief ontological vice of the unvirtuous physician is uncaring and the two ontic derivatives are carelessness and incompetence, while the compound vice is imprudent lovelessness. In a penultimate chapter, I analyze two medical stories in terms of caring, care and competence, prudent love, and the other virtues, as well as uncaring, carelessness and incompetence, imprudent lovelessness, and other vices. In a concluding chapter, I discuss the role the notion of virtuous physician plays in addressing the crises of quality-of-care and professionalism in modern medicine.

2.1.1 Traditional Virtue Theory

The ancient Greeks propose various definitions for the nature of virtue, which forms the foundation of how people should behave and comport themselves (Irwin, 1998; Pence, 1991; Peterson and Seligman, 2004; Prior, 1991). They define virtue as an excellence (*arête*), which is necessary for living a good or fulfilled life (*eudaimonia*). For the Greeks, virtue refers not simply to the moral or ethical but also to a more expansive meaning and usage for daily life. Importantly, it is not simply a theoretical notion per se but it also represents the concrete characteristics or traits of an individual person or object. "When the ancient Greeks thought about virtue," explains William Prior, "they were more likely to think of examples than of abstract principles or rules of virtuous conduct" (1991, p. 5).

Initially, the ancient Greeks take their examples of the virtuous not from philosophy but rather from poetry. For example, Homer in the *Iliad* refers to virtues as the excellences of warriors, horses, and implements of warfare. "In Homer," notes Gary Ferngren and Darrel Amundsen, "*arete* is not used with reference to moral behavior but to describe efficiency or fulfillment of a natural function" (1985, p. 3). Moreover, for Homer virtues qua excellences refer broadly to many different qualities. Thus, no specific list of virtues is available to identify a particular trait as virtuous. Importantly, virtues function in a heroic society not only for the good of the individual but also—and even more importantly—for the good of that society (MacIntyre, 1984). Finally, virtues are gifts from the gods.

Socrates is one of the first ancient Greeks—if not the first—to struggle with the philosophical notion of virtue, although he still tethers the notion to its particular or common-sense usage (Irwin, 1995). "The virtues are directed towards action; their point is to cause the agent," claims Platonist commentator Terence Irwin, "to do the actions that we count (for whatever reason) as desirable" (1998, p. 39). In Plato's early dialogues, Socrates examines notions or definitions of a variety of virtues, e.g. the pious or courageous. He wants to know what it means for a person to be pious or courageous, in terms of being virtuous. He is especially interested in the common sense or practical notion of virtue. In the *Meno*, for example, Plato's Socrates describes the practical nature of virtues necessary for performing the different functions of daily activities, at the various stages of life (youth and old age), or with respect to gender (male or female) or social status (free or slave). In this dialogue, he also discusses the very practical issue of whether the virtues are teachable.

The nature of virtue is plural for Plato's Socrates and no inherent unity or single definition exists (Day, 1994). However, in one of the last early Platonic dialogues—the *Protagoras*—Socrates asserts that the many virtues are really one in nature. Terry Penner (1973) identifies two types of unity for Socratic or Platonic virtues: extension and form. The latter refers to the identity of various virtues to a universal form, a general form within all virtues share to some extent its fundamental properties. The former unity involves "the identity of the extensions of the different species of virtue" (Penner, 1973, p. 66). In other words, an overlap with respect to the characteristics of the various virtues unites them into a single whole. For Penner, then, the unity of the Platonic virtues represents an identity of specific virtues with each other, such that courageous people act bravely in such a manner that courage is identical ontologically to virtue writ large, i.e. "Virtue = bravery = wisdom = knowledge = temperance = and so forth" (1973, p. 66).

Plato works out a philosophical or theoretical notion of virtue in his middle dialogues, especially in the *Republic*, with respect to the form of the good. Plato identifies what tradition later calls the four cardinal virtues, which are fundamental for the health or wellbeing not only of an individual but also of a society. The first virtue is prudence, which is the only rational or intellectual virtue. It represents the chief virtue among the four and is associated with the rational part of a person's soul or the ruling class of society. Prudence is the ability to make a sound judgment or to give good counsel, which is the result of knowledge and not ignorance. The next virtue is courage (*andreia*), which is a character trait of the emotive part of a person's soul or of the guardian class. According to Plato,

> Courage is a preservative. Strengthened by education, it preserves conviction about the things that are legitimately to be feared and those that are not. Courage makes a man hold fast to these convictions no matter whether he is threatened by danger or lured by desire. Neither pain nor pleasure will move him (1996, p. 125).

The third virtue is temperance (*sôphrosyne*), which involves mastering oneself vis-à-vis desires or pleasures. This virtue is a character trait of the appetitive part of a person's soul and of the lower classes.[2] The final virtue is justice (*dikaiosunê*), which is the focal point of discussion in the *Republic*. Whereas Plato associates the other three virtues with a specific part of a person's soul or a class' status within a society, justice pervades a person or a society in terms of the harmony it produces among the other virtues. Specifically, it "sustains and perfects the other three [virtues]; justice is the ultimate cause and condition of their existence" (Plato, 1996, p. 128). Justice pertains to an inner order, either of a person or of a society, which allows either to govern itself well.[3]

---

[2] Plato distinguishes between the lower and ruling classes, in that the latter rules their desires via reason (431c).

[3] Interestingly, Plato utilizes medicine to explain justice: "Justice, like health, depends upon the presence of the natural order governing the soul in the relations of its parts and in the conduct of the whole" (1996, p. 138).

Plato's student, Aristotle, also develops a virtue theory in both the *Eudemian ethics* and *Nichomachean ethics*.[4] However, the pupil's theory, although initially reliant upon the teacher's theory, is more systematic than the teacher's. Aristotle's focus on virtue is also in terms of excellence (*arête*), which is responsible for achieving *eudaimonia* or the good life based on right action. To that end, Aristotle distinguishes between intellectual and moral or practical virtues. According to Aristotle, an "intellectual virtue in the main owes both its birth and its growth to teaching (for which reason it requires experience and time)" (1998, p. 28). He identifies five intellectual virtues, which are excellences for knowing the truth: theoretical or philosophical wisdom (*sophia*), practical wisdom (*phronêsis*), empirical or scientific knowing (*epistêmê*), art or technical knowing (*téchnê*), and intuitive reasoning (*noûs*). Aristotle's intellectual virtues concerned with necessary knowledge are theoretical wisdom, scientific knowing, and intuitive reasoning, while those concerned with contingent knowledge are practical wisdom and technical knowing (Ross, 1995). The virtuous person, with respect to theoretical wisdom, has the capacity to ascertain the universal and the necessary, whereas practical wisdom concerns the particular, with respect to the "capacity to act with regard to human goods" (Aristotle, 1998, p. 143). The virtuous person, in terms of scientific knowing, involves the ability to demonstrate certain or sure knowledge empirically, while technical knowing involves the ability to manufacture objects. Finally, the virtuous person with respect to intuitive reasoning involves a direct insight into the intelligibility of universals (Hill, 1995; Ross, 1995).

Aristotle's moral or practical virtues, in contrast to the intellectual virtues, are a result of habit (*éthos*). According to Aristotle, "none of the moral virtues arises in us by nature; for nothing that exists by nature can form a habit contrary to its nature" (1998, p. 28). Aristotle specifically situates practical virtue as a mean between two extremes or vices. "Virtue, then," as Aristotle defines it, "is a state of character concerned with choice, lying in a mean, i.e. the mean relative to us, this being determined by a rational principle, and by that principle by which the man of practical wisdom would determine it" (1998, p. 39). According to the definition of practical virtue, Aristotle equates virtue with a state or condition (*hexis*) of the agent's character. Importantly, virtue is not simply a feeling or emotional disposition but a way of doing. That way of doing involves choice (*prohairesis*), which results in a decision with respect to a mean between deficiency and excess. For a virtuous person, the mean, which is not a mathematical mean, is relative to the person's context. Importantly, the mean does not represent moral relativity, however, for Aristotle holds that extremes lead to a vicious life. For example, the virtue of courage is a mean between the excess of rashness and the deficiency of cowardice. "For the man who flies from and fears everything and does not stand his ground against anything becomes a coward," claims Aristotle, "and the man who fears nothing at all but goes

[4] Traditionally, commentators consider *The Nicomachean ethics* a revision of *The Eudemian ethics*—although not all commentators agree.

to meet every danger becomes rash" (1998, p. 31).[5] Finally, practical reason is critical for living a virtuous life, for it allows a virtuous person to reason wisely in terms of choosing among various options available for achieving *eudaimonia*.

The Roman orator and philosopher, Marcus Tullius Cicero, develops a virtue theory based on the ancient Greeks. In the treatise *De invention*, Cicero defines virtue (*virtus*) "as a habit of the mind in harmony with reason and the order of nature" (1949, p. 327). In agreement with his Greek intellectual antecedents, Cicero also recognizes four chief virtues: prudence, justice, courage, and temperance. Prudence represents knowledge of good and bad things through the faculties of memory by which one recalls knowledge, intelligence by which one comprehends and understands, and prescience by which one presages what is to happen. "Justice," according to Cicero, "is a habit of the mind which gives every man his desert while preserving the common advantage" (1949, p. 329). In other words, the virtue represents the proper dignity of everything while preserving a right regard for general welfare. Its origin is a natural inclination, although social and religious laws support a natural inclination to be just. Courage, for Cicero, involves consciously engaging danger and hardship, with magnificence or vividness, confidence or assurance, patience or endurance, and perseverance or persistence. The final virtue is temperance, which involves a balance between reason and desire such that improper desires do not overwhelm right conduct. For Cicero, the virtues are avenues for overcoming the passions that often lead to a debased life. For example, in a letter to his son, Marcus, which later he published as *On duties*, Cicero outlines not only the duties but also the virtues for living a well-ordered and productive life.

Besides intellectual and ethical virtues, theological or transcendental virtues are another category of virtues important for the development of virtue theory in western civilization. According to Christian theologians, the cardinal or natural virtues are necessary but insufficient for living ultimately the good life. The theological or supernatural virtues are also required for living the good life to its fullest, although a life ultimately not realized until after physical death and bodily resurrection. Traditionally, these virtues are faith, hope, and love. They originate in Paul's first letter to the Corinthians (1 *Corinthians* 13:13). After discussing the nature of love, Paul lists the three theological virtues but gives pride of place to love. In the Christian scriptures, the goal of the theological virtues is the completion or perfection of humans, vis-à-vis divine grace, for eventual union with God in the new heaven and earth. Importantly, God infuses the believer with the virtues. Paul lists faith (*pistis*) first. "Faith," the author of Hebrews claims, "makes us sure of what we hope for and gives us proof of what we cannot see" (*Hebrews* 11:1, CEV). Faith functions as a virtue that sustains human reason and assists the believer in discerning the will of God, as revealed in scripture. Hope (*eipis*), the next virtue listed by Paul, pertains to belief in the fulfillment of expectations, with the result that humans will align themselves with God's will. Love (*agape*) is the final virtue. In the first letter to the Corinthians, Paul presents a comprehensive description of love

---

[5] In the *Eudemian ethics*, Aristotle (1992) provides a representative list of moral virtues.

in both positive and contrapositive terms, including kindness, patience, forgiving, magnanimous, forbearing, believing, hoping, etc. Finally, faith and hope are temporary virtues in that they function to correct human fallen nature in this life. Love, however, is eternal because "God is love" (1 *John* 4:16, CEV); and, as redeemed, Christians participate in that love eternally.

Aurelius Augustine's virtue theory, if not his entire theology, intimately depends on the Christian scriptures in his attempt to reform pagan—particularly Plato's—philosophy (Wetzel, 1992).[6] Like Plato before him, Augustine identifies both rational and volitional dimensions of virtues. However, Augustine utilizes the theological virtues to reform the cardinal virtues. Moreover, he, as Paul before him, gives pride of place to love among the theological virtues. "For when we ask whether someone is a good man," queries Augustine, "we are not asking what he believes, or hopes, but what he loves" (1955, p. 409). Augustine goes so far as to identify or equate virtue per se with love, especially love of God. In addition, he makes love the focal point or root for unifying the cardinal virtues (Langan, 1979). For Augustine, the cardinal virtues are manifestations of love itself: "temperance is love giving itself entirely to that which is loved; fortitude is love readily bearing all things for the sake of the loved object; justice is love serving only the loved object, and therefore ruling rightly; prudence is love distinguishing with sagacity between what hinders it and what helps it" (1872, p. 17). Overall, the aim of virtue or love is to provide a sure path to communion with God. Contrary to the ancients, Augustine does not believe that virtue leads to perfection unaided by God's assistance. "Virtue is a threshold," notes Augustinian commentator Bonnie Kent, "not the end of the road of moral development, so that we are justified in considering people virtuous if they are only moving in the right direction, are steadily *trying*, and have already made noteworthy progress" (2001, p. 229).

Thomas Aquinas weds Augustine's theology with Aristotle's philosophy to produce a highly systematic virtue theory that influenced philosophers throughout the centuries until even today (Spohn, 1992).[7] "Virtue," according to Thomas, "denotes a determinate perfection of power (*potentiae*)" (1969a, p. 5). The end of that power can be towards either being or acting. For Thomas, the power vis-à-vis virtue is the capacity or ability to act rightly or to do the good. Virtue, then, is an "operative" habit of the soul or an agent that results in good works and a life lived well. Thomas identifies two avenues by which habit leads to such an end. The first is an "aptness for a good act," while the second "ensures that this is brought to bear rightly" (1969a, p. 25). For example, the virtue of courage not only imparts an ability to a person to act courageously but also makes that person act courageously. Thomas, like his predecessors, divides the virtues into the four cardinal and three theological virtues. The intellectual virtue is prudence, which conforms or leads to its good mean or rule of measure—the truth. The moral virtues, according to Thomas, are

---

[6] Cicero's *Hortensius* is also important in Augustine's conversion to Christianity.

[7] Thomas O'Meara (1997) claims that contemporary virtue theorists fail to ground the virtues, particularly the theological virtues in God's grace, as Thomas did.

the appetitive virtues—justice, temperance, and courage or fortitude. These virtues, following Aristotle, conform not to a real or mathematical mean but to a rational one, i.e. between deficiency and excess.[8]

Thomas, following his ancestors, also acknowledges three theological virtues—faith, hope, and love or charity. As articulated by Thomas, the rule for the mean of theological virtues, in conformity with Augustine, is God: "his truth for our faith, his goodness for our charity, his sheer omnipotence and loving kindness for our hope" (1969a, p. 175). Thomas, however, makes a distinction between living or formed faith and lifeless or formless faith, with the former a virtue and the latter not. Living faith requires both an act of the intellect, as well as an act of the will; lifeless faith is deficient in the latter. "Whereas its act does have the quality called for in terms of mind," notes Aquinas, "the act does not have the quality called for in terms of the will" (1974, p. 133). Finally, Thomas, as true of Augustine, envisions love as the most excellent of the virtues, since "faith and hope attain to God according as from him comes knowledge of truth or possession of good, but charity attains God himself so as to rest in him without looking for any gain" (1975, p. 25).

2.1.2 Eclipse of Virtue Theory

Beginning with the Renaissance and continuing into the Enlightenment, deontological and consequential/utilitarian theories eclipse traditional virtue theory. Virtue theory is simply inadequate in a world rapidly becoming scientific and empirical, i.e. analytically action-based. Kant's deontological theory in particular is an attempt to remove the agent, whether divine or human, from the development and justification of morality. Kant (1993) wants to derive morality strictly from human reason. For Kant, the moral person acts according to categorical imperatives that are independent of any personal or subjective desires or needs. A categorical imperative asserts, "The moral person ought to do *X*," where *X* is an ethical law or rule.[9] A person obeys the ethical law not because of personal gain or excellence but because obedience is the right thing to do. A person's moral duty then is to obey ethical laws, which, according to Kant, accounts for a person's good will. Kant recognizes three formulations of the categorical imperative. The first, known as the law of nature, claims that ethical laws must be universal and apply to people at all times and places. "I should never act," notes Kant, "except in such a way that I can also will that my maxim should become a universal law" (1993, p. 14). The next, known as the end in itself, states that humans must treat each other as ends and not merely or simply as means. For Kant, humans are not instruments to one's own gains. The final, known as the kingdom of ends, insists that humans must comport themselves

[8] Thomas does acknowledge that at times the mean of moral virtues can also be real, as in the case of justice.

[9] In contrast, for Kant a hypothetical imperative states, "If a person wants *Y*, then the person must do *X*."

as if they live in a world guided by rational ends. In other words, humans compose a rational community in which every person contributes to the community's ethical laws.

As deontological theories focus on rules and laws to determine moral worth, utilitarian theories concentrate on consequences or outcomes. The British are particular champions of utilitarianism, especially with Jeremy Bentham's dictum that people respond almost exclusively to pleasure and pain. John Stuart Mill, one of the best-known British utilitarians and raised by his father as a Bentham devotee, develops a robust theory of utilitarianism, which is influential on subsequent utilitarian and consequentialist theorists. He bases his theory on the notion of utility. "The creed which accepts as the foundation of morals, Utility, or the Greatest Happiness Principle," writes Mills, "holds that actions are right in proportion as they tend to promote happiness, wrong as they tend to produce the reverse of happiness" (1871, pp. 9–10). Happiness, according to Mill, represents pleasure and the absence of pain; while unhappiness involves pain and the absence of pleasure. The greatest good, then, is the utility that is responsible for the greatest amount of good for the greatest number of people. However, Mill breaks with Bentham's animalistic approach to utility by recognizing grades of utility. Mill holds that physical pleasure is not as refined or morally worthy as intellectual or spiritual happiness. "It is better," Mill maintains, "to be a human being dissatisfied than a pig satisfied; better to be Socrates dissatisfied than a fool satisfied" (1871, p. 14).

2.1.3 Contemporary Virtue Theory

Contemporary virtue theorists credit Elizabeth Anscombe with the revival, in the middle of the twentieth century, of virtue-based theory for ethics and morality. In a 1958 landmark paper, "Modern moral philosophy," she criticizes both the deontological and utilitarian/consequential theories. For example, Anscombe finds deontological theories defective, "That legislation can be 'for oneself' I reject as absurd; whatever you do 'for yourself' may be admirable; but is not legislating" (1958, p. 13). For, laws derived from personal experience are revisable as better laws appear. She goes on to argue that modern analytic philosophy is unable to account satisfactorily for morality.

> In present-day philosophy an explanation is required how an unjust man is a bad man, or an unjust action a bad one; to give such an explanation belongs to ethics; but it cannot even be begun until we are equipped with a sound philosophy of psychology. For the proof that unjust man is a bad man would require a positive account of justice as a "virtue" (Anscombe, 1958, pp. 4–5).

Although she does not develop a virtue theory, Anscombe realizes that virtue could resolve "a huge gap, at present unfillable as far as we are concerned, which needs to be filled by an account of human nature" (1958, p. 18). Following her lead, contemporary virtue theorists denigrate deontological and utilitarian theories in order to promote their particularly virtue theory (Pojman and Fieser, 2008). Martha

Nussbaum (1999) divides virtue theorists into two camps: anti-Kantians and anti-Utilitarians. The former emphasize the emotions in their virtue theories, while the latter reason.[10]

Since Anscombe's 1958 paper, contemporary virtue theories developed steadily (Foot, 1978; Geach, 1977; Pence, 1984). Within the past several decades, these theories proliferated, along with various proposed definitions for the nature of virtue. Robert Adams (2006) divides virtue theories into two camps. The first consists of excellence-based theories, which define virtues in terms of intrinsic excellences. According to proponents of these theories, virtues and their associated moral goodness are inherently worthy or valuable apart from any benefit derived from the virtues per se. Examples of these theories abound in the literature. For instance, Robert Roberts and Jay Wood define virtue as "an acquired base of excellent functioning in some generically human sphere of activity that is challenging and important" (2007, p. 59). Benefit-based theories of virtue, which define virtues in instrumental terms, represent the second camp. For these theories, the virtues provide the agent with tangible advantages and gains. According to Adams, "moral goodness is not worth having for its own sake, but only for the sake of its extrinsic benefits" (2006, p. 15). For example, one of the more influential virtue theorists, Alasdair MacIntyre, defines virtue as "an acquired human quality the possession and exercise of which tends to enable us to achieve those goods which are internal to practices and the lack of which effectively prevents us from achieving any such goods" (1984, p. 191).[11] In another example, Thomas Hurka defines virtue recursively by "equat[ing] the virtues with appropriate attitudes to intrinsic goods and evils; namely, loving the former and hating the latter for themselves" (2001, p. 20). In other words, the virtues undergird the appropriate stance towards striving for the good and avoiding the bad or evil.

Adams also provides a definition of virtue: "persisting excellence in being for the good" (2006, p. 14). He notes that his definition falls squarely within the excellence-based category. He expounds upon three components of his definition. The first is that virtue is "being for" something. He contrasts "being for" with simply "doing." "Being for" is an intentional stance or attitude, according to Adams, which covers a lot of ground, including "loving it, liking it, respecting it, wanting it, wishing for it, appreciating it, thinking highly of it, speaking in favor of it and otherwise intentionally standing for it symbolically, acting to promote or protect it, and being disposed to do such things" (2006, pp. 15–16). As intentional, "being for" also involves a virtuous agent's will, especially with respect to psychological states. The next component is the goods that virtues are "for." As the notion of "being for" is expansive, so is the notion of the goods. "I hold a very broad view of the goods that virtue is

---

[10] Nussbaum argues that an overall theory of virtue is often misleading.

[11] Edmund Pincoffs takes issue with the reductive sound of MacIntyre's definition, specifically with restriction of virtues to goods associated with internal practices. For Pincoffs, virtues must also include external practices and the qualities associated with the virtuous life that "are mutually irreducible one to another" (1986, p. 97).

for," claims Adams, "including any goods that human beings can exemplify excellence in caring about" (2006, p. 19). These goods can include not only the moral but also the intellectual and even the spiritual. The final component is persisting excellence. Persisting refers to a period longer than a day, although the exact length is unclear.[12] As Adams admits, exhibiting a virtue for five years is certainly long enough to call that person virtuous with respect to that virtue. Excellence pertains to a quality of goodness that is worthy because it is intrinsically good and not simply useful. For Adams, "what is excellent is good as an end in itself and not merely as a means to some ulterior end" (2006, p. 25).

Recently, Christine Swanton (2003) proposes a neutral definition of virtue to overcome problems associated with the difference approaches to defining it. "A virtue," writes Swanton, "is a good quality of character, more specifically a disposition to respond to, or acknowledge, items within its field or fields in an excellent or good enough way" (2003, p. 19). She claims her definition is acceptable both to eudaimonistic virtue theorists and to non-eudaimonistic virtue theorists. Her definition is acceptable to the former theorists because virtue contributes some benefit to the moral agent (Adams' benefit-based virtue theories). On the other hand, her definition of virtue is also acceptable to the latter theorists because virtue is simply an admirable trait that does not necessarily contribute to the agent's benefit (Adams' excellence-based virtue theories).[13]

Besides the two traditional categories for classifying virtues, i.e. cardinal virtues and theological or transcendental virtues, contemporary virtue theorists provide alternative classifications. For example, Bernard Gert (1985) divides virtues into personal and moral. Personal virtues are character traits that rational persons desire for themselves, while moral virtues are "traits of character that all impartial rational persons want everyone to have" (Gert, 1985, p. 95). The former virtues include courage, prudence, and temperance, while the latter justice, kindness, and trustfulness. Adams (2006) also proposes a classification of virtues. He makes a distinction between a capital "V" virtue and small "v" virtues. The former Virtue represents a "holistic property" of the virtuous person vis-à-vis "persisting excellence in being for the good" (Adams, 2006, p. 33). To be a Virtuous person, claims Adams, "one must not only have a number of excellent traits. One must also have them excellently composed into a whole" (2006, p. 32). Small "v" virtues represent particular character traits. Adams makes a distinction between motivational and structural virtues. The former are "defined by motives which in turn are defined by goods that one is for in having them" (Adams, 2006, p. 33). Benevolence is an example of a motivational virtue. Structural virtues represent the organization of motivational virtues. "The excellence of structural virtues," maintains Adams, "is a manner of personal psychic strength—of ability and willingness to govern one's behavior in accordance with

---

[12] Instead of persisting exclusively in virtuous behavior for a given period, a relaxed notion is preserving in large part in virtuous behavior for the same period.

[13] Adams (2006) would quibble about the finer details of their two classifications, e.g. restraining virtues to admirable character traits.

values, commitments, and ends one is for" (2006, pp. 33–34). Courage is an example of a structural virtue. Adams admits that this classification is not exhaustive, but does capture the holistic nature of virtue(s).

Edmund Pincoffs (1986) proposes one of the more comprehensive taxonomies for classifying virtues. He divides them into instrumental and noninstrumental virtues. Whether in an individual agent or a group of agents, the instrumental virtues, according to Pincoffs, "are those that, in relatively direct fashion, make it more probable that he will successfully pursue goals and ends, or objectives" (1986, p. 84). For the single agent these virtues include prudence, courage, carefulness, and persistence, among others, while for the group of agents they include cooperativity, practical wisdom, and other virtues of leaders and their followers. In contrast to instrumental virtues, noninstrumental virtues make the attainment of goals or objectives likely in a direct fashion. Pincoffs divides them into aesthetic, meliorating, and moral virtues. The aesthetic virtues are valued not for their instrumental effect but simply for themselves. They include two subcategories. The first are the noble virtues such as dignity and magnanimity, while the second are the charming virtues such as gracefulness and imaginativeness. The meliorating virtues pertain to "the making of common life, whether or not it is structured for common endeavor, more tolerable" (Pincoffs, 1986, p. 87). Pincoffs divides these virtues into mediating virtues, such as reasonableness and tactfulness; temperamental virtues, such as amiability and openness; and formal virtues, such as civility and modestly. Lastly, the moral virtues pertain to an agent's "concern" for the interests of either one's own or others. They can be mandatory, such as honesty and trustworthiness, or nonmandatory, such as altruism and forgivingness.

Finally, since the reintroduction of virtues into mainstream contemporary philosophy as a viable option for conducting ethical inquiry, philosophers revise both deontological and consequential/utilitarian theories. Kant's rule-based theory is now a virtue-based theory. In *Fundamental principles of the metaphysics of morals*, Kant writes, "Nothing can possibly be conceived in the world, or even out of it, which can be called good, without qualification, except a good will" (2008, p. 12). Robert Louden claims that Kant's notion of good will qualifies the early modern philosopher as a *quasi* virtue theorist. According to Louden, "Kant's virtuous agent is a human approximation of good will who through strength of character of mind continually acts out of respect for the moral law while still feeling the presence of natural inclinations which could tempt him to act from other motives" (1986, p. 478). The moral or ethical law, i.e. the categorical imperative, is what the virtuous person strives to keep. Indeed, Kant's notion of virtue presumes an agent's willing or desiring the ethical law. Kant defines virtue as "fortitude in relation to the forces opposing a moral attitude of will *in us*" (1964, p. 38).[14] Although virtue is certainly at the center of Kant's moral theory, still the ethical law subordinates virtue to its ends or duties. Louden concludes that the moral theory of Kant does not

[14] Kant defines fortitude as "the power and deliberate resolve to withstand a strong but unjust opponent" (1964, p. 37).

exclude virtue but does not place it at center stage; and, others note this is also true for utilitarians such as Mill.

In conclusion, Adams provides a good foundational or working definition for explicating the general nature of virtues, especially moral virtues with a capital V; however, his definition needs expanding in order to explicate the notion of virtuous physician. That notion requires not simply moral virtues for grasping the moral or ethical dimensions of the virtuous physician but also the intellectual or cognitive virtues and the theological or transcendental virtues. To that end, we need a more expansive or inclusive definition of virtue. Christine McKinnon provides such a definition of virtue. According to McKinnon, "a virtue is a chosen settled disposition to think, feel, and act in ways which are beneficial to oneself and/or to others and to do so for reasons having to do with ways in which one thinks humans ought to think, feel, and act" (1999, p. 37). To think requires epistemic or intellectual virtues, to feel theological or transcendental virtues, and to act moral or ethical virtues. All three sets of virtues are necessary to describe and explain fully the notion of virtuous physician. Moreover, her definition—like Swanton's definition—is neutral with respect to eudaimonistic aims, i.e. it includes eudaimonistic aims but it also takes the virtues to be intrinsically important. Reflecting this expansive and comprehensive definition of virtue, the types of virtues include epistemic or intellectual, theological or transcendental, and moral or ethical categories. However, the number of virtues exceeds simply the four cardinal virtues and three theological virtues and includes virtues such as honesty, creativity, humility, curiosity, tolerance, reasonableness, gratitude, and integrity—to name but a handful. Finally, virtues can serve double duty in terms of their categorization or classification. For example, faith is not only a theological virtue—such as belief in the divine—but it can also function as an intellectual virtue, i.e. an epistemic agent must have faith in an intellectual community and its members to report accurate information.

2.2 Vice

Almost all virtue theories include a definition of vice, which often reflects a contraposition of virtue. For example, Plato conceives of vice as the opposite of virtue in terms of a disharmony (see *Republic* 4). For him, courage is a virtue and cowardliness is a vice, or justice a virtue and injustice a vice, or temperance a virtue and intemperance a vice, or prudence a virtue and imprudence a vice. In other words, vice is the antithesis of virtue. Moreover, as virtue pulls a person in one direction, especially toward the happy or fulfilled life, vice pulls in the opposite direction, particularly to the reprobate or degenerate life. Plato uses the metaphor of a puppet to illustrate the relationship between virtue and vice, "like tendons or cords, drawing us and pulling against one another toward opposite directions toward opposing deeds, struggling in the region where virtue and vice lie separated from one another" (1988, p. 25). For a person to live a happy and fulfilled life, according to Plato, virtue must replace vice—just as knowledge must replace ignorance for a person to live an informed life. As Charles Young articulates Plato's position, the "path to happiness is the removal of ignorance and vice from our souls and their replacement with virtue and knowledge" (2006, p. 55). As Plato summarizes his position concerning virtue and vice, "the life that possesses virtue, of body or also of soul, is more pleasant than the life possessing vice, and that in the other respects. . .it is far superior to the life of vice. Thus it makes the one who possesses it [virtue] live more happily than his opposite [vice], in every way and on the whole" (1988, p. 121).

Aristotle agrees with Plato that vice is the opposite of virtue: "vice, too, makes choice, a choice with a view to the opposite thing" (1992, p. 34). However, he transcends his teacher's analysis of vice. Aristotle characterizes vice in terms of his definition for virtue, which, as discussed earlier, is a mean between two extremes. Those extremes or vices are excess and deficiency. As Aristotle claims in the *Nicomachean ethics*, "excess and defect are characteristic of vice" (1998, p. 38). In the *Eudemian ethics*, as noted above, Aristotle provides a chart of virtues and their corresponding excesses and deficiencies or vices. For the virtue of courage, for example, the vice of excess is rashness or over-confidence and the vice of deficiency is cowardice or lack of confidence. In other words, the virtue of courage moderates between an excessive fear in the face of danger (cowardice) and an inability to fear when it is appropriate to do so (rashness). To quote Aristotle, "the man who exceeds in confidence is rash, and he who exceeds in fear and falls short in confidence is a coward" (1998, p. 40). Moreover, vice, for Aristotle, is responsible for blinding a person to the "originating cause of action" necessary to make sound judgments for how best to act (1998, p. 143). A life of vice leads not only to an unfulfilled life but it also leads to a life of pain. In the case of courage then, the vice of cowardice may lead to shame while the vice of rashness to personal injury or even death.

Just as vices to the cardinal virtues exist; so vices—or the alternative term sin—exist to the theological virtues.[15] For example, in the letter to the Galatians, Paul lists a number of vices or sins that the Galatians must avoid if they are to follow Christ whole-heartedly.[16] The chief vice is faithlessness or unbelief in Christ as savior, for the Galatians were reverting to the Judaic law for salvation. As Paul explains in his letter, "The Law isn't based on faith. It promises life only to people who obey its commandments" (*Galatians* 3:12, CEV). Thus, the Galatians' hope is also misplaced and the object of their love is not Christ or the spirit of God but themselves or their flesh. The vices or sins include, e.g. anger, envy, hate, idolatry, impatience, jealousy, malice, and selfishness (*Galatians* 5:19–21).[17] These vices are in contrast to the virtues needed to live a robust and abundant Christian life, which traditionally commentators base on the four cardinal and three theological virtues.[18] For example, self-control represents temperance while compassion charity. Paul

---

[15] Solomon Schimmel argues that vice is not equivalent to sin but rather it is the foundation of sin. According to Schimmel, vices "are basic, perhaps universal human tendencies, from which sins result" (1997, p. 14).

[16] Paul also lists vices and virtues in a letter to the Colossians.

[17] Traditionally, the vices correspond to seven deadly sins (Schimmel, 1997).

[18] Paul's list of virtues or fruits of God's spirit are, "loving, happy, peaceful, patient, kind, good, faithful, gentle, and self-controlled" (*Galatians* 5:22–23, CEV).

embeds the discussion in terms of a spiritual warfare in which the Galatians—and all Christians—participate. According to Louis Martyn, "having identified them [Galatians] as soldiers, Paul now lists in vv. 19–23 the *vices* soldiers *should avoid* and the *virtues* they *should cultivate*" (2005, p. 261). Paul's discussion provides the foundation for further exposition on vice by Christian theologians, especially Augustine and Thomas.

Augustine's notion of vice is deeply personal, in that the struggle over the godly or virtuous life and the sinful or vicious life ravaged Augustine throughout his early life (Wetzel, 1992). In *Confessions*, he acknowledges that sin and vice clouded his ability to reason clearly and separated him from God's love. Whereas virtue pulled him towards peace and life, vice pulled in the opposite direction towards discord and death. According to Augustine, sin originates in vice and often exhibits the "false and shadowy beauty which attends the deceptions of vice" (1955, p. 56). For Augustine, as for Plato, vice is in opposition to virtue: "There are...certain vices forming contraries to the virtues by a clear distinction, as imprudence to prudence" (Langan, 1979, p. 87). In addition, some vices have a deceptive likeness to virtues, such as craftiness, which mimics prudence but is certainly not a virtue. Thus, at times a person might be unable to determine whether an action arises from virtue or vice. However, according to Augustine, the solution to this problem involves charity or love. Thus, a person who acts out of charity cannot act out of vice or sin. As Augustine writes, "where charity is full and perfect there will be no remains for vice" (Langan, 1979, p. 91). In other words, the person who is truly charitable or loving has all the virtues and "if all the virtues [are] present, there would be no vice; if no vice, absolutely no sin" (Kent, 2001, p. 228). Virtue, then, for Augustine, allows the believer to serve God who is love, while vice fleshly pleasures for the unbeliever.

In the *Summa Theologiæ*, Thomas (1969b) identifies three contraries to virtue. The first is sin, in which the contrary represents an immoderate act that does not result in a good deed—which derives from virtue—but in an evil deed. The next is malice, in which the contrary results in wickedness while virtue in goodness. The final is vice, which is contrary to virtue per se, i.e. "a suitable disposition of a given thing to its nature" (1969b, p. 5). Thomas consequently defines vice—with respect to virtue—as "any disposition which is not suitable to a given nature" (1969b, p. 9). Besides being contrary to virtue per se, vice is also contrary to the nature of what it means to be human—especially the rational dimension. Thomas summarizes his position on vice *contra* virtue accordingly, "human virtue, *which makes both man himself and his work good*, is in accord with human nature only to the extent that it is in accord with reason; and vice is contrary to human nature to the extent that it is contrary to the order of reason" (1969b, p. 9). Thomas categorizes the various vices with respect to the four cardinal and three theological virtues. The chief vices with respect to the cardinal virtues are imprudence and negligence versus prudence; injustice versus justice; fear, fearlessness, and daring versus courage or fortitude; and intemperance and insensibility versus temperance. The chief vices with respect to the theological virtues are unbelief versus faith; despair and presumption versus hope; and hatred versus charity.

The revised Kant, whom contemporary virtue ethicists portray now as a virtue theorist, also addresses the nature of vices. However, the original Kant is anything but a traditional or even a contemporary virtue theorist for a number of reasons. For one, he rejects Aristotle's notion of vice as a unitary idea and as the extreme and deficiency that bracket the virtuous mean. According to Kant, vice is not a matter of degree or weakness of virtue but a transgression of ethical law or duty. For Kant, the degree or weakness of virtue (*Tugend*) is not vice "but rather only lack of virtue [*Untugend*], a want of moral strength (defectus morlais)" (1948, p. 22). He defines vice as a "deliberate transgression" of moral or ethical law or duty, whether to oneself or others. "Kantian vices," according to Lara Denis, "can be understood as perversions of natural human tendencies—that is, as vices grafted onto animality and humanity" (2006, p. 522). For example, Kant identifies defamation as a vice that runs contrary to the virtue of respect for others. This vice represents not simply a lack of respect for others but an active effort to defame others. In addition, malice is also an example of a vice opposed to the virtue of love, since it inhibits an agent's duty to promote the happiness of others. Thus, for Kant, an agent's moral character depends on ethical laws or duties and not vice versa, as for virtue theory, since the vices oppose these laws or duties and thereby run contrary to human moral nature (Atwell, 1986).

Just as contemporary philosophers propose a number of definitions for virtue, so they do for vice.[19] For example, McKinnon (1999) defines vice in terms of functional deficits. These deficits can be purely cognitive in nature, i.e. "the agent may not know that part of her function is to construct a self, or the agent may not judge correctly the content of the desires of this self or the ways in which they may coexist" (McKinnon, 1999, p. 49). Thus, she bases the deficiencies of vices on defects in practical reasoning. "The bad person's life is bad," argues McKinnon, "because his reason has let him down: it has failed to exercise its proper authority over his desires or it has misconstrued the true nature of humans and the range of candidate good human lives" (1999, p. 50). Besides reason, volition in terms of willfulness or carelessness also ground vices. Other deficits can be both cognitive and volitional in nature, i.e. "the agent may not succeed in identifying with those desires she values, or she may have a very weak will and not succeed in overcoming contrary impulses" (McKinnon, 1999, p. 49). According to McKinnon, bad persons and their vices exists along a continuum, ranging from persons who knowingly make decisions that are bad to those who fail to deliberate at all about the choices they make.

Hurka (2001) also categorizes vices along a continuum, based on a recursive account of vices. He defines vices as "those attitudes to goods and evils that are intrinsically evil" (2001, p. 20). In other words, vice undergirds an inappropriate stance towards the good and the bad or evil in that the vicious person strives for the bad and avoids the good. On the one end of the continuum are the pure vices, which "involve attitudes that are inappropriately oriented towards their objects, either love

[19] As a number of virtue ethicists note, however, ethicists often ignore the role of vices in ethical theory. So, few theories of vices are available.

of an evil or hatred of a good" (Hurka, 2001, p. 92). Malice is an example of such a vice, since the malicious person intentionally inflicts pain on another. Such malice is a lower-level pure vice. A higher-level pure malice involves the vicious person taking pleasure in inflicting pain on others. The next category of vices represents an intermediate category on the spectrum, the vices of indifference. These vices "involve not a positively inappropriate orientation to a good or evil, but the absence, at least to a minimum threshold intensity, of an appropriate one" (Hurka, 2001, p. 94). For example, a person who is unmoved by the pain of another is uncaring and callous. On the other end of the spectrum of vices are the vices of disproportion. These vices "involve two or more attitudes both of which are appropriately orientated and above the threshold intensity, so that on their own they are good. But the intensities of these attitudes," claims Hurka, "are so out of proportion to their objects' values that their combination is evil, not just as a combination, as in some shortfalls in virtue, but on balance" (2001, p. 96). Selfishness is an example of this vice, in which a person cares for his or her lesser good more than for another's greater good.

Adams proposes a definition for vice that mirrors his definition for virtue: "a trait that counts against the overall excellence of the way you are for and against goods and evils" (2006, p. 36). In other words, a vice negates a virtue, especially virtue with a capital V, i.e. Vice is a bad character trait. For Adams, the notion of vice involves not simply an absence of virtue or goodness. A vice can be parasitic on or in opposition or even hostile to a virtue or the good. Like Aristotle, Adams divides vices into two types. The first are vices of weakness or deficiency, which he further divides into two types: motivational and structural vices. Motivational vices are often "vices of opposition or indifference to actual or potential goods" (2006, p. 37). An example of a motivational vice is malevolence. Structural vices, on the other hand, "consist not in opposition or indifference to specific goods, but in deficiency in strengths of self-government" (2006, p. 37). Cowardliness is an example of a structural vice. The second major type of vices involves excess. According to Adams, this type exhibits "concern for some good or type of good [that] is badly swollen in some way" (2006, p. 38). Examples of these vices include avarice, chauvinism, and sensuality. Importantly, for Adams, vices of excess in contrast to vices of weakness lead to wickedness, since a "wicked person is someone whose heart is in a bad place" (2006, p. 38). For example, a person who cannot effectively plan towards a good end exhibits folly as a vice of weakness, which does not lead to wickedness but simply to foolishness, while a person who plans evil knowingly is wicked.

In conclusion, vice is a bad character trait or disposition that results in diminution of what it means to be human. "Vices corrupt, vitiate, and destroy;" according to Paul Wadell, "they disfigure us morally and spiritually" (2008, p. 51). Vice robs its possessor of a full and good life, i.e. *eudaimonia*, and substitutes a truncated and depraved life for it. Vicious or unvirtuous persons never really fulfill their potential as human beings; rather, they live lives not only harmful to themselves but also to others. Vice acts as a myopic blinder that prevents vicious people from seeing clearly the damage they cause to themselves and others. It distorts the world, and the people who live in it. In fact, vicious people think that the vicious or unvirtuous life is really the happy and pleasurable life; however, the truth is that vice leads to the very opposite. Vice has the appearance of leading to a fulfilled and productive life, but it leads to a dysfunctional, tortured, and deformed life. Rather than flourishing, the vicious or unvirtuous person flounders.

2.3 Contemporary Virtue Ethics

Now that we have discussed virtue theory—and a workable definition of virtue (and vice) is in hand—we next turn our attention to modern virtue ethics, i.e. post-Anscombe (Crisp and Slote, 1997; Darwall, 2003; Statman, 1997; Swanton, 2003; Taylor, 2002). Unfortunately, the question, What is virtue ethics?, is a difficult question to answer for several reasons. Contemporary virtue ethicists define virtue ethics in contrast to and defend it against deontological and consequential or utilitarian ethics, i.e. virtue ethicists define virtue ethics negatively rather than positively. Gregory Trianosky (1990), for example, identifies nine tenets of neo-Kantian deontological ethics, which virtue ethicists either disagree with strongly or reject completely. The tenets range from "what is it right or obligatory to do?" to "virtuousness of a trait is always derivative from some relationship it displays to what is antecedently specified as right action" (Trianosky, 1990, p. 335). Trianosky notes that almost all virtue ethicists reject out of hand the tenet concerning the conjunction of obligation or duty and ethical rightness of an action, along with at least one or even more of the other tenets. He cites the example of Foot, who not only rejects the tenet of obligation but also the tenet claiming that ethical rightness of an action derives from the categorical imperative and not from the moral agent's desires or interests. For Foot, virtue ethics provides the proper motivation to act morally since the impetus for right action is always conditional or hypothetical and depends on an agent's desires or interests. Thus, an agent must be genuinely motivated via virtue to be moral or runs the risk of possibly being immoral.

The contention between virtue ethicists and deontological and consequential or utilitarian ethicists led to a niggly debate, which still engages ethicists today. An instructive articulation of the debate is found in the book, *Three methods of ethics: a debate*, representing the combined efforts of Marcia Baron, who champions Kantian or deontological ethics, Philip Pettit, who advocates consequential ethics, and Michael Slote, who defends virtue ethics. The authors make several important assumptions to frame the debate, especially that the method of any "substantive ethics should proceed analytically: by argument, example, and distinction-making" (Baron et al., 1997, p. 2). Given this assumption, each ethicist promotes and defends his or her chosen ethics vis-à-vis the other two competing ethics, often in traditional terms. Baron articulates Kantian ethics in terms of universal rules or laws and respect for others—not as means towards certain ends but as ends in themselves. Next, Pettit advocates consequential ethics in terms of maximizing an action's utility or good results. Finally, Slote presents a version of virtue ethics that analyzes morality from the perspective of the agent's character or disposition. The authors certainly raise important issues in ethical theory and methodology; but, as they freely admit, they do not provide the resources to resolve them.

David Solomon (2003) also examines the debate and frames it in two astute ways, especially for understanding the confusion and frustration surrounding expositions on the nature of virtue ethics. The first focuses on the chief terms or ideas of the debate. For deontological ethics, the chief idea is duty or rules; for consequential ethics, it is the outcome of an act; and, for virtue ethics, it is character excellence. The issue is which of these three chief ideas should function as the main idea for developing a robust ethics. In other words, which idea has pride of place in the moral life? The second way Solomon frames the debate represents greater nuance and reflects better, he claims, the complex nature of the debate. To that end, he identifies ten major themes that run through the debate among these ethicists. These themes include wariness of rules and duties as adequate for justifying morality, preference of concrete over abstract ethical terms, centrality of community in the moral life, the moral agent's narrative, and the role of mentoring in moral education. According to Solomon, the list of themes is not meant to be exhaustive or definitive but rather to "remind us how diverse and rich are the differences between many contemporary advocates of the virtues and their neo-Kantian and consequentialist opponents" (2003, p. 69). Obviously, the debate among these ethicists is not going to be resolved anytime soon.

Another reason why the question concerning the nature of virtue ethics is difficult to answer revolves around its different types, which is also true for deontological and consequential ethics. Justin Oakley (1996), for example, divides virtue ethicists into two major types. The first is consequentialist virtue ethics, which "treats rightness as ultimately a function of the value(s) an agent *promotes* (or the values an agent's rules would promote, when followed by people generally)" (Oakley, 1996, p. 144). He divides this type of virtue ethics into character-utilitarianism, Aristotelian perfectionism, and satisfying perfectionism. For character-utilitarian virtue ethics, agents promote virtues, which represent the best means by which to maximize a consequence's utility. Next, for Aristotelian perfectionism, agents promote virtues, which are objective and neutral values for achieving a virtuous life. Finally, for satisfying perfectionism, which is Oakley's preferred type of virtue ethics, agents promote virtues, which reflect the relative values adequate for living a virtuous life. The second type is nonconsequentialist virtue ethics, which "regards rightness as ultimately a function of the value(s) an agent *honors* or *exemplifies*" (Oakley, 1996, p. 144). In other words, virtuous agents exemplify the virtues since virtues are necessary for achieving the good life by motivating agents to conform to particular rules or to perform certain duties.

Like Oakley, Solomon also divides contemporary virtue ethicists into two types. The question animating his division is, "has the turn to virtue in ethics involved a genuine revolution in ethics or have we simply been undergoing a slight course correction in ethical theory?" (Solomon, 2003, p. 61). Advocates of the first type answer the initial part of the above question affirmatively, and Solomon claims its proponents advocate what he calls radical virtue ethics. According to its proponents, virtue ethics is incommensurable with traditional deontological and consequential or utilitarian ethics. Solomon cites the following example, "virtue theorists think that action should be guided not by action guiding devices entailed by principles in the theory—but rather by the virtues themselves" (2003, p. 75). Proponents of the second type answer the second part of the above question affirmatively and subscribe to what Solomon calls routine virtue ethics. Advocates of this type of virtue ethics attempt to situate virtues vis-à-vis deontological and consequential ethics. For Solomon, for instance, Oakley's consequentialist and nonconsequentialist virtue ethics are routine, since both locate virtue within these traditional ethical theories. Finally, Solomon notes that these two types of virtue ethics are responsible for compounding the debate. Specifically, routine virtue ethicists encompass a circumscribed conflict about locating virtues in the ethical life, while radial virtue ethicists a wider conflict in which the nature of ethics itself is at stake.

As noted earlier, contemporary deontologists and consequentialists or utilitarians attempt to incorporate virtues into their ethics; and, for good reason, since virtuous character is not foreign to these ethics. However, contemporary virtue ethicists, especially in terms of Solomon's notion of routine virtue ethics, turn the tables and attempt to incorporate these latter ethics into virtue ethics. Solomon differentiates these attempts into three categories. The first is assimilation, in which virtue ethicists integrate virtues into the traditional ethical theories to achieve synchronization. Solomon illustrates this category with William Frankena's dictum, "rules without virtues are impotent; virtues without rules are blind" (2003, p. 71). The next category is subordination, in which virtue ethicists subordinate either rules or consequences to virtues. He recognizes two patterns of subordination. The first is master subordination in which virtue ethicists identify one virtue as predominant and then subordinate rules or consequences to it. The second is distributed subordination in which virtue ethicists recognize a set of virtues as necessary for virtuous behavior and subordinate rules or consequences to it. The final category is condescension, in which virtue ethicists debase rules or consequences vis-à-vis virtues but still require them for moral behavior.

A number of virtue ethicists attempt to bring consensus or order to the nature of virtue ethics, given the diversity of contemporary virtue ethics, and to define it in positive rather than simply negative terms. For example, Trianosky identifies two features of what he calls a "pure ethics of virtue." The first is "at least some judgments about virtue can be validated independently of any special judgments about rightness of actions" (1990, p. 336). In other words, some virtues of a moral agent are justifiable irrespective of the consequences of the agent's action. He gives an example of Plato, who claims in the *Republic* that the moral goodness of the act originates from the congruent nature of a person's psychology rather than from the moral goodness of the act per se. The first feature of a pure virtue ethics leads to its second, which asserts that a virtue's fundamental goodness is responsible for an act's moral goodness. Again, citing Plato, he argues that a person's congruent psychology is responsible for that person's acts being good rather than ethical correctness of the acts per se. As Trianosky concludes, "for the pure ethic of virtue the moral goodness of traits is always both independent of the rightness of actions and in some way originative of it as well" (1990, p. 336).

In another positive and probably the most comprehensive exposition to date, Oakley (1996) identifies six key tenets of virtue ethics. The first is what he calls the "primacy of character." It claims that an agent's character or disposition justifies an action's morality. In other words, for example, a particular action is just because a just person performs it. The next tenet states, according to Oakley, "Goodness is prior to rightness" (1996, p. 138). Thus, goodness resulting from virtuous behavior defines an action's rightness or ethical nature and not vice versa. The third tenet claims that the virtues represent a spectrum of intrinsic goods, which are not reducible to one another, and are necessary for *eudaimonia* or a flourishing life. The objective goodness of the virtues is the next tenet. As such, a virtue's goodness is independent of an agent's desire, i.e. the virtue's good is agent-neutral. However, according to the penultimate tenet, the good of some virtues is agent-relative in the sense that the virtue acquires additional moral significance from a particular virtuous agent possessing the virtue. Oakley gives the example of the virtue of integrity, in contrast to the agent-neutral virtue of justice. The final tenet rejects the idea that virtues maximize the good whether in terms of quality or quantity. Rather, virtues refer to an action's excellence. These six tenets, according to Oakley, define virtue ethics in contrast to deontological and consequential or utilitarian ethics.

In response to virtue ethicists, both deontological and consequential ethicists levy several criticisms against virtue ethics. Probably the most common criticism assails Oakley's first tenet. Briefly, critics "raise doubts about whether the notion of virtue is clear or detailed enough to serve as a basis of a criterion of rightness" (Oakley and Cocking, 2001, p. 31). In other words, virtue ethics is too ambiguous to provide a robust system for directing or, more importantly, justifying moral action. Solomon (2003) calls this criticism the "action-guiding objection." Its foundation is circularity in the argument for justifying virtue ethics, i.e. a virtuous action is what a virtuous agent does, which in turn defines a virtuous action. The assumption behind this criticism is that any ethics must provide a direct means for motivating and guiding a moral agent. Virtue ethicists defend their reliance on virtue as the basis of moral action by citing the complexity of the moral life. No algorithm exists for specifying how a moral agent should use virtues to behave morally, other than how a virtue ultimately aids such an agent to achieve human *eudaimonia*. Moreover, the criticism is unfair, according to virtue ethicists, since no objective ground for justifying moral action supports either deontological or consequential ethics (Solomon, 2003). For example, what system-independent, objective criteria can a deontologist give for justifying a particular rule or duty?[20] Consequently, why are virtue ethicists held to a different standard than deontological or consequential ethicists?

Solomon (2003) identifies two additional criticisms of virtue ethics. The first he calls the "self-centeredness objection." According to this criticism, virtue ethics inverts the traditional trajectory of ethics because it promotes primarily one's own good, e.g. in terms of personal *eudaimonia*, and not the other's good. "Instead of my needing to be good in order to benefit others," as Solomon articulates the criticism,

[20] A deontologist might invoke, for example, the criterion of the golden rule to prohibit murder.

"I am required to be the sort of person who benefits others in order to be fulfilled myself" (2003, p. 74). The assumption here is that ethics must supply a direct channel from self to another to circumvent egoism and to promote altruism. Virtue ethics, according to critics, provides an indirect channel, at best, in which another's good is secondary or derivative of the agent's good. The second criticism Solomon calls the "conscientiousness objection." Critics claim that virtue ethics is unable to provide proper motivation for the moral act. Rather than acting in accord with the demands peculiar to the limits of moral behavior, the virtuous agent apparently achieves the morally good without effort. The assumption here is that moral motivation is distinct from non-moral motivation. Solomon concludes that these criticisms are problematic for routine versions of virtue ethics but not for radical versions, since the latter seek to replace the moral landscape by focusing primarily on the agent rather than on rules or consequences.

2.4 Virtue Epistemology

Recently the role of virtues in the acquisition, acceptance, and transmission of knowledge has gained prominence in philosophy, in a sub-discipline called virtue epistemology (Code, 1987; DePaul and Zagzebski, 2003; Kelp, 2011; Kvanvig, 1992; Montmarquet, 1993; Plantinga, 1993; Pritchard, 2005; Roberts and Wood, 2007; Sosa, 1991; Zagzebski, 1996; Zagzebski and Fairweather, 2000). Virtue epistemology utilizes traditional virtue theory, in which ethicists evaluate actions in terms of an agent's normative character traits exemplified in those actions rather than simply with reference to either the agent's motives or commitment to duty or the consequences of the acts themselves. In like manner, virtue epistemologists are interested in the normative or properly functioning epistemic faculties of an agent rather than just in the knowledge itself or its justification. Intellectual virtues, in general terms, are the innate or "acquired bases of excellent intellectual functioning" (Roberts and Wood, 2007, p. 60). Whereas traditional epistemologists focus on the discovery and justification of knowledge in terms of the evidence or methods used to produce it, virtue epistemologists focus on the intellectual virtues of an epistemic agent required to deliver knowledge or the epistemic goods, as well as on the vices that hinder such delivery.[21]

Virtue epistemologists generally divide intellectual virtues into two types (Greco, 2002). The first pertains to intellectual virtues as reliable or sound sensory or perceptual and cognitive or conceptual faculties, powers, or processes of an epistemic agent. The virtues are necessary for obtaining and ensuring knowledge, given an appropriate intellectual environment or context. They are innate and include, for

[21] According to Roberts and Wood, the epistemic goods represent more than simply the notion of justified true beliefs, if such goods are attainable, but rather a richer or broader notion that includes "warranted true belief, acquaintance, and understanding" (Roberts and Wood, 2007, p. 33). In addition, transmission or communication of the epistemic goods from teacher to pupil is an important epistemological issue.

example, sight or hearing for sensory or perceptual faculties and memory, intuition, inferential reasoning, insight, or introspection for cognitive or conceptual faculties.[22] Importantly, they are truth-promoting. This kind of virtue epistemology is generally called "reliabilist" virtue epistemology, since knowledge is based on the dependability of perceptual and conceptual faculties and on the causal role of a dependable belief-producing mechanism. The second type of intellectual virtue pertains to the virtuous character traits of epistemic agents. These agents acquire and develop these intellectual virtues over their lifetime. These virtues include, for example, honesty, courage, open-mindedness, humility, fairness, curiosity, tenacity, generosity, and integrity. This kind of virtue epistemology is often called "responsibilist" virtue epistemology, since knowledge is based on an epistemic agent's desire and motivation to know the truth. In large part, epistemically virtuous agents are accountable for delivering true epistemic goods (analogous to the responsibility of ethically or morally virtuous agents to perform right ethical or good moral actions, respectively), whether those goods are acquaintance or propositional knowledge or understanding.

For reliabilist virtue epistemologists, intellectual virtues are the proper functioning of intellectual faculties, powers, or processes, given an appropriate intellectual environmental context (Plantinga, 1993; Sosa, 1991). These faculties pertain to innate sensory or perceptual and cognitive or conceptual skills, which can be developed through training. The perceptual faculties are composed of somatic senses, such as sight or hearing, while the conceptual faculties consist of cognitive processes involved in memory, intuition, inferential reasoning, insight, or introspection.[23] The proper functioning of these perceptual and conceptual faculties provides the warrant necessary for the acquisition, acceptance, and communication of true beliefs and the avoidance of false beliefs, under appropriate environmental conditions. The reliabilist intellectual virtues, especially as reliable processes, are critical for the acquisition and justification of knowledge. Accordingly, as John Greco articulates reliabilist virtue epistemology, "A belief B(p) is epistemically justified for S if and only if B(p) is the outcome of a sufficiently reliable cognitive process, i.e., a process that is sufficiently truth-conducive" (2002, p. 291). In other words, the reliabilist intellectual virtues as properly functioning perceptual and conceptual faculties warrant beliefs as true since they are adequate for discovering a belief's veracity.

The perceptual virtues consist of properly functioning somatic senses, especially sight and hearing. These virtues are tied to the epistemic agent's embodiment. "A materially embodied human person," as Charles Taliaferro so ably describes these virtues, "feels with his skin, sees with his eyes, hears with his ears, smells with his

[22] Although the faculties are innate, in that such faculties as sight and hearing are not learned per se, this does not mean that their use cannot be developed through training or learning. Thus, the faculty of sight can be sharpened though learning to use it under specific conditions. For example, a clinician can learn to use sight effectively in observing certain clinical signs that are indicative of specific illnesses.

[23] Of course, perception can be influenced by concepts, which have been shown to determine what a person observes under specific conditions.

nose, and tastes with his mouth" (2001, p. 116). They are the faculties or powers that allow an epistemically virtuous agent to perceive the world. "In a word," claims Thomas Reid, "perception is most properly applied to the evidence which we have of external objects by our senses" (1790, p. 10).[24] The perceptual virtues are innate skills, which an agent can develop through training, and permit access to the physical world.[25] Without these skills, an epistemic agent cannot obtain the necessary sensory evidence or experience to formulate ideas or notions about the world, especially through the conceptual intellectual virtues. Importantly, sight or vision plays a predominant or unique role epistemologically, as compared to the other perceptual virtues.[26]

The conceptual virtues consist of the properly functioning cognitive faculties. Just as the perceptual virtues allow epistemic agents to perceive the world, so the conceptual virtues allow them to think about or to conceive it, i.e. to form concepts, ideas, or notions about the perceived world. "To think a thing, and to have a thought of it . . . to conceive a thing, and to have a conception, notion, or idea of it," as Reid writes, "are phrases perfectly synonymous" (1790, p.174). The conceptual virtues are also innate skills, which epistemic agents can develop through training; but rather than giving these agents direct access to the physical world, they permit indirect access through theories about it or access to a conceptual world that may not be tethered to the physical world. The conceptual virtues include faculties such as memory, intuition, inferential reasoning, insight, or introspection. Memory and inferential reasoning are two of the more important conceptual virtues. Properly functioning memory is the process of recollecting or recalling the relevant information necessary for conceiving the world, while properly functioning inferential reasoning is the process of drawing a valid conclusion—deductively, inductively, or abductively—also essential for conceiving the world.

For responsibilist virtue epistemologists, intellectual virtues are the character traits that are necessary for the best or excellent use of the epistemic faculties (Roberts and Wood, 2007; Zagzebski, 1996). These virtues consist of traits such as curiosity, courage, generosity, honesty, humility, and tenacity. Typically, epistemic agents—who exhibit intellectual virtues—are warranted in the acquisition, acceptance, and transmission of beliefs (although some virtue epistemologists do not consider the justification of belief to be the most or only appropriate goal for

[24] Reid (1790) also recognizes that perceptions are related to conceptions through prior beliefs.

[25] It is important to note that the perceptual virtues can be divided into the physical and the mental. The physical perceptual virtues are part of the process that does not necessarily involve conscious awareness, while the mental perceptual virtues do. In other words, the physical dimension of the perceptual virtues gives an epistemic agent contact with the world, while the mental dimension allows such an agent to mediate consciously that contact.

[26] "The unique distinction of sight," claims Hans Jonas, "consists in what we may provisionally call the *image*-performance, where 'image' implies these three characteristics: (1) *simultaneity* in the presentation of the manifold, (2) *neutralization* of the causality of sense-affection, (3) *distance* in the spatial and mental senses" (1966, p. 136). The consequence of sight's uniqueness is that the mind often goes where sight leads (Jonas, 1966, p. 152). Of course other senses, like hearing or touch, also function to lead the mind but sight is considered predominant or preeminent.

epistemic agents).[27] Responsibilist virtue epistemologists claim that the intellectual virtues are acquired habits rather than innate skills. Indeed, Robert Roberts and Jay Wood (2007) go so far as to describe these virtues as habits of the "heart" and argue that these virtues must mature over the lifetime of an epistemically virtuous agent. Importantly, these virtues also include a notion of the will in motivating such an agent to deliver the epistemic goods, with the virtues functioning in both a truth-conducive and truth-desiring manner. An epistemic agent then is largely responsible for delivering the truth. Consequently, the agent is to avoid the intellectual vices that hinder epistemic goods or truth.

A short exposition on four representative intellectual virtues and their associated vices provides an able introduction to how responsibilist virtue epistemologists utilize virtues for their epistemological purposes. The intellectual virtue of curiosity is an epistemic agent's desire or disposition to investigate or explore what is intellectually interesting but unknown to the agent. It does not allow an agent to avoid or turn away from the unknown or to acquiesce to the vice of intellectual indifference, but curiosity motivates the agent's attraction to or willingness to engage it until the unknown becomes known, even possibly at great cost to the agent. It also involves epistemic openness or receptivity to the unknown and does not close it off as being completely unknowable or mysterious. Although the epistemically virtuous agent may admit that there are mysteries that cannot be known exhaustively, and as such must be respected, the epistemically curious agent forges ahead and examines the boundaries of the unknown or mysterious so what is truly knowable can be known. Intellectual curiosity does not allow the epistemic agent to be satisfied with the status quo of epistemic goods; rather, it drives the agent to extend those goods in terms of epistemic goals. "Merely having at one's disposal a plethora of true beliefs," notes Wayne Riggs, "does not satisfy our natural curiosity about *why* things are the way they are" (2003, p. 221).[28] The epistemically curious agent thrives in a world of the unknown; and, if epistemically benevolent, this agent strives to know it for the benefit of all.

The intellectual virtue of courage mimics closely its ethical cousin. Just as an ethically courageous agent does not shrink from doing what is right simply because of personal danger, so an epistemically courageous agent does not shrink from believing what is true or from communicating it. For instance, when such an agent is in danger of being ostracized by an epistemic community for proposing novel beliefs that may challenge the community's consensus beliefs or dogmas, the epistemically courageous agent does not cower before the fear that threatens to disrupt

[27] Virtue epistemologists, like Roberts and Wood (2007), claim that the goal of virtue epistemology is not simply justified true belief, although such belief is important, but the maturation of a robust epistemic agent.

[28] Intellectual curiosity involves a strong drive or desire to know. According to Neil Cooper, it "is the capacity and the willingness to be interested and involved, even obsessed, with the object of inquiry" (1994, p. 461).

his or her cognitive functioning.[29] In other words, the agent does not succumb to the vice of intellectual cowardice. On the other hand, the epistemically courageous agent does not exhibit the vice of intellectual recklessness. According to Roberts and Wood (2007) intellectual virtues generally come in pairs, which for courage often includes caution. Caution as an intellectual virtue is a proper fear of epistemic danger, which keeps the epistemic agent from acting recklessly. "In general," claim Roberts and Wood, "courage and caution enable us to find our way among the threats, real and apparent, that we encounter in the course of our practices, sometimes circumventing these threats, sometimes facing them, and sometimes paying their price" (2007, p. 216). Without courage and caution, an epistemic agent could either acquiesce under pressure and from fear to conform to epistemic beliefs or dogmas that the agent believes are false, or propose reckless ideas that hinder acquiring or communicating the epistemic goods.

The intellectual virtue of honesty is a disposition to be straightforward or forthright, i.e. intellectually upright in acquiring and accepting, as well as in communicating and transmitting, epistemic goods like knowledge or truth. In other words, epistemically honest agents are frank or candid about what they know or believe, or about what they do not know or believe.[30] Such agents are fastidious in ensuring that what is known is, to the best of their ability, what is believed to be true. Linda Zagzebski says the intellectually honest agent "respects [the truth] and does her best to find it out, to preserve it, and to communicate it in a way that permits the hearer to believe the truth justifiably and with understanding" (Zagzebski, 1996, p. 158). In other words, intellectually honest agents do not fall prey to "immaculate perception" (Rivers, 2004, p. 254). Intellectual honesty is often defined in negative terms, as "a disposition or dispositions such that notwithstanding contrary to incentives, the agent refuses, in respect of assertion or other means of communication, to gain an unfair advantage, to indulge laziness diminishing the quality of the impression left or to indulge in exaggeration" (Guenin, 2005, p. 218).[31] The vice opposed to the epistemic virtue of intellectual honesty is intellectual dishonesty, where epistemically dishonest agents lie or cheat in an attempt to distort the truth or the known and to deceive other epistemic agents.

Finally, the intellectual virtue of humility is a disposition to make an unpretentious or a realistic assessment of one's knowledge and intellectual faculties or powers. Virtue epistemologists define intellectual humility, in part, in negative terms as not ascribing to oneself more intellectual excellence or ability than one actually

---

[29] Intellectual courage often involves leaving an epistemic comfort zone to forge new notions of reality. "Serious exploring of ideas," notes Thomas Rivers of this virtue, "risks shattering our preconceived notions, our images of the world" (2004, p. 251).

[30] "Intellectual honesty," as Louis Guenin articulates it, "assures that forthrightness dominates, delivering candor when it counts" (2005, p. 218).

[31] Pellegrino and Thomasma define intellectual honesty, with respect to medical practice, also in negative terms as "the habitual disposition not to deceive, or to move positively to reveal what we know and do not know about the clinical situation—the diagnosis, treatment, prognosis, and so on" (1993, pp. 25–26).

possesses. Epistemically humble agents realize that they do not know everything and can thereby benefit through instructions from others (even patients). As such, the epistemic virtue of intellectual humility contrasts with such intellectual vices as pride or arrogance. “As the opposite of intellectual arrogance,” according to Roberts and Wood, “humility is a disposition not to make unwarranted intellectual entitlement claims on the basis of one’s (supposed) superiority, out of either a concern for self-exaltation, or some other vicious concern, or no vicious concern at all” (2007, pp. 250–251). Importantly, epistemically humble agents are willing to acknowledge openly and self-effacingly that they make mistakes vis-à-vis the epistemic goods and that they can change their mind concerning them.

2.5 Summary

Virtue represents an attitude, a disposition, a character trait, a quality, a property, or an excellence that empowers and enables its bearer to live a fulfilled, flourishing, and vibrant life, not only morally or ethically but also intellectually and even transcendentally or spiritually. However, it not only benefits the bearer’s life but also the lives of others who live in community with the virtuous person. A virtuous community is a highly functioning community that ensures the welfare of its members. However, virtue is valuable not only because of its benefits, both individually and collectively, but it is also valuable in and of itself. In other words, virtue possesses an inherent worth, which makes the virtuous person attractive. That person is attractive because virtue makes him or her so, in that virtue disposes goodness and wellbeing onto its bearer who, in turn, disposes goodness and wellbeing onto others. The virtuous person, then, is one who not only knows how to live life well but who is also able to live life well.

Vice, on the other hand, represents not only the absence of virtue but also the contraposition of virtue. In other words, unvirtuous or vicious people not only lack requisite virtue but they also possess vice. For example, the imprudent agent not only lacks prudence but that agent is also hasty and rash in judgment. Vice robs its bearer of a fulfilled and productive life, substituting a depraved and distorted one of viciousness and rancor. As the virtuous person is attractive because of virtue, the vicious person is unattractive because of vice. Vice harms not only the person who possesses it but also those who interact with the vicious. The end-result is not only dysfunctional community members but also a dysfunctional community. Unvirtuous or vicious people, then, are those who do not know how to live life well but even if they knew how to live it well they would be incapable of doing so.

The aim of this chapter is to provide a background and to establish a framework for examining the ontologically prior virtue of caring, the ontically derived virtues of care and competence, and the compound virtue of prudent love for explicating the notion of virtuous physician, as well as the ontologically prior vice of uncaring, the ontically derived vices of carelessness and incompetence, and the compound vice of imprudent lovelessness for the unvirtuous physician. But, before we can examine the chief ontological virtue and vice, the ontically derived virtues and vices, and the compound virtue and vice, completion of the metaphysical analysis of virtue and vice requires a fuller discussion of specific virtues and vices than undertaken in this chapter. I complete that analysis in the next chapter. Part of the reason for the succeeding chapter on the specific virtues and vices is that I agree with Solomon's claim that radical virtue ethics is the more promising type of virtue ethics. Such a position requires at least a modestly detailed discussion of the individual virtues and vices.

References

Adams, R.A. 2006. *A theory of virtue: Excellence in being for the good*. Oxford, UK: Claredon Press.

Anscombe, G.E.M. 1958. Modern moral philosophy. *Philosophy* 33:1–19.

Aquinas, Thomas. 1969a. *Summa theologiæ, volume 23, Virtue (1a2æ. 55–67)* (trans: Hughes, W.D.). New York, NY: McGraw-Hill.

Aquinas, Thomas. 1969b. *Summa theologiæ, volume 25, Sin (1a2æ. 71–80)* (trans: Fearon, J.). New York, NY: McGraw-Hill.

Aquinas, Thomas. 1974. *Summa theologiæ, volume 36, Prudence (2a2æ. 47–56)* (trans: Gilby, T.). New York, NY: McGraw-Hill.

Aquinas, Thomas. 1975. *Summa theologiæ, volume 34, Charity (2a2æ. 23–33)* (trans: Batten, R.J.). New York, NY: McGraw-Hill.

Aristotle. 1992. *Eudemian ethics*, 2nd edition (trans: Woods, M.J.). New York, NY: Oxford University Press.

Aristotle. 1998. *The Nicomachean ethics* (trans: Ross, W.D.). New York, NY: Oxford University Press.

Atwell, J.E. 1986. *Ends and principles in Kant's moral thought*. New York, NY: Springer.

Augustine, A.1872. Writings in connection with the Manichaean heresy, volume V. In *The works of Aurelius Augustine, Bishop of Hippo: A new translation*, ed. M. Dods. Edinburgh: T. & T. Clark.

Augustine, A. 1955. *Confessions and enchiridion* (trans: Outler, A.C.). Philadelphia, PA: Westminster.

Baron, M.W., P. Pettit, and M. Slote. 1997. *Three method of ethics*. Oxford, UK: Blackwell.

Cicero, M.T. 1949. *De invention* (trans: Hubbell, H.M.). Cambridge, MA: Harvard University Press.

Code, L. 1987. *Epistemic responsibility*. Hanover, NH: University Press of New England.

Cooper, N. 1994. The intellectual virtues. *Philosophy* 69:459–469.

Crisp, R., and M. Slote (eds.) 1997. *Virtue ethics*. New York, NY: Oxford University Press.

Darwall, S. (ed.) 2003. *Virtue ethics*. Oxford, UK: Blackwell.

Day, J.M. 1994. *Plato's Meno in focus*. New York, NY: Routledge.

Denis, L. 2006. Kant's conception of virtue. In *The Cambridge companion to Kant and modern philosophy*, ed. P. Guyer, 505–537. Cambridge, UK: Cambridge University Press.

DePaul, M., and L. Zagzebski (eds.) 2003. *Intellectual virtues: Perspectives from ethics and epistemology*. New York, NY: Oxford University Press.

Ferngren, G.B., and D.W. Amundsen. 1985. Virtue and health/medicine in pre-Christian antiquity. In *Virtue and medicine: Explorations in the character of medicine*, ed. E.E. Shelp, 3–22. Dordrecht: Reidel.

Foot, P. 1978. *Virtues and vices and other essays in moral philosophy*. Berkeley, CA: University of California Press.

Geach, P.T. 1977. *The virtues*. Cambridge, UK: Cambridge University Press.

Gert, B. 1985. Virtue and vice. In *Virtue and medicine: Explorations in the character of medicine*, ed. E.E. Shelp, 95–109. Dordrecht: Reidel.
Greco, J. 2002. Virtues in epistemology. In *The Oxford handbook of epistemology*, ed. P. Moser, 287–315. New York, NY: Oxford University Press.
Guenin, L.M. 2005. Intellectual honesty. *Synthese* 145:177–232.
Hill, S. 1995. Two perspectives on the ultimate end. In *The crossroads of norm and nature: Essays on Aristotle's ethics and metaphysics*, ed. M. Sims, 99–114. Lanham, MD: Rowman & Littlefield.
Hurka, T. 2001. *Virtue, vice, and value*. New York, NY: Oxford University Press.
Irwin, T. 1995. *Plato's ethics*. New York, NY: Oxford University Press.
Irwin, T.H. 1998. The virtues: Theory and common sense in Greek philosophy. In *How should one live? Essays on the virtues*, ed. R. Crisp, 37–55. New York, NY: Oxford University Press.
Jonas, H. 1966. *The phenomenon of life: Toward a philosophy of biology*. Chicago, IL: University of Chicago Press.
Kant, I. 1948. *Metaphysical elements of ethics* (trans: Abbott, T.K.). Lexington, KY: Forgotten Books.
Kant, I. 1964. *The doctrine of virtue: Part II of the metaphysics of morals* (trans: Gregor, M.J.). Philadelphia, PA: University of Pennsylvania Press.
Kant, I. 1993. *Grounding for the metaphysics of morals*, 3rd edition (trans: Ellington, J.W.). Indianapolis, IN: Hackett Publishing.
Kant, I. 2008. *Fundamental principles of the metaphysics of morals* (trans: Abbott, T.K.). Radford, VA: Wilder Publications.
Kelp, C. 2011. In defense of virtue epistemology. *Synthese* 179:409–433.
Kent, B. 2001. Augustine's ethics. In *The Cambridge companion to Augustine*, eds. E. Stump and N. Kretzman, 205–233. New York, NY: Cambridge University Press.
Kvanvig, J.L. 1992. *The intellectual virtues and the life of the mind: On the place of the virtues in epistemology*. Savage, MD: Rowman & Littlefield.
Langan, J.P. 1979. Augustine on the unity and the interconnection of the virtues. *Harvard Theological Review* 72:85–91.
Louden, R.B. 1986. Kant's virtue ethics. *Philosophy* 61:473–489.
MacIntyre, A. 1984. *After virtue*, 2nd edition. Notre Dame, IN: University of Notre Dame Press.
Martyn, J.L. 2005. *Theological issues in the letters of Paul*. London: Continuum.
McKinnon, C. 1999. *Character, virtue theories, and the vices*. Orchard Park, NY: Broadview Press.
Mill, J.S. 1871. *Utilitarianism*, 4th edition. London: Longmans, Green, Reader, and Dyer.
Montmarquet, J. 1993. *Epistemic virtue and doxastic responsibility*. Lanham, MD: Rowman & Littlefield.
Nussbaum, M.C. 1999. Virtue ethics: A misleading category. *Journal of Ethics* 3:163–201.
Oakley, J. 1996. Varieties of virtue ethics. *Ratio* 9:128–152.
Oakley, J., and D. Cocking 2001. *Virtue ethics and professional roles*. Cambridge, UK: Cambridge University Press.
O'Meara, T. 1997. Virtues in the theology of Thomas Aquinas. *Theological Studies* 58:254–285.
Pellegrino, E.D., and D.C. Thomasma. 1993. *The virtues in medical practice*. New York, NY: Oxford University Press.
Pence, G.E. 1984. Recent work on virtues. *American Philosophical Quarterly* 21:281–297.
Pence, G. 1991. Virtue theory. In *A companion to ethics*, ed. P. Singer, 249–258. Oxford, UK: Blackwell.
Penner, T. 1973. The unity of virtue. *Philosophical Review* 82:35–68.
Peterson, C., and M.E.P. Seligman. 2004. *Character strengths and virtues: A handbook and classification*. New York, NY: Oxford University Press.
Pincoffs, E.L. 1986. *Quandaries and virtues: Against reductivism in ethics*. Lawrence, KS: University Press of Kansas.
Plantinga, A. 1993. *Warrant and proper function*. New York, NY: Oxford University Press.
Plato. 1988. *The laws of Plato* (trans: Pangle, T.L.). Chicago, IL: University of Chicago Press.

Plato. 1996. *The republic* (trans: Sterling, R.W. and Scott, W.C.). New York, NY: Norton.

Pojman, L.P., and J. Fieser. 2008. *Ethics: Discovering right and wrong*, 6th edition. Belmont, CA: Wadsworth.

Prior, W.J. 1991. *Virtue and knowledge: Introduction to ancient Greek ethics*. London: Routledge.

Pritchard, D. 2005. Virtue epistemology and the acquisition of knowledge. *Philosophical Explorations* 8:229–243.

Reid, T. 1790. *Essays on the intellectual and active powers of man*, volume 1. Dublin: Byrne & Milliken.

Richter, D. 1999. Virtue without theory. *Journal of Value Inquiry* 33:353–369.

Riggs, W.D. 2003. Understanding 'virtue' and the virtue of understanding. In *Intellectual virtues: Perspectives from ethics and epistemology*, eds. M. DePaul and L. Zagzebski, 203–226. New York, NY: Oxford University Press.

Rivers, T.M. 2004. Ten essentials for character education. *Journal of General Education* 53:247–260.

Roberts, R.C., and W.J. Wood. 2007. *Intellectual virtues: An essay in regulative epistemology*. New York, NY: Oxford University Press.

Ross, D. 1995. *Aristotle*. New York, NY: Routledge.

Schimmel, S. 1997. *The seven deadly sins: Jewish, Christian, and classical reflections on human psychology*. New York, NY: Oxford University Press.

Solomon, D. 2003. Virtue ethics: Radical or routine? In *Intellectual virtues: Perspectives from ethics and epistemology*, eds. M. DePaul and L. Zagzebski, 57–80. New York, NY: Oxford University Press.

Sosa, E. 1991. *Knowledge in perspective: Selected essays in epistemology*. New York, NY: Cambridge University Press.

Spohn, W.C. 1992. The return of virtue ethics. *Theological Studies* 53:60–75.

Statman, D. (ed.) 1997. *Virtue ethics*. Edinburgh: Edinburgh University Press.

Swanton, C. 2003. *Virtue ethics: A pluralistic view*. New York, NY: Oxford University Press.

Taliaferro, C. 2001. The virtues of embodiment. *Philosophy* 76:111–125.

Taylor, C. 2002. *Sympathy: A philosophical analysis*. New York, NY: Palgrave Macmillan.

Trianosky, G. 1990. What is virtue ethics all about? *American Philosophical Quarterly* 27:335–344.

Wadell, P.J. 2008. *Happiness and the Christian moral life: An introduction to Christian ethics*. Lanham, MD: Rowman & Littlefield.

Wetzel, J. 1992. *Augustine and the limits of virtue*. New York, NY: Cambridge University Press.

Young, C.M. 2006. The Socratic Elenchus. In *A companion to Plato*, ed. H.H. Benson, 55–69. Oxford, UK: Blackwell.

Zagzebski, L.T. 1996. *Virtues of the mind: An inquiry into the nature of virtue and the ethical foundations of knowledge*. New York, NY: Cambridge University Press.

Zagzebski, L., and A. Fairweather (eds.) 2000. *Virtue epistemology: Essays on epistemic virtue and responsibility*. New York, NY: Oxford University Press.

Chapter 3
Virtues and Vices

I extend the philosophical analysis of the previous chapter by examining specific traditional virtues and vices, which constitute an agent's character and provide a basis for understanding that agent's actions ethically or morally, intellectually or epistemically, and theologically or transcendentally. To that end, I examine various virtues and vices in terms of the four cardinal and three theological virtues and vices, as initially introduced in the last chapter. Consequently, I categorize or group the intellectual or epistemic virtues and vices under prudence and imprudence, the ethical or moral virtues and vices under courage and cowardice, temperance and intemperance, and justice and injustice, and the theological or transcendental virtues and vices under faith and faithlessness, hope and hopelessness, and love and lovelessness. These particular categories and their specific virtues and vices serve as readily general headings for grouping and discussing other similarly associated or cognate virtues and vices. For example, compassion or sympathy, empathy, humility, forgiveness and mercy, and loyalty are virtues akin or cognate to the theological or transcendental virtue of love. I must stress, however, the list of cognate virtues and vices is not exhaustive but rather illustrative of the main traditional virtues and vices.

Moreover, the virtues and vices of one grouping, as noted in the previous chapter, can function in another grouping. For example, the virtue of love can operate not only theologically or transcendentally as love of God but it can also operate ethically as love of one's neighbor or epistemically as love of knowledge. Thus, this taxonomy of the traditional virtues and vices as cardinal and theological virtues is fluid or flexible and serves as a convenient means to examine and discuss not only the nature of these virtues and vices but also the nature of cognate virtues and vices associated with them. Finally, I must reiterate that this categorization and list of virtues and vices is not complete or comprehensive. Although I attempt to cover as many of them as possible, the attempt here is representative in order to understand and explain the notions of virtuous and unvirtuous physician.

J.A. Marcum, *The Virtuous Physician*, Philosophy and Medicine 114,
DOI 10.1007/978-94-007-2706-9_3, © Springer Science+Business Media B.V. 2012

3.1 The Intellectual Virtues and Vices

Although prudence is but one example or form of wisdom, i.e. practical wisdom, traditionally the ancient Greek and Roman philosophers recognized prudence as *the* intellectual virtue among the four cardinal virtues. Some of these philosophers even considered it the first among equals of the cardinal virtues. As a virtue, prudence equips an epistemic agent to make good quality judgments and decisions about the validity of arguments and about the soundness of epistemic claims concerning reality and the world. And, in a certain sense, it permits an agent to foresee the results of certain actions, prior to their occurrence. Prudence derives from the Greek word *phronesis*, which means to have understanding or to be wise (Carson, 2006). The Romans translated the Greek word as *prudentia*, from which we derive our current English word prudence (Comte-Sponville, 2001).

In the western tradition, Aristotle identifies prudence in terms of contingent truth and making practical decisions or judgments in contrast to *episteme*, which concerns theoretical knowledge and necessary truth. Prudence, according to Aristotle, represents an ability to deliberate well and is critical for human flourishing or *eudaimonia*. Aristotle, as William Gerhard articulates the Greek philosopher's notion of the virtue, identifies prudence as "a virtue of the practical intellect which has as its particular function to make an exact determination of what is to be done in a particular situation" (1945, p. 438). Aristotle definition of *phronesis* or prudence involves "knowing the right thing to do in a particular circumstance through understanding the circumstance rightly, knowing what matters, and effective means-end reasoning to bring about what matters" (Dekkers and Gordijn, 2007, p. 231).

Prudence, or being able to discern an appropriate or even the best course of action, then, according to Aristotle's definition, is composed of three steps (Bryan, 2006). The first is the prudent agent's capacity to determine accurately the conditions and context surrounding a set of facts that requires or demands a judgment or decision. The next step involves the agent's ability to distinguish significant or germane facts from those that are insignificant or superfluous vis-à-vis a prudent decision, in order to comprehend the intelligibility or meaning of facts. The last step is the capability to reason in terms of means-ends analysis, in order to negotiate the best possible plan of action for carrying out a prudent decision effectively and efficiently.[1] Finally, prudence is the chief virtue that governs the other three cardinal virtues (Comte-Sponville, 2001). Without prudence, "justice endorses cruelty, temperance promotes fanaticism, courage invites foolhardiness, faith fosters intolerance, hope lacks guidance, and love courts disaster" (Bryan, 2006, p. 4).[2]

[1] Anoré Comte-Sponville offers a contemporary definition of prudence based on Aristotle's definition but which only captures two of the three steps, "prudence is the disposition that makes it possible to deliberate correctly on what is good or bad for man (not in itself but in the world as it is, and not in general but in specific situations) and through such deliberations to act appropriately" (2001, p. 32).

[2] According to Peter Geach (1977), the prudent agent develops a strategy for right and effective action through wisely following ethical rules or precepts.

Thomas (1974a) follows Aristotle's lead in explicating prudence as an ability to deliberate well. This deliberation is a cognitive activity or faculty of reason that has consequences for an agent's moral character. As such, prudence is an intellectual virtue, in that it pertains to right reason, but it also has an impact upon an agent's moral life, in that it allows the agent to choose the best ethical course of action. Thomas defines the virtue, then, as "right reason applied to human conduct" (1974a, p. 27). Accordingly, prudence is the judicious application of theoretical or universal knowledge to particular or concrete situations. Like Aristotle before him, Thomas identifies three steps in prudent action. The first he calls "taking counsel" or deliberation and pertains to a discovery process in which the prudent agent determines the facts and ends of a concrete situation. The next stage involves the formulation of a judgment concerning the discovered facts and ends. Up to this point, Thomas acknowledges that theoretical reason prevails in the process. However, with the final stage practical reason enters as the prudent agent commands "execution [of] what has been thought out and decided on" (Thomas, 1974a, pp. 27–28). Thus, prudence is a function not only of the rational faculty but also of the volitional faculty. Finally, like Aristotle, Thomas distinguishes prudence from other intellectual virtues, e.g. *sapientia* or *scientia*, which pertain to universal or necessary and not contingent truth. However, prudence also differs from the ethical virtues, e.g. justice or temperance, which arise from the soul's affective not cognitive faculties or powers, and pertain to the good and not to the true.

Contemporary virtue ethicists, in reviving the virtue of prudence, preserve the Aristotlelian-Thomistic notion, particularly with respect to practical deliberation. For example, Hurka defines prudence "as the ability to select effective means to good ends" (2001, p. 107). Some contemporary ethicists, however, introduce a postmodern twist in the revival. Their notion of prudence expands the virtue beyond the traditional realm of the intellectual and even the moral to include the performative, especially with respect to social practices. To understand prudence, one must turn to exemplars of the prudent. "Indeed," notes Robert Hariman, "rather than see the wise person as a specific case of prudence—although that certainly is a sensible definition—we might consider how the prudent decision maker is so because he or she has become a linking mechanism for joining rules and cases, universal precepts and particular situations. And by examining how prudence is articulated through such a person," continues Hariman, "we can not only recognize how it is nested into the idiosyncratic nooks and crannies of the individual personality but also discern the outline of its more general, personal operations in the flow of gestures, expressions, movements, persons, and events that make up the world of action" (2003, p. 7). As such, prudence is a virtue that mediates the intellectual and ethical on a larger stage for both intellectual and ethical ends, but not outstripping the exemplary prudent agent. Moreover, as McKinnon (1999) argues, prudence leads not simply to the ethical good but also to the truly good. Thus, prudence qua virtue is essential for guiding the intellectual and ethical direction of actions of a social group and its members to achieve true and good moral ends.

Besides being the chief governing virtue, prudence is also a central virtue, around which coalescence closely associated or cognate intellectual virtues. These virtues include curiosity, open-mindedness, creativity, insightfulness, thoughtfulness, understanding or knowledgeableness, rationality, truthfulness, wisdom, reasonableness, and judiciousness (Peterson and Seligman, 2004; Roberts and Wood, 2007; Zagzebski, 1996). The intellectual virtue of curiosity involves an agent's genuine interest and inquisitiveness about the world and a desire to ask questions about it. Closely associated with curiosity is open-mindedness, in which the epistemic agent exhibits a cognitive willingness and expansiveness to investigate the world and to entertain questions about it. These intellectual virtues often lead to love of learning over the agent's lifetime. Creativity is the next intellectual virtue associated with prudence. And, it pertains to the imaginativeness or ingenuity to generate or invent original facts. This intellectual virtue often depends on another intellectual virtue, insightfulness. The insightful epistemic agent is capable of evaluating pertinent data and making the necessary epistemic associations among them. This virtue involves epistemic skills in which the epistemically virtuous agent grasps the intelligibility or meaning of events and phenomena (Lonergan, 1978). Insightfulness is important for predicting future events (foresight) or for gaining understanding of past events (hindsight).

The next intellectual cognate virtue of prudence is thoughtfulness. This virtue is essential for an agent's introspective and reflective cognitive activities. Introspection involves the ability or capacity to think about what is occurring under or behind the surface of a phenomenon, while reflection pertains to contemplation of a phenomenon in a deeply meditative manner. The next cognate virtue is understanding, which is an agent's ability or capacity to apprehend and comprehend the intelligibility of an event or a phenomenon and to judge whether that intelligibility is the case. Apprehension involves the proper functioning of the sensory faculties, while comprehension the cognitive faculties. An important goal of the epistemic agent's understanding is knowledgeableness, which is the ability or capacity to perceive and conceive the intelligibility of events and phenomena as facts. The next intellectual virtue is rationality, which is the ability to reason coherently and logically. The rational epistemic agent is skilled at inferring what is valid from an argument. Truthfulness is also a cognate intellectual virtue to prudence. The truthful epistemic agent is able not only to assent to whether one's knowledge of something corresponds or coheres to the thing itself but also to communicate truth instead of falsity.

The final, but certainly not the least, virtue associated with prudence is wisdom. Wisdom is the ability or capacity to grasp not only the intelligibility of events or phenomena but also their meaning or significance. Traditionally, as noted earlier for Aristotle in particular, wisdom is either theoretical or practical. Theoretical wisdom is concerned with the fundamental or universal principles or laws that govern the world. The theoretically wise person is someone who grasps the eternal or necessary truths or intelligibility, since the goal of theoretical wisdom is nothing less than truth itself. Practical wisdom, however, concerns the specific meaning or significance of a person's actions. Thus, the practically wise individual is someone who grasps the meaning not just of the eternal but also of the here and now and knows how to act in a common-sense manner.[3] According to William Prior, "just as theoretical wisdom included both a knowledge of first principles and an ability to demonstrate, practical wisdom includes both a correct desire for the ultimate end of conduct and an ability to calculate the proper means to that end" (1991, p. 179). Two important intellectual virtues associated with wisdom are reasonableness and judiciousness. Reasonableness is the ability or capacity to think or reason in a sensible and sober way, while judiciousness is the ability or capacity to judge or decide in a sound and discrete way.

Imprudence, as a vice, involves rashness and incautiousness in coming to judgments and decisions. The imprudent person is one who makes poor judgments about the validity of arguments and the soundness of epistemic claims about the world. According to Bernard Gert, that person permits "present conditions to lead her to act unreasonably by neglecting the long-term consequences of her action for herself and those for whom she is concerned when these may result in significant evils or failure to achieve significant goods" (2005, p. 295). Importantly, imprudent people may be unable to foresee the results of their actions or, if they do foresee the results, are incapable of doing anything about them. The first is an epistemic issue, while the latter an ethical one. Irrespective of the underlying pathology of this vice, imprudent people may lead a life that appears to be flourishing but that flourishing, if resembling the real, is only transitory and non-substantive. The imprudent life is one in which a person makes decisions that lead to an unhappy and eventually unfulfilled end.

Whereas prudence is the chief virtue around which different cognate intellectual virtues coalescence, imprudence is the chief vice around which various cognate intellectual vices coalescence. These cognate vices include indifference and close-mindedness or narrow-mindedness, unoriginality, stupidity, thoughtlessness, ignorance, irrationality, untruthfulness, foolishness, unreasonableness, and injudiciousness. Indifference, as an intellectual vice cognate to imprudence, involves incapacity to ask meaningful questions about the world. Associated with this vice is close-mindedness or narrow-mindedness in which the imprudent epistemic agent allows intellectual biases and prejudices to limit attempts to ask questions about the world. Importantly, this vice leads to fear of learning. The next cognate intellectual vice is unoriginality, which refers to an inability to generate or create novel data and facts about the world. The end-result of these intellectual vices is another vice, stupidity, in which the epistemic agent is unable to grasp insightfully the intelligibility of evidence and observations. Often, this vice leads to an imprudent person whose behavior is foolish or even childish.

The next cognate vice of the imprudent person is thoughtlessness, in which the epistemic agent lacks the introspective and reflective cognitive capacities or fails to use them effectively. The thoughtless person often does not take the time to

[3] Prudence, as *the* intellectual virtue, differs from traditional Aristotelian practical wisdom in that the former involves both the universal nature of knowledge and the meaning or significance of that knowledge for the individual, as well as its community, while the latter pertains only to the meaning or significance of knowledge qua solution of a practical problem.

think deeply or meditatively about something but rather acts quickly and without a second thought. The outcome of such thoughtlessness generally is profound ignorance, which is an intellectual vice that plagues the imprudent person. Moreover, the imprudent agent may also exhibit the intellectual vice of irrationality, in which, rather than reasoning logically, the agent is prone to logical fallacies and false inferences. Without the ability to reason logically, the imprudent epistemic agent engages in the intellectual vice of untruthfulness. Untruthfulness yields epistemic falsehoods or lies. Overall, the imprudent agent is foolish or unwise in that the agent fails to recognize or even to acknowledge the true meaning and significance of phenomena; rather, the meaningless and insignificant are mistaken as the meaningful and significant. The foolish or unwise epistemic agent is both unreasonable, i.e. unable to think in a sensible or sober manner, and injudicious, i.e. incapable of making discrete decisions.

Prudence (imprudence) is an important virtue (vice) in medical practice. For example, Pellegrino and Thomasma (1981, 1993) claim prudence or *phronesis* is critical for achieving the goals of medicine. "Clinical judgment," according to these authors, "is essentially an exercise of prudence ... in a complex situation fraught with uncertainties. It is here," Pelegrino and Thomasma observe, "that the clinician must discern what means are most appropriate to the ends, how to balance the benefits and harms in clinical interventions, and how to put the moral and the technical issues in a proper relationship with each other" (1993, p. 86). Unfortunately, the uncertainties of medical practice often overwhelm the imprudent physician, who is thereby unable to determine the requisite means for treating patients, to square the risks and benefits of treatment plans, or to align technical and ethical demands. In addition, prudence is required, according to Pelegrino and Thomasma, to balance both the affective and technological dimensions of medical practice successfully and effectively. The prudent physician can navigate these dimensions of both the heart and the head. The imprudent physician, however, cannot navigate them, often with deleterious consequences for the patient.

For Pelegrino and Thomasma, the prudent physician includes—in a wise and caring manner—the patient's physical and existential needs in a treatment plan, which generally meets these needs. The imprudent physician is unable to meet them, often to the patient's detriment. Moreover, the prudent physician's clinical judgments include a decision tree, according to Pelegrino and Thomasma, with algorithmic processes, which include not only clinical observations and laboratory results but also the existential concerns of patients and their personal contexts. The imprudent physician often dispenses with such algorithmic processes, substituting the most expedient means by which to treat patients. Importantly, the prudent physician not only achieves the goals of medicine, which is to meet the clinical needs of patients sanctioned by society; but, in meeting those needs through prudence, that physician also practices a type of medicine that results in a fulfilled clinical practice for the physician. The imprudent physician generally does not achieve those goals, which often results in not meeting the medical needs of patients sanctioned by society and in an unfulfilled practice from which the physician generally experiences burnout.

Pelegrino and Thomasma's role for prudence or *phronesis* in medical practice has found adherents in the literature. For example, Ineke Widdershoven-Heerding (1987) agrees with Pelegrino and Thomasma in that reasoning in medicine requires inclusion not only the technical but also the practical—especially in terms of the patient's existential needs. Both of these aspects of reasoning converge to form neither just a technical medicine nor simply a practical medicine but rather a tertium quid. In other words, medicine is neither science nor art only; rather, it must negotiate between these two attributes of medical practice, especially in terms of prudence, to fashion a discipline that achieves the goals of medicine with respect to aiding the suffering patient. As Widdershoven-Heerding concludes, in contrast to other professions, "an inspection of medical reasoning, of clinical reasoning, indeed serves to set off medicine as a special discipline" (1987, p. 184). As a profession, what makes medicine special or unique, according to Widdershoven-Heerding, is the application of theoretical medical knowledge to the individual clinical case.

The above role for prudence or *phronesis* in medicine, however, does have its critics. For example, Bjørn Hofman (2002) argues that Hippocratic *techne* provides a better basis for undergirding medical practice, especially the moral and ethical challenges that face contemporary medicine, than Aristotelian *phronesis*. In fact, Hofman contends that Pelegrino and Thomasma's notion of *phronesis* corresponds more closely to the notion of *techne*, since "medicine, as a Hippocratic *téchnê*, is a practical activity aimed at healing the particular patient" (2002, p. 146). The notion of *phronesis*, especially as Aristotle conceived it, is simply restricted to the rational to support the goals of medical practice Pelegrino and Thomasma seek to achieve with the virtue. Also, *phronesis* is too ambiguous and thereby applicable to any profession, which prohibits its specific application to medicine as demarcating it from other professions. *Techne*, according to Hofmann, avoids these problems. Thus, *techne*—not *phronesis*—serves to integrate both the art and science of medicine at a fundamental level. For Hofman, "the truly virtuous physician is one who acts according to *téchnê*, that is, as a *technites*" (2002, p. 148). In other words, Hippocratic *techne* guides both the theoretical and practical activities of medical practice, which *phronesis* cannot do.

Besides prudence, its cognate virtues are also important attributes of the virtuous physician. For example, curiosity or open-mindedness is essential for gathering pertinent and salient clinical information in order to make an accurate diagnosis and to develop an effective treatment plan. The curious physician is open to exploring important leads that surface during the medical interview or physical exam. By pursuing these leads, that physician aids the patient's recovery. The close-minded physician, however, ignores or is unable to recognize such critical leads and to pursue them adequately, generally following minimal standard protocols too closely. Part of the closed-minded physician's inability to stretch standard protocols is fear of recrimination if clinical problems arise. Hence, the closed-minded physician often engages in defensive medicine.

Another important cognate virtue for a robust medical practice is insightfulness. The insightful physician is able to grasp the intelligibility of the clinical data to arrive at the proper diagnosis and therapeutic protocol. The stupid physician is unable to grasp the clinical data's intelligibility, often leading to additional pain and suffering for the patient. One last cognate virtue suffices to assist the reader in appreciating the importance of these virtues in medicine. Wisdom is central to any medical practice. Not only must the intellectually virtuous physician be knowledgeable and well-trained, but such a physician must be able to use that knowledge and training reasonably and judiciously. The foolish physician, on the other hand, fails to acquire such knowledge and training or uses them unreasonably and injudiciously—often causing the patient irreparable harm or injury.

3.2 The Ethical Virtues and Vices

In this section, I introduce and discuss the three traditional ethical virtues—courage, temperance, and justice—and the corresponding vices—cowardice, intemperance, and injustice. Whereas the intellectual virtues and vices pertain to the life of the mind, the ethical virtues pertain to the life of the will or the volitional life. Throughout the centuries, commentators on ethics maintained that these virtues are important for living a morally upright and fulfilled life, while the vices lead to a morally depraved and unfulfilled life. According to contemporary virtue theorists and ethicists, the moral life then depends on and is an outcome of these virtues or good traits and the immoral life results from the corresponding ethical vices or bad traits. "Such good and bad traits," claims Adams, "are a major factor in how well your life (and not just your day) is going morally. Indeed, they constitute what is called *moral character*, and are commonly seen as determining the extent to which one is *a morally good person*" (2006, p. 3). Finally, just as for prudence, cognate ethical virtues and vices are associated with each of these traditional virtues and vices. I not only discuss and elaborate upon these virtues and vices, in this section, but I also explore the application of the traditional ethical virtues and vices to medical practice in terms of the virtuous and unvirtuous physician.

3.2.1 Courage and Cowardice

Ancient commentators attributed courage to heroes, especially military heroes. For example, in discussing the warrior's role in the Trojan War Homer—in the epic poem *The Iliad*—identifies the virtue as essential for motivating warriors to engage, often times, valiantly in combat. Given these conditions, courage is a masculine virtue that enables a man to overcome his fear of dying in battle. To these ends, then, courage is a laudable virtue that allows the military hero to act bravely in the face of death. Plato addresses this notion of courage in the dialogue, *Laches*. To Socrates' query concerning the nature of courage, the military general Laches responds that courage empowers a soldier not to flee because of his fear of danger or harm but rather to stand his ground and fight. Courage is a sort of wise endurance of the soul. Socrates challenges Laches' notion of courage by asking him who is braver, the soldier who knows that he has the upper hand in a battle or his enemy. Laches chooses the enemy and Socrates reminds him that he had associated courage with wise deliberation in the face of dangerous odds. Thus, the enemy is not so much courageous as foolish. Plato leaves the reader with the conundrum of how is the courageous agent to deliberate wisely in the face of impending danger. In other words, how is an agent to determine whether a particular danger is truly worthy of courageous action? Plato's student, Aristotle, takes up this perplexing question in his exposition on courage.

In the *Nicomachean ethics*, Aristotle acknowledges as courage a noble trait that allows a person to be brave and to act valiantly. To resolve the Platonic conundrum, he defines courage in terms of his standard definition of virtue as a mean between two extremes. According to Aristotle, "courage is a mean with respect to things that inspire confidence or fear, in [particular] circumstances . . . and it chooses or endures things because it is noble to do so, or because it is base not to do so" (1998, p. 67). Thus, an apparently courageous agent who fears nothing and is overly confident in the face of real danger is simply foolish, since the agent is incapable of reasonably deliberating over or evaluating the nature of how dangerous a person, thing, or situation truly is. On the other hand, the agent who is overly fearful or has little or no confidence in the face of that which is—or even in the face of that which is not—dangerous is a coward. "To be truly courageous," according to Douglas Walton's articulation of Aristotle's notion of courage, "an act must be reasoned out and executed thoughtfully and carefully by its agent" (1986, p. 59). Walton identifies three important components of Aristotle's notion of courage: fearing what is rightly fearful, exhibiting the appropriate level of fear, and fearing something for the best end or outcome.

Utilizing Aristotle's notion of courage, Thomas Aquinas (1966a) defines courage in terms of removing impediments or obstacles to a right course of action. According to Thomas, two components constitute courage. The first is removing a volitional impediment through proper reasoning, i.e. "firmness of mind," which allows a courageous agent to distinguish real dangers that require courageous acts (Thomas, 1966a, p. 9). The second component of courage is the removal of physical or bodily obstacles that impede the agent from performing a courageous act. In other words, a courageous person must have "fortitude of body" to carry out an intended brave feat. For Aquinas then, courage permits a virtuous person to follow his or her will to perform the right action in the face of danger, as determined by proper reasoning. Finally, Aquinas defends the cardinal nature of courage because courage is the principle virtue by which a person remains steadfast in a course of action. Moreover, he relates courage to the other cardinal virtues in that courage removes the impediments to prudent, just, and temperate actions and thereby safeguards those actions.

Most modern definitions of courage certainly utilize an Aristotelian-Thomistic definition of courage, especially in terms of practical reasoning in the face of danger. For example, James Wallace defines courage as "the ability to weigh up correctly the pros and cons of various alternative courses of action when some courses involve danger and the ability to face dangers" (1978, p. 77).[4] Walton takes a similar approach to define courage "as an excellence of practical action both through skills

[4] Wallace predicates this definition upon courage as the absence of cowardice.

of execution and through deliberation that enter into human action" (1986, p. 9). For him, courage is a supererogatory outcome of practical reasoning in the face of danger that results in good not just for the individual agent but also for the community. Courage for many modern commentators involves endurance, as Peter Geach articulates the virtue, "courage is the virtue of the end: what makes a man endure to the end and in the extremity of evil" (1977, p. 150). Finally, as Comte-Sponville notes, "The ancients saw courage as a mark of virility (the word *andreia*, which means courage in Greek, and the word *virtus* in Latin comes from *anêr* and *vir*, respectively, root words that denote man, not in a general sense but man as opposed to woman) and many people would still agree today" (2001, p. 49). One contemporary pundit who does not agree is Richard White (2008), who argues for a notion of courage that transcends gender roles and that is not based on military aggression but on passivism.

Earl Shelp provides a rather comprehensive definition for courage that incorporates many of the essential features of the virtue. According to Shelp, "courage is the disposition to voluntarily act, perhaps fearfully, in a dangerous circumstance, where the relevant risks are reasonably appraised, in an effort to obtain or preserve some perceived good for one self or others recognizing the desired perceived good may not be realized" (1984, p. 354). From this definition, he identifies several important key features of courage. The first is the volitional nature of courage in the face of genuine fear over a dangerous and risky situation that impedes one from acting. The courageous agent is one who acts voluntarily, with an understanding of the risk. The next feature is risk, which "refers to the vulnerability as the condition of the agent relevant to courage. Without vulnerability, without risk," claims Shelp, "courage has no place" (1984, p. 355). The third key feature of courage is a worthy end, in which the courageous agent is willing to risk one good, such as the agent's life, for another equally or greater good, such as the life of a child or the lives of family or community members. The final feature is uncertainty, which demarcates courage from confidence. The courageous agent has no guarantee that the outcome's value is commensurate with the risk, but still that agent acts as if it was commensurate. As Shelp concludes, "courage is a moral virtue the expression of which benefits the agent or others and expresses one's solidarity with and concern for the goods of personal and social existence" (1984, p. 356).

The virtues cognate to courage include fortitude, bravery, daringness, fearlessness, valor, and endurance (Peterson and Seligman, 2004). These virtues are simply more than synonyms for courage but reflect important characteristics of a virtuous agent who acts in the face of danger. Traditionally, fortitude represents physical or moral strength or potency.[5] It affords the virtuous agent power to overcome the fear associated with either a physically or morally dangerous thing or situation. Next, bravery pertains to the quality of boldness in the face of danger. It allows the brave agent to confront the fear associated with danger and to prevail. Associated with this

[5] Utilizing Thomas Aquinas, Pellegrino and Thomasma (1993) associate courage with physical strength or endurance and fortitude with moral courage.

virtue is daringness, which represents heroicness in the face of danger. Daringness pertains to a willingness to take on risk for greater good. Fearlessness is not an unmeasured action in the face of danger, but rather it is a measured one that takes into consideration the risk of harm to the virtuous agent relative to the potential good. The cognate virtue of valor refers to the nobleness of character, such that the valorous agent appears larger than life or superhuman in terms of facing danger. Finally, endurance is a cognate virtue to courage that endows an agent to persist in the face of danger. It also refers to an agent's patience to deliberate about the most effective strategy to overcome danger or to the agent's capacity to wait until an opportune time to achieve a goal.

Traditionally, ethicists identify cowardice as the main vice contra the virtue of courage. Upon reflection, however, two vices are associated with fear of dangerous events or persons. The first is an inability to deliberate properly upon the danger in terms of the risk involved in acting. In forging ahead without deliberating in terms of the most appropriate strategy for obtaining the best possible outcome, the apparently courageous agent acts in a rash or foolish manner. For example, a person who blindly jumps without second thought into a middle of gunfire to rescue someone would appear rash to the person who deliberates first as to the best course of action in order to enact the rescue. The second vice is an inability or a refusal to act in the face of danger even though deliberation over the ratio of risk versus benefit represents a courageous act. According to most, this represents the vice of cowardice. The coward is someone incapacitated or overwhelmed by the fear or risk associated with danger. "A coward," according to Hurka, "cares about his comfort or safety, which is good in itself or as a means, but he cares more about his comfort or safety than about some greater good he could achieve by risking them" (2001, p. 85).

Wallace (1978) identifies four necessary conditions of the cowardly act. The first involves an agent who performs act B instead of act A. In other words, the agent has an option to do a particular act but chooses to do another instead. The next condition is that the cowardly person believes that he or she has good reason to do act A but instead chooses to do act B for another reason not related to the reason for doing act A. The third condition involves fear and danger. The reason animating the cowardly agent's performance of act B is that the agent perceives some danger or harm in performing act A and consequently does act B because of the fear of that danger or harm. The final condition revolves around the risk to benefit ratio for performing an act. The cowardly agent believes that the ratio for doing act A is too high compared to that for act B.[6] In other words, the agent is not willing or is afraid to sustain the perceived risk of harm associated with doing act A and opts to do act B. For Wallace, then, the cowardly act represents a conflict between reason and fear. The coward is someone who is unable to deliberate adequately about the potential risk of an action and allows fear to overwhelm the decision making process.

---

[6] Wallace makes a distinction between two types of cowardly acts based on a modification of the fourth condition.

The cognate vices of cowardice are weakness, pusillanimity, timidity, fearfulness, faintheartedness, and frailty. As for the virtues associated with courage, these vices are simply more than synonyms for cowardice but reflect important features of an agent who fails to act or who acts unvirtuously in the face of danger. As a cognate vice, weakness produces a powerless agent with little or no physical or moral strength or potency. In other words, the weak person lacks the strength to overcome the fear associated with a particular danger and is thereby ineffectual to act courageously or even to act at all. The next cognate vice, pusillanimity, refers to the agent's quality of being craven or spineless in the face of danger. It prevents a person from confronting fear associated with danger and acting appropriately. Timidity is the next cognate vice. The timid agent is a person who exhibits apprehensiveness or lacks boldness in the face of danger. The outcome is a person who is unwilling to take risks for a greater good and who scurries from danger. Related to timidity is the next cognate vice, fearfulness. The fearful agent is a person who is simply scared when danger appears and is so terrified as to act in an unvirtuous manner. Next, faintheartedness is a vice often associated with the fearful agent. The fainthearted agent is someone whose actions are ignoble and anti-heroic. The last cognate vice is frailty. The frail agent is fragile or feeble in terms of enduring in the face of danger and consequently yields to fear.

According to Shelp (1984), the virtue of courage is important in medical practice, especially in terms of the patient-physician relationship. Courage, for Shelp, enables the physician to be a "sustaining presence" in the midst of the patient's pain and suffering, as well as the clinically unknown and mysterious vis-à-vis death and dying. "For physicians," stresses Shelp, "it is assisting patients in not letting fear overtake them so that the opportunities present in sickness and dying are not lost to them and others" (1984, p. 359). In other words, courage allows the virtuous physician to "encourage" the patient when plagued by illness and dying; otherwise, both illness and dying can overwhelm the patient and cut off options for healing—even when the physician is unable to do anything in terms of prescribing drugs or performing surgical procedures. For Shelp, courage is a critical virtue for a profession that defines itself in terms of "care and concern" for the patient's welfare. The virtuous physician then assists the patient to live courageously in the face of illness and dying. "The patient-physician relationship," concludes Shelp, "can be a context for courage and encouragement within which the nature of the human condition is learned and the capacities necessary for its negotiations are developed" (1984, p. 359). The cowardly physician, on the other hand, flees from the patient's medical challenges. Rather than stand with the patient to face the full impact of the illness, the cowardly physician often abandons the patient.

Pellegrino and David Thomasma (1993) also acknowledge the importance of courage or fortitude, their preferred term, in the practice of medicine, especially within medicine's contemporary corporate setting.[7] They fear that the patient's

[7] As noted earlier, Pellegrino and Thomasma define fortitude as moral courage.

"interests are at risk of being supplanted by interests of the doctor's own self-interest, those of the hospital, the managed health care system, or society in general" (1993, p. 113). Although being a team player is important in modern medicine to meet the patient's physical and existential needs, it can also lead to cowardice on the part of physicians when challenged to curb costs of healthcare or to fill quotas in terms of patient contact. Charles Bryan (2006) identifies this unvirtuous or cowardly behavior of physicians with the social phenomenon of "groupthink" in which physicians allow corporate or at times personal biases or prejudices instead of objective criteria to determine clinical practice. As Pellegrino and Thomasma conclude, "medical fortitude [is] the virtue that inspires confidence that physicians will resist the temptation to diminish the patient's good through their own fears or through social and bureaucratic pressure, and that they will use their time and training resourcefully to accomplish good in society" (1993, p. 114). Unfortunately, in today's medical climate ruled by a cooperate culture of maximizing revenue, cowardly physicians may acquiesce to the temptation to trump the patient's welfare with an institution's profitability.

The cognate virtues of courage and the cognate vices of cowardice also play a critical role in virtuous and unvirtuous medical practice, especially given the legal vulnerability of physicians. Unfortunately, fear looms large in the medical decisions and judgments of many physicians. Fearful physicians worry that one wrong step or action may result in a malpractice suit. Rather than boldly practice medicine that is good for the patient, fearful physicians lose heart at the possibility of a legal suit and practice defensive medicine. Such fearful or defensive medicine often cuts off a physician from attempting a daring procedure that might benefit or help the patient. The fainthearted physician cowers in the face of uncertainties that are common in medicine. Rather than bravery when faced with such uncertainties, the pusillanimous physician hides behind the complexities of medical technology to cover any eventuality. Often, such behavior not only adds to the cost of medicine for both the patient and society but it generally reveals little if any useful information for an accurate diagnosis and effective therapy.

3.2.2 Temperance and Intemperance

For the ancient Greeks, just as courage is associated with the heroic so temperance is associated with the tragic (North, 1966; White, 2008; Young, 1988). Indeed, the heroic or the courageous and the temperate are often opposed to one another. For the Greek tragedian, temperance or *sophrosyne* represents an ideal, which involves not only self-knowledge—knowing one's strength and weakness—but also self-restraint—being capable of limiting or controlling oneself. In her monumental work on the genealogy of *sophrosyne*, Helen North argues, "observance of limits is the essence of Aceschylean sophrosyne" (1966, p. 35). For example, in Aeschylus' writings the gods—particularly Zeus—punish the person who exhibits *hybris* or arrogance. A person's desires must be in balance, such that they are neither deficient nor excessive. Thus, in the *Persians*, Aeschylus has Darius condemning his son, who incurred the gods' wrath, because of the son's *hybris* over the victory at the Battle of Salamis. However, temperance is not only a personal virtue but it is also an important virtue of the polis. For example, in the *Seven against Thebes*, Thebes "is saved because its cause is just and its champions are as *sôphrones* as the attackers are hybristic" (North, 1966, p. 40). The polis and temperance are inseparable in that the polis requires moral order and restraint of its citizens for its continued survival and success.

Plato inherits this literary tradition and expands upon it philosophically (North, 1966; White, 2008). In the Platonic dialogue *Charmides*, a dialogue proleptic to the *Republic* and later dialogues, Socrates engages his interlocutors on a journey to define temperance by rejecting the common definitions of minding one's own business, modesty, doing one's own work, and knowing oneself (Kahn, 1988; North, 1966). Although Plato arrives at no single definition, he paves the way for incorporating *sophrosyne* into understanding the nature of humanity, the polis, and the cosmos—from the microcosm of the human soul to the macrocosm of the universe. For example, in the *Symposium* Plato contrasts Socrates, who is able to control his desires and is thereby admired, with Alcibiades, who is unable to control his desires and is therefore pitied. According to White, Socrates' "temperance involved the achievement of personal sovereignty and the harmonious arrangement of all the different aspects of his life" (2008, p. 48). In the *Republic*, Plato not only contrasts Socrates to the tyrant who lacks any capacity for self-control and personal restraint; but, he also locates temperance with the appetitive part of the soul in terms of human nature and with the lower social classes with respect to the nature of the polis.[8] As for the macrocosm, in the *Timaeus* Plato's notion of *sophrosyne* is critical for the order and harmony of the universe. Finally, the two main features of Platonic *sophrosyne* are, according to North, "the control of appetite by reason and the harmonious agreement within the soul [whether of the individual human, polis, or universe] that this control should be exercised" (1966, p. 176).

Although Aristotle rejects the notion that only the four cardinal virtues are adequate to account for virtuous behavior, he does explicate a notion of *sophrosyne* even though it is different from his predecessors (North, 1966; White, 2008; Young, 1988). Using the doctrine of the mean, he defines temperance—in the *Nicomachean ethics*—as a mean between the deficiency (abstinence) and excess (indulgence) of pleasure (and pain).[9] Pleasure for Aristotle refers specifically to the bodily activities associated with touch, especially eating and drinking, as well as sexual intercourse. "Temperance," as Charles Young interprets Aristotle, "is a virtue that regulates

---

[8] North (1966, p. 173) cautions that *sophrosyne* is not critical simply for the functioning of the appetitive soul but also for the other two parts of the soul. Given the Platonic unity of the soul, temperance is an important means to that unity. "Temperance," as White articulates the relationship between the virtue and the soul, "involves the proper harmony, balance, and order of the soul" (2008, p. 50).

[9] Charles Young (1988) maintains that Aristotle's notion of *sophrosyne* is not the result of a mean but of regulating a person's animality.

appetites occasioned by physical needs" (1988, p. 532). The *sophron* or temperate person engages in bodily pleasure in moderation. Although *sophrosyne* refers to a person's irrational parts, the virtue itself is subject to a person's reason. In other words, the desire for bodily pleasure must conform to good sense and not to extreme. According to Aristotle, "the appetitive element in a temperate man should harmonize with the rational principle ... and the temperate man craves for the things he ought, as he ought, and when he ought" (1998, p. 78). Thus, the *sophron* person submits bodily pleasure to rational scrutiny in order to live a healthful and eudaimonic life.

As a concept, *sophrosyne* presented a knotty challenge to the Romans in terms of finding an adequate translation from Greek into Latin. The Romans had a different set of values from the Greeks, "sophrosyne was so intensely Hellenistic that in its totality it always remained as exotic in Rome" (North, 1966, p. 258). Cicero was not the first to attempt to translate the term but his translation of it as *temperantia* still carries contemporary resonance. In *De inventione*, Cicero defines temperance as "a firm and well-considered control exercised by the reason over lust and other improper impulses of the mind" (1949, p. 331). He then identifies three components of *temperantia*: *continentia*, *clementia*, and *modestia*. First, *continentia* or continence involves the intentional or wise regulation of desires through right reason or sound judgment. Next, *clementia* or clemency is a type of gentleness that modulates a quick temper towards violence or hatred of others, especially those considered to be of lesser social status. Finally, *modestia* or modesty is the outcome of maintaining one's honor in the face of temptation to act shamefully. Importantly for Cicero, a person should seek temperance and its various components not so much for their utility as for their inherent value.

Early Christianity appropriated temperance, especially in terms of the seven deadly sins (Bryan, 2006; Geach, 1977; North, 1966; White, 2008). In the next to last chapter in the letter to the Galatians, for example, Paul lists a number of sins that represent a violation of temperance, particularly sexual promiscuity and drunkenness, and contrasts these with the gifts of the Spirit, especially self-control. "For many of the early Christians," according to White, "the goal was to achieve supremacy over this world by renouncing physical pleasure, especially sexual pleasure, and becoming indifferent to physical pain" (2008, p. 54). The early Church Fathers also struggled with earthly pleasure in terms of achieving spiritual purity. For example, Tertullian renounces all conjugal rights with his wife to pursue the path to the divine.[10] Later Church Fathers follow in the footsteps of the earlier Fathers, especially in terms of chastity. For example, Augustine focuses upon the physical elements of temperance in terms of eating and drinking, as well as sex; however, he distinguishes his notion of the virtue by embedding it in his theology of love. "Its particular function," as North articulates Augustine's notion of temperance, "lies in

[10] Of course, the early Christian church sharply criticized and condemned Tertullian's approach to marriage. In his writings, Paul exhorts Christians not to deny their conjugal rights, unless for a brief time. Paul advises followers to be content with their physical situation, whether in want or in need. This advice parallels Socrates' life in which he was not fazed by its vicissitudes.

restraining and quieting the passions by which we lust after those things that separate us from the laws of God and from the enjoyment of his goodness" (1966, p. 373). Finally, Aquinas endorses the tradition of focusing on the physical dimension of temperance and the need to restrain the desire to overindulge in earthly pleasures, especially in sexual desire and lust.

Given this tradition and especially the temperance movement of the early twentieth century, the contemporary notion of temperance is generally associated with the denial of physical or bodily pleasure, especially in terms of eating, drinking, and sex (one must also include drugs). For the modern person, temperance is an "ambivalent virtue today: Of course we know that it *should* be a virtue because there is something wrong with self-abandonment, but at the same time we sense that the meaning it has acquired (as self-control and even as self-denial) is quite impoverished and says nothing about our place in the world around us" (White, 2008, p. 58). In an attempt to restore temperance as a virtue for contemporary culture, Mark Carr (2001) divides traditional approaches to the virtue into broad and narrow conceptions. The broad conception includes the Greek *sophrosyne* as mental health, Aristotle's notion of the mean as moderation, the Latin *temperantia* as proper mixture or balance, and the Greek *prepon* and the Latin *decorum* as social propriety and decorum, respectively. The narrow conception of temperance includes self-restraint or suppression of desires and self-control or management of desires.

Based on the above broad and narrow conceptions, Carr proposes a normative notion of temperance for contemporary culture that attempts to combine both the physical and psychological dimensions of the moral life. He defines temperance normatively as "self-management of both sensate and intellectual desires" (2001, p. 69). Importantly, according to Carr, temperance as self-management reflects a specific goal: "the settled state where no conflict arises between reason and passion in the human experience of emotion" (2001, p. 69). The virtue then is critical for expressing one's "current self" in the midst of situations that without the virtue of temperance one would elicit bad or undesirable passions or reactions. Besides negating or modulating adverse passions, the temperate person also relies on advantageous passions to deliberate about the proper moral course of action. Quoting Nicholas Dent, Carr invokes unity of personhood in which "desires carry, now, our conscious and deliberately sought scheme of values for living and action" (2001, p. 70). Temperance, instead of being irrelevant for modern lifestyles, is important for living a robust virtuous life.[11]

The cognate virtues for temperance include moderation, discipline, self-control or self-restraint, patience, civility, and modesty (Peterson and Seligman, 2004). The first three virtues relate to one another in terms of an active promotion of limits or boundaries. Moderation, discipline, and self-control or self-restraint generally indicate an agent's ability to curb or check behavior. The virtue of moderation typically

[11] Comte-Sponville also articulates a similar notion for temperance in contemporary life, "the voluntary regulation of life force, a healthy affirmation of our power to exist... an affirmation especially of the power of the mind over irrational impulses of our affects or appetites" (2001, p. 43).

refers to an ability to moderate or balance excesses or extremes, such that the agent is not extreme in quantity—whether the quantity refers, e.g. to consuming a substance like alcohol or to expressing an emotion like anger. Discipline, as a virtue, reflects a capacity to regulate behavior through extensive training and education. A disciplined agent is able to rule or govern actions and passions as an outcome of intentional instruction. The virtue of self-control or self-restraint involves an ability to modulate one's behavior volitionally and rationally, i.e. it entails a person who is in command of himself or herself. The self-controlled or self-restrained agent is a person motivated to achieve a behavioral goal or end, such as resisting an overreaction to a minor nuisance. Finally, self-control or self-restraint involves restricting one's behavior or conduct to conform to particular standards or criteria.

The cognate virtues of patience, civility, and modesty, do not share the common element of limits or boundaries directly with the virtue of temperance, as do the other three associated virtues, but they do so only indirectly. In other words, the notion of limits or boundaries informs these virtues in terms of providing a foundation for their expression. For example, patience assumes a limit to an agent's behavior but does not involve an active controlling or restraining; rather, the patient agent is someone who is able to endure pain and hardship without over reacting. Indeed, such a person can tolerate another's limitations or extremes with aplomb. In like manner, the civil agent is one who is well-bred and behaves in a polite and respectful way towards others, even when others are rude or uncivil. Such a person is humane, gentle, caring, and kind, and treats others as the civil agent would like to be treated. Civility is certainly a major root that supports a society, a virtue that defines a society. Finally, modesty refers to the state of decorum and propriety in an agent's behavior. The modest agent is someone who exercises a sober evaluation of oneself so as not to be proud or arrogant but rather humble or unassuming.

Traditionally, intemperance is the vice counterpoised to the virtue of temperance. Just as temperance refers to the propitious management of desires such that a moral agent does not behave in the extreme, so intemperance entails the mismanagement of desires such that the agent behaves excessively and thereby immorally. In other words, intemperate people indulge themselves in either excessive pleasure or appetitive desires to their detriment and/or to those around them. "The intemperate person is like a slave," analogizes Comte-Sponville, "all the more subjugated in that his master—the monkey on his back—is with him wherever he goes. He is the prisoner," he goes on to explain, "of his body, of his desires or habits, of their strength or weakness" (2001, p. 39). In a narrow sense of the vice, intemperate people indulge in eating, drinking, and sexual activity, in excess or to the extreme. Such behavior often destroys these people and those around them.

The cognate vices of intemperance include immoderation, indiscipline, self-indulgent and self-excess, impatience, incivility, and immodesty. The cognate vice of immoderation pertains to disproportional behavior in terms of what is reasonable. In other words, the immoderate agent is unable to moderate or balance excesses or extremes, especially in terms of quantity. The undisciplined agent is incapable of regulating behavior because of lack of or improper training or education. The self-indulgent agent is someone who is unable to control or modulate volitionally desires and is unmotivated to achieve behavioral goals. The vice of self-excess pertains to an inability to restrict one's behavior or conduct, especially with respect to specific standards or criteria. Impatience, as a cognate vice of intemperance, involves an inability to endure pain and hardship without over reacting. The uncivil agent is one who is ill-bred and behaves badly, treating others with disrespect. Finally, the vice of immodesty refers to an unvirtuous agent's rudeness and impropriety in behavior, especially in a social setting. Such an agent is generally arrogant, conceited, and proud.

Temperance plays a significant role in virtuous medical practice (Bryan, 2006; Carr, 2001; Pellegrino and Thomasma, 1993; Telfer, 1990). According to Bryan (2006), for example, the virtue is important for the practice of clinical medicine, especially in terms of promoting reasonable constraints on behavioral extremes to avoid patient harm and to facilitate the patient's (as well as the physician's) best outcome. For Bryan, cultivation of temperance goes hand-in-hand with the main goals of medicine—service to and healing of patients. The temperate physician exhibits the discipline necessary to focus on patients and their medical needs, while the intemperate physician is inadequately disciplined to concentrate on patients but is often too self-adsorbed—generally leading to impatience on the part of the physician with patients. The temperate physician treats patients with civility, realizing that medicine's chief aim ultimately is to heal and not simply to cure patients. The intemperate physician treats patients with incivility, trying to hurry them along towards a cure with little regard for their emotional or existential condition.

One of the behavioral extremes that plagues modern medicine, according to Pellegrino and Thomasma (1993), is a physician "playing God." This extreme is obviously a consequence of the vice of immodesty. Modern technology equips today's physicians with what seems like divine power to intervene in the disease process and to "save" patients. Consequently, such apparent power often tempts the medical practitioner to act like god. In addition, according to Pellegrino and Thomasma, the "temptation" for the physician is "to employ technology rather than to give oneself as a person in the process of healing" (1993, p. 124). The temperate physician does not succumb to this temptation to allow technology to intervene between the physician and patient, thereby separating them. Besides the physician, temperance is also an important virtue for the patient. Elizabeth Telfer (1990) examines for the patient's welfare the health advantages of the virtue and the disadvantages of the vice. Temperance is a particular important virtue in terms of preventive medicine. For example, Telfer claims that if "people can learn to like best the kinds and amounts of food and drink which are healthiest . . . they can have a long life *and* a merry one" (1990, p. 158). Physicians, she counsels, must cultivate this virtue and discourage the vice of intemperance in their patients to ensure the most healthful and happy lives for them.

On May 1, 1889, William Osler gave the Valedictory Address—entitled *Aequanimitas*—to the graduating medical class at the University of Pennsylvania. In that address, he defines *aequanimitas* in terms of imperturbability and equanimity. The former refers to an assured coolness and the latter to a steady composure in the face of the patient's plight. Critics cite Osler's notion of *aequanimitas* as the foundation of a dehumanized, detached medicine (Bryan, 2006). Carr (2001), however, appropriates Osler's notion of *aequanimitas* for his normative notion of temperance for medical practice. According to Carr, Osler emphasizes *aequanimitas* to maintain a physician's physical condition in order to facilitate mental functioning. Carr then goes on, vis-à-vis his normative definition of temperance, to equate Osler's notion of imperturbability with a notion of physical temperance. Next, he equates Osler's notion of equanimity with a notion of mental or emotional temperance. Carr stresses that temperance provides a means to integrate both the physical and mental dimensions for the physician to achieve "a sympathetic attunement toward the patient" that authentically connects the physician to the patient. "This focus on connection," concludes Carr, "does not need to be understood in opposition to Osler's equanimity. Osler need not be seen as an enemy of the care ethic nor, conversely, need the care ethic be seen as an enemy of Osler's supposed detachment" (2001, p. 153). Rather than inhibiting a physician's connection to the patient to forge a therapeutic bond, Osler's notion of *aequanimitas* promotes it.

3.2.3 Justice and Injustice

Although generally listed last among the cardinal virtues, justice is certainly not least (MacIntyre, 1988). For Plato, it was the chief virtue making possible the other virtues—especially the moral virtues.[12] According to White, "justice is often considered the highest of all the virtues, or the master virtue that animates the other virtues and makes them virtues to begin with" (2008, p. 75). In other words, it directs the other cardinal virtues so that, e.g. courage is not just foolhardiness but rather bravery. Besides being a master virtue, justice is also a "cold virtue" (Lucas, 1972). It exhibits this feature because justice is impartial and dispassionate, taking into consideration a person only within a larger social context. As such, justice is not simply a personal virtue but it is also a virtue of communities and societies. The nature of this complex virtue has not remained static over the millennia but has exhibited considerable development. We now turn to its history and genealogy briefly to set the stage for the contemporary understanding of justice.

For the ancient Greeks, especially the Presocratics, justice (δίκη) pertains to the cosmos (κόσμοσ) in terms of a well-ordered and harmonious universe or existence (Havelock, 1978). In other words, the virtue of justice is an attribute of the universe or a person such that disharmony, i.e. injustice, is quickly rectified and harmony restored. For these Greeks, the basis of this virtue is the regularities found in nature (Alexander, 1987; Gagarin, 1974). Just as the sun courses through the heavens on a daily basis, providing light by its presence and darkness by its absence, so justice maintains the regularity in human life. When a person wrongs another, for example, justice demands restitution. "What then is 'justice' except again," concludes

[12] Aquinas lists justice first among the moral virtues but second to prudence among the cardinal virtues.

Eric Havelock on an analysis of the virtue in early Greek literature, "the rule of reciprocity required to keep 'regularity' in affairs both cosmic and human?" (1978, p. 267). Without justice, chaos rules in both the cosmos and polis, as well as for the individual. Justice demands regard for the well-structured nature of existence. "To respect the nature of anyone or anything," according to Gregory Vlastos, "is to be 'just' to them" (1947, p. 156). Thus, the just person is one who honors the balance inherent in social and natural relations.

Although the early Greeks define justice in terms of harmony, they do not provide a fuller explication of justice. Plato extends the notion of justice qua harmony with respect to both the structure of the human soul and the polis.[13] In the *Republic*, Plato defines justice in terms of harmony and defends the definition with respect to the harmony and order found within both the soul and the polis. According to Plato, the soul consists of three parts. The first is appetitive, which pertains to a person's biological functions, and is associated with the virtue of temperance. The next part of the soul is rational, which involves a person's mental functions, and is ruled by the virtue of wisdom. The final part is spirited, which is associated with the emotions, and is driven by the virtue of courage. Justice as a virtue at the personal level consists of the harmonious interactions of these three parts of the soul, with each accomplishing its specific function or duty and without interfering with the others.

As for the just person, so too Plato defines in similar terms justice as a virtue for the polis or state. The polis consists of three classes of people, each with its own virtue. For the ruling class that virtue is wisdom, for the guardian class courage, and for the worker class temperance. Each of these classes serves a useful function and only that function. The rulers are to rule, the guardians to guard, and the workers to provide goods and services. At no time, should once class aspire to discharge the function of another class. Plato articulates his position on justice vis-à-vis the polis accordingly, "You remember the original principle we laid down at the founding of the city: each citizen should perform that work or function for which his nature best suits him. This is the principle, or some variation of it," he concludes, "we may properly call justice" (1996, p. 128). In other words, as each class performs its function in a collegial and harmonious fashion, the good is realized for each citizen individually and for the polis collectively. Not to perform one's proper class function or to perform another class' function is tantamount to disrupting the state's harmony and to acting unjustly.

Whereas Plato searches for justice in terms of the ideal polis, Aristotle looks for it with respect to the practical person and state. To that end, in the *Nicomachean Ethics* he divides justice into two classes.[14] The first is universal or general justice,

---

[13] For Plato, the distinction between justice and temperance is generally indistinguishable such that Plato often uses them synonymously (Larson, 1951).

[14] Based on this dual nature of Aristotle's conception of justice, Xianzhong Huang denotes justice in terms of a "non-individual individual ethical virtue": "an individual virtue rooted within the character of justice, but which shows through just actions and its relationship to the eudaemonia of the whole city-state" (2007, p. 271).

in which the just person follows the laws of the just state. It is legal justice, based on the "character" of either an individual person or a society in which that individual resides. "We see that all men mean by justice," claims Aristotle, "[is] that kind of state of character which makes all people disposed to do what is just and makes them act justly and wish for what is just" (1998, p. 106). The second class is particular or special justice, which pertains to what is equitable with respect to goods and services. "Of particular justice," claims Aristotle, "and that which is just in the corresponding sense, (A) one kind is that which is manifested in distributions of honour or money or other things that fail to be divided among those who have a share in the constitution (for in these it is possible for one man to have a share either unequal or equal to that of another), and (B) one is that which plays a rectifying part in a transaction between man and man" (1998, p. 111).[15] He then goes on to distinguish two kinds of the latter type of justice: voluntary in which the agent is free to enter into an exchange, and involuntary in which the agent is not.

Aristotle sets the stage for the contemporary discussion of justice in terms of distributive, retributive, and restorative justice. First, retributive or legal justice, probably the oldest form of justice and common to all civilizations, refers to appropriate punishment for violation of the law (Griset, 1991; Weiler, 1978). The foundation of such justice is proportionality. In other words, the punishment must fit the crime, e.g. tooth for tooth, and must not exceed it. Next, restorative or compensatory justice pertains to restoring the loss a person incurs at the hands of another person (Johnstone, 2002). Whereas retributive justice focuses on the perpetrator of the injustice, restorative justice focuses on the victim to restore the loss suffered in the course of the injustice (Zehr, 1997). Lastly, distributive or social justice concerns the equitable distribution of goods and services (Barry, 1989; Fleischacker, 2004). Simply put, it involves "how a society or group should allocate its scarce resources or products among individuals with competing needs or claims" (Roemer, 1996, p. 1). That "how" often reflects normative rules and principles that allot resources based on necessities and merit.

John Rawls' theory of justice is certainly the most widely discussed notion of distributive justice and deserves closer inspection.[16] Rawls begins his defense of distributive justice by identifying its social utility: "Justice is the first virtue of social institutions" (2005, p. 3). To that end, he posits an abstract concept—the original position. This position refers to a state in which bargaining parties find themselves unaware of each other's particular situation or circumstances, such as socio-economic status; what he calls the veil of ignorance. For Rawls, this position is "a purely hypothetical situation characterized so as to lead to a certain conception of justice" (2005, p. 12). Consequently, he derives two principles of justice from the original position and maintains people would choose them given this prior position.

[15] Howard Curzer (1995) argues that general justice is a second order virtue derived from particular justice, which he identifies as a primary virtue.

[16] For an able introduction to Rawls's theory of justice, see Jon Mandle (2009).

"They are the principles," according to Rawls, "that free and rational persons concerned to further their own interests would accept in an initial position of equality as defining the fundamentals terms of their association" (2005, p. 11).

With respect to the first principle of justice, "each person is to have an equal right to the most extensive scheme of basic liberty compatible with a similar liberty for others" (Rawls, 2005, p. 60). He gives priority to this principle, known as the equal liberty principle, unless a greater system of liberties is negotiable in the event that liberties conflict. "Social and economic inequalities," according to the second principle, "are to be arranged so that they are both (a) to the greatest benefit of the least advantaged and (b) attached to offices and positions open to all under conditions of fair equality of opportunity" (Rawls, 2005, p. 83). This principle's first component, known as the difference principle, asserts that inequality is just only if it protects the liberties of the least advantaged from exploitation. In other words, inequalities must benefit the least advantaged and not the privileged. Its second component pertains to the equity of social advantages. Based on these two principles, Rawls concludes that society through its procedural system should protect the liberties of its citizenry and ensure just allocation or distribution of scarce resources.[17]

The cognate virtues to justice include fairness, impartiality, honesty, integrity, trustworthiness, and reliability (Peterson and Seligman, 2004). They are similar to the virtue of justice in terms of mimicking a particular dimension of it but not its entirety. For example, fairness pertains to a person's evenhandedness but not necessarily to honesty, as would justice. These virtues are divisible into three sets of pairs. The first pair consists of fairness and impartiality. Both virtues involve a sense of evenhandedness and neutrality in that fair or impartial people do not allow extraneous factors to influence unduly their actions to benefit or harm another person (Callan, 1994; Gert, 1995). However, fairness is not equivalent to impartiality in that the former pertains to parity while the latter to objectivity. The next pair is honesty and integrity, which are similar to each other in that they maintain the unity or fabric of the moral life and do not fragment or divide it. But, honesty is grounded metaphysically in identity, while integrity on wholeness (McFall, 1987; Smith, 2003). Honest people always tell the truth because the truth is what it is, while people of integrity tell the truth because of who they are.[18] The final pair is trustworthiness and reliability, which share dependability as a common feature associated with justice (Hardin, 2002). In other words, trustworthy and reliable people are as predictable as just people. However, trustworthiness pertains to fidelity while the latter to dependability. In other words, trustworthy people are faithful while the reliable people are constant.

---

[17] Rawls' theory of justice provoked considerable discussion and criticism (Daniels, 1989; Kukathas and Pettit, 1990; Nozick, 1974; Sen, 2009).

[18] Stephen Carter also acknowledges the difference between honesty and integrity: integrity demands much more work than honesty. For Carter, the work of integrity includes "discerning what is right and what is wrong; acting on what you have discerned, even at personal cost; and saying openly that you are acting on your understanding of right and wrong" (1996, p. 74).

Injustice is the chief vice contra the virtue of justice. Since justice (δίκη) for the Presocratics pertains to balance and harmony, injustice (αδίκη) represents an imbalance or disharmony (Alexander, 1987; Gagarin, 1974; Vlastos, 1947). Injustice is not just negation of justice, but rather it is intent to disrupt harmony and to violate the law, i.e. it results in harm. For the Presocratics, injustice is the obverse of justice. "Justice," according to Eric Havelock, "subsists in virtue of its antithesis to the nonjust: neither can be known without the other" (1978, p. 268). For Plato, injustice pertains to the rule of the appetitive or spirited rather than to the rational part of the person's soul. "The soul whose reason does not rule," according to Nickolas Pappas, "is the soul that does least whatever benefits it 'as a whole'" (2003, p. 172). Likewise, for the polis, injustice is rule by the workers or warriors rather than by the ruling class. Thus, in the *Republic*, for Plato the greatest injustice occurs in forms of governments like democracy or tyranny. Finally, according to Aristotle, injustice represents an extreme: "the core of injustice is selfishness and involves self-assertion and an insistence on our own desires even when this means disregarding others and their legitimate claims" (White, 2008, p. 81). In other words, unjust persons are greedy and receive more than merit would warrant. Injustice, for Aristotle then, refers to that which is "excess or defect, contrary to proportion" (1998, p. 121).

For contemporary western society, injustice is a vice that also refers to the excessive or the defective, which deprives others of what is rightfully theirs. It can reflect the inequitable distribution of physical and material, as well as epistemic, goods (Fricker, 2007). Rawls' theory of justice best illustrates such distributive injustice. For Rawls, injustice derives from an inability of an individual or of a society to conform to the principles of justice for the equal distribution of goods and services. "Injustice, then," concludes Rawls, "is simply inequalities that are not to the benefit of all" (2005, p. 62). For example, rather than negotiating from an original position, one party may have access to privy information of a negotiating party's particular socio-economic position and then uses that information to deprive that party of specific goods and services rightfully owed to it. Moreover, a society may enact a law to entitle a particular social group access to such information. How then should citizens respond to other citizens who avail themselves to such a law? One course of action is vigilante justice, in which citizens judge such citizens in kangaroo courts. But such recourse, according to Rawls, would be unjust. Rather, citizens must either repeal or amend the law through legally or socially sanctioned actions.

The above example is rather straightforward, since it involves active injustice that is often plain for all to see. Judith Shklar identifies an insidious type of injustice that creates not only vicious citizens but also a vicious society in which atrocities may and do occur. Shklar identifies this type of injustice as passive, which she claims Cicero originally described. According to Shklar, passive injustice is "the refusal of both officials and of private citizens to prevent acts of wrongdoing when they could and should do so" (1990, p. 5). The actively unjust person is culpable for legal infractions that cause others unwarranted harm and injury. The passively unjust person, on the other hand, "is simply indifferent to what goes on around him, especially when he sees fraud and violence" (Shklar, 1989, p. 1142). She gives the example of the Joshua DeShaney case in which a father brutally beat his four-year old son, even though social and healthcare workers knew of the danger to the child. She recognizes the right of families to privacy but also argues that such rights may lead to passive injustice with unintended harm to innocent victims. According to Shklar, paths to resolving the problem of passive justice, unfortunately, are indistinct.

The cognate vices of injustice include unfairness, prejudice, dishonesty, corruptness, untrustworthiness, and unreliability, and are divisible as well into three sets of pair (Fricker, 2007; Shklar, 1990). The first pair, unfairness and prejudice, involve a sense of favoritism or bias in that unfair or prejudiced people allow extraneous factors to influence unduly their actions thereby causing harm to others. However, unfairness is not equivalent to prejudice in that the former pertains to disparity while the latter to subjectivity. Dishonesty and corruptness are the next pair and are similar to each other in that they damage or even destroy the unity or fabric of a person's moral life by fragmenting or dividing it. However, they differ from each other in that dishonesty is rooted metaphysically in deceit while corruptness in distortion. The dishonest agent lies because the agent hides the truth but still knows it, while the corrupt agent distorts the truth to such an extent that this agent is no longer able to recognize it. Untrustworthiness and unreliability are the final pair of vices (Hardin, 2002). They both share unpredictability as a common feature. As such, one can never predict what an untrustworthy and unreliable person will do under given circumstances. However, untrustworthiness pertains to infidelity while unreliability to fickleness. In other words, the untrustworthy person is treacherous while the unreliable person is capricious.

Justice is certainly an important virtue for good medical practice, especially in terms of the patient-physician relationship. For example, Pellegrino and Thomasma (1993) discuss the role of what they call "loving justice" in terms of the healing relationship.[19] From a negative perspective, loving justice requires that a physician does not take advantage of a patient's vulnerability in order to promote the physician's personal agenda or good. Such a physician would obviously be unjust and far from loving. From a positive perspective, however, loving justice demands that "the physician becomes committed to some suppression of self-interest, comfort, and preferences in order to serve the patient" (Pellegrino and Thomasma, 1993, p. 105). In other words, the physician who practices the virtue of loving justice places the patient first—but not in a way that hampers the physician's ability to care for other patients properly. Loving justice does not include self-destruction. In a similar vein, Bryan links what he calls reciprocal justice, with respect to the patient-physician relationship, to the Golden (he prefers Platinum) Rule: "do unto others as they themselves would like to be treated" (2005, p. 389). Thus, reciprocally just physicians seek to do good rather than harm to patients while unjust physicians do not. Unjust physicians seek not to give to patients but only to receive, whether economically or socially.

[19] According to Pellegrino and Thomasma, justice has its origins in love and is transformed by love from "the legalistic justice of a chess-game approach to our duties to one another...[to] the loving concern of the community of care itself" (1993, p. 95).

The cognate virtues of justice certainly aid in the task of just healthcare, while the cognate vices of injustice do not but rather they promote unjust healthcare. For example, trustworthiness is critical for establishing caring and functional patient-physician relationships in order to practice effective medicine. According to Daniel Sulmasy, "Trustworthiness may very well be the central professional virtue of health care" (2000, p. 514). Without the bond of trust between the healthcare provider and the patient, the former is hindered from assisting the latter in terms of medical treatment. Generally, distrust between the patient and physician—often the fruit of untrustworthiness—leads to patient noncompliance or the physician abandoning the patient. Another essential cognate virtue for virtuous clinical practice is honesty. Although honesty is essential, it is problematic. Moral and ethical issues arise as to how much a healthcare provider should disclose to a patient who is suffering from a terminal disease (Begley, 2008; Da Silva et al., 2003). Is full or no disclosure the only options? Or, is partial disclosure justifiable? Such questions are not easily answered in the daily world of healthcare practice. Finally, the cognate virtue of impartiality is necessary for fair distribution of healthcare resources. Prejudiced physicians often cause not only patient harm but also harm to the healthcare system in terms of unfair resource allocation.

The notion of justice in medicine is not only important as a virtue but also as an ethical principle, especially with respect to examining issues in biomedical ethics. Justice, along with patient autonomy, beneficence, and non-maleficence, are the four ethical principles championed by Tom Beauchamp and James Childress (2001). For Beauchamp and Childress, justice is a principle to adjudicate ethical issues concerning the fair distribution of healthcare goods and services. Such use of the notion of justice is certainly a significant and common one in healthcare. For example, James Drane encourages the development of a just healthcare system that is in harmony with the needs of patients and a system that can meet those needs. "A badly imbalanced medical delivery system," according to Drane, "dehumanizes both the patients and the doctors" (1995, p. 109). Unfortunately, the resolution of unjust allocation of healthcare is not an easy task given the scarce resources available and the immense demands for them. As Rosamond Rhodes notes, "it is difficult to achieve justice in medical and public health policy because there is neither a single ideal governing principle nor a simple formula for success" (2005, p. 24). Certainly, the just distribution of healthcare is one of the greatest challenges facing contemporary medicine.

3.3 The Theological Virtues and Vices

Whereas the intellectual or epistemic virtues involve a person's correct knowing or understanding and the ethical or moral virtues a person's right action or doing, the theological or transcendental virtues pertain to a person's complete or essential being or existence (Kreeft, 1992; Peterson and Seligman, 2004; Robinson, 2004). While the intellectual virtues involve the life of the mind and the moral virtues the life of the will, the theological virtues pertain to the life of the spirit or soul, i.e.

one's very essence. The notion of theological virtue reflects not so much one's spiritual character, although it certainly does reflect that, but the transcendent character of a person in that it goes beyond one's physical and rational or logical, as well as emotional or affective. Without these virtues, the virtuous person lives a truncated or possibly a trivial life in terms of meaning. "Theological virtue," according to Josef Pieper, "is an ennobling of man's nature that entirely surpasses what he 'can be' of himself" (1986, p. 99). As noted in the last chapter, Paul of Tarsus identifies in his first letter to the Corinthians the theological virtues as faith, hope, and love. Although Paul does not use the term virtue, Christian tradition identifies them as such. Importantly, the theological or transcendental virtues transform and make possible the other virtues, especially the cardinal virtues (Keenan, 1995). For example, courage without hope might simply be despair, or prudence without faith mere foolishness (Kreeft, 1992, p. 73). From a religious perspective, theological virtues play a critical role in a person's life, especially in the sight of God, by helping one to live a life pleasing to God and others and preparing a person for life eternal. From a secular perspective, the transcendental virtues allow a person to rise above merely achieving one's own good to achieving the good of others—even at the expense of one's own good.

3.3.1 Faith and Faithlessness

Christian scripture describes faith (πίστις) as "the assurance of things hoped for, the conviction of things not seen" (Hebrews 11:1, RSV). In New Testament koine Greek, the verse reads, Ἔστιν δὲ πίστις ἐλπιζομένων ὑπόστασις, πραγμάτων ἔλεγχος οὐ βλεπομένων. The first phrase describing faith is composed of two Greek words. The first is ἐλπιζομένων, a present passive genitive plural masculine participle, which can be translated literally as of (something or someone) being expected or hoped for. The second word is ὑπόστασις, a nominative singular feminine noun, which can be translated literally either as confidence, in terms of a supporting structure or source, or as substance or essence of something. The second phrase is composed of four words. The first is the πραγμάτων, a genitive plural neuter noun that can be translated as of things or events. The next word is ἔλεγχος, a nominative singular masculine noun that can be translated literally as conviction or proof. The last words are οὐ and βλεπομένων, which go together grammatically and represent a negative particle and a present passive genitive plural neuter participle, respectively. They can be translated literally as of not being seen or observed. Thus, a literal translation of the verse is, "Now faith is confidence of [things] being expected, conviction of things not being seen." Given this translation, faith broadly construed is a virtue in the sense that it represents an excellence for living a fulfilled life, which transcends corporeal limits—i.e. not dependent solely on the senses or even on the mind—to include the spirit.

The author of Hebrews gives a number of powerful examples of faith, as a virtue, particularly from the Old Testament. Abraham, for instance, had faith in God's promise that his progeny would inherit a land in which he was but a vagrant, even though he had no progeny to fulfill that promise at the time God made it. His faith was the source of confidence that what God promised could be expected to come true. Again, Moses is another example of someone who had faith in divine promises—as are a host of others, including Noah, Rahab, Samson, and David—in the sense that life is not restricted just to what is observed but also includes what is expected vis-à-vis God's promises. What is important about these examples is that they illustrate the nature of faith as a virtue, particularly with respect to knowing and understanding the divine. In other words, these examples focus on the epistemic, i.e. knowing the promises God makes and keeps. Thus, faith serves as a precondition for gaining knowledge not just about the natural but also about the supernatural or transcendent world. Without this knowledge, no one can truly know or please God.

Augustine (1961), accepting the scriptural description of faith in *Hebrews*, refines the relationship between the virtue of faith and knowing vis-à-vis intellectual virtues (Wetzel, 1992). For him, faith is a precondition for knowing and begins, as do the other theological virtues, with the infusion of God's grace. Through a life devoted to God, a person's faith matures until its consummation in the beatific vision. "Faith," as Robert Cushman articulates Augustine's position, "is the lowly door by which the 'heart,' bowing to enter, is cleansed in order that at length the whole mind may apprehend the universal abiding Truth—may see God" (1950, p. 273). In other words, faith precedes reason and, as a virtue, makes reason robust enough to investigate and eventually to understand not only the apparent mysteries of the natural world but also not to shy away or eliminate the subtle mysteries of the supernatural world. While natural reason is limited in terms of knowing the supernatural, a person infused with faith is able to assent confidently to the truths associated with the divine. For Augustine, natural reason is not inherently defective vis-à-vis the divine or even natural mysteries; but, it is untrustworthy in the sense that it is often—though not always—fallen or corrupted. "To suppose that we can attain happiness by means of intellectual contemplation or moral virtue," as Elaine Robinson summarizes Augustine's stance on faith, "denies the reality of our human nature and our need for God's gracious assistance" (2004, p. 46). Without faith, we are strictly dependent upon an unreliable *ratio* or reason that may lead us astray.

Thomas Aquinas (1974b) extends Augustine's appropriation of the virtue of faith to knowing and the intellectual virtues. For Thomas, faith is a virtue since it is a habit that assents to the good, particularly the good in terms of both truth about God and God's truth. However, it is not only a virtue of the intellect, in terms of perfecting the human intellect vis-à-vis the divine intellect and knowing God, but also of the will, with respect to conforming the human will to the divine will (Brown, 2002; Garrigou-Langrange, 1965; Penelhum, 1977). "Faith," as John Lamont summarizes Thomas' position, "can be said to be a virtue in the intellect, because it subjects the intellect to the will when the will commands it to achieve the good of eternal life through believing, and the good of the intellect lies in its being subject to the will adhering to God" (2004, p. 58). Moreover, Thomas (1974b) divides the virtue of faith into an inner and outer act. The inner act of faith (*actus interior fidei*) represents the cognitive process by which one assents to the truth concerning the beatific vision and one's need for the divine, i.e. "the process of the mind searching before reaching its term in the full vision of a truth" (Thomas, 1974b, p. 61). The outer act of faith (*actus exterior fidei*), which consists in confessing publicly Jesus as the Christ and other creedal articles, complements in turn the inner act. In sum, for Thomas, faith qua theological virtue "directs its possessor to the supernatural end of perfect beatitude" (Jenkins, 1997, p. 162).

Contemporary notions of faith maintain its relationship to knowing, particularly as articulated by Augustine, but secularize that relationship.[20] Probably one of the most celebrated proponents of this secularized faith is the scientist-turned-philosopher Michael Polanyi. Polanyi stresses the importance of faith in the modern setting, especially for scientists. Faith operates in the lives of scientists in two senses vis-à-vis scientific objectivity. According to Polanyi, in his Riddell Memorial Lecture, "there can be no way of aiming at the truth unless you believe in it. And furthermore there is no purpose in arguing with others unless you believe that they also believe in the truth and are seeking it" (1946, p. 56). Thus, faith operates both at the personal level, in terms of believing that truth exists independent of the individual, and at the social level, with respect to a community in which members believe that collectively truth exists. In other words, "all human knowing takes place through a framework of tacitly held, formally unprovable commitments (a faith-structure) that motivate and guide the knower in the acquisition of knowledge" (Neidhardt, 1984, p. 42). However, such faith, as Polanyi stresses, is necessary for knowing the truth, in terms of providing—but is not sufficient for guaranteeing—it.[21] For such a guarantee, if possible, scientists must engage in exploring and testing their hypotheses and theories about the world. This secularized faith represents the contemporary notion of faith as a virtue, especially as an epistemic virtue (Comte-Sponville, 2001).

The cognate virtues of faith include belief, trust, commitment, confidence, assurance, and conviction. They can be divided—in terms of knowing—into two classes, with the first representing behavioral dimensions while the second attitudinal dimensions. The first consists of belief, trust, and commitment, since each represents a way of conducting oneself vis-à-vis knowing, in that a person may be believing, trusting, or committed, especially to some authority or source of knowledge or truth. But, each of these cognate virtues stresses a particular dimension of faith. Belief emphasizes acceptance of authority as a justifiable means in terms of knowing (Geach, 1977), while trust dependence on authority and commitment allegiance to authority. Confidence, assurance, and conviction compose the second class of cognate virtues of faith. Each represents an attitude that a person of faith exhibits towards what is knowable. And, each shares a passion or sincerity for knowing the knowable. The confident person, however, has an attitude of certainty, the assured person one of reliance, and the convicted person one of persuasion.

---

[20] Robert Audi (1991) points out the danger of secularizing faith, especially if the purpose is to reconcile science and religion. Audi argues that certain religious beliefs are not reducible rationally.

[21] The inability of faith to guarantee truth has led to a discussion of whether faith is a virtue. See, e.g. Chappell (1996) and Perkins (2010).

Faithlessness is the vice counterpoised to the virtue of faith. It involves both a negative and positive dimension. The negative dimension is the absence of faith. In other words, the faithless person is someone devoid of faith. From a religious perspective, such a person denies divine existence and does not assent to religious dogmas and doctrines because such faith is immiscible with the empirical, especially with the empirical evidence of the natural sciences. From a secular perspective, such a person denies the veracity of empirically verified epistemic statements because of preconceived ideas. Although chiefly defined in terms of the absence of faith, faithlessness also includes a positive dimension. That dimension involves a bad faith as opposed to no faith, i.e. bad faith is still faith (Haynes-Curtis, 1988). Bad faith pertains to people who think they have faith but such faith corrupts, distorts, and deceives. For example, prior to Copernicus people had faith or believed in a flat world based on specific empirical evidence. Once that evidence changed to support a different worldview, faith in the previous evidence as robust for supporting a flat earth view became bad faith. Such faith involves a disingenuous search for the truth that leads to falsity and inauthenticity, as opposed to good faith (Santoni, 1995; Sartre, 1956).

Disbelief, mistrust, doubt, timidity, indifference, and skepticism, are the cognate vices of faithlessness. Like the cognate virtues of faith, these vices can be divided—with respect to knowing—into two classes, with the first being behavioral and the second attitudinal. The first consists of disbelief, mistrust, and doubt, since each represents a way in which a person conducts himself or herself towards knowing or the knowable. In other words, such a person might be disbelieving, mistrusting, or doubting, especially to a particular authority or source of knowledge or truth. But, each of these cognate vices stresses a particular dimension of faithlessness. Disbelief emphasizes rejection of authority as a justifiable means with respect to knowing (Salmon, 1995), while mistrust independence of authority and doubt apprehension of authority. The second class of cognate vices of faithlessness consists of timidity, indifference, and skepticism. Each depicts an attitude that a person displays towards what is knowable. And, they each share insincerity towards knowing the knowable. The timid person, however, has an attitude of fearfulness, the indifferent person one of disinterest, and the skeptic one of cynicism.[22]

Faith is a critical virtue for robust practice of contemporary medicine, while faithlessness a vice for ineffectual practice. From a religious perspective, Pellegrino and Thomasma claim the virtue is central to the notion of a Christian physician. Faith in the healing power emanating from God is at the heart of such a physician to care for the sick and dying. As they point out, the gospels illustrate this power in myriad healing miracles attributed to Jesus, whom God anointed to heal the sick and dying (Luke 5:17). However, such faith is not limited simply to curing a patient's disease, for all patients must die eventually. According to Pellegrino and Thomasma,

---

[22] Finally, these vices at times may function like virtues, especially in terms of suspending belief in an epistemic claim that lacks sufficient evidential support. Here, suspension of belief is an outcome of a skeptical attitude that leads to a doubting stance. Such an apparent virtuous outcome is the result of the principle of disbelief (Stubenberg, 1988).

"faith calls to mind that the purpose of human existence is union with God, not immortality or freedom from pain and suffering" (1996, p. 52). Moreover, faith transforms the natural cardinal virtues for medical practice. For example, Pellegrino and Thomasma maintain that benevolence vis-à-vis faith is more than just benefiting the patient; rather, "[i]t requires doing good at some sacrifice to oneself" (1996, p. 52). From a secular perspective, the virtue of faith is what holds the medical system together (Bryan, 2006). Not only must patients have faith in their healthcare providers but also in the healthcare system. Physicians must have faith in themselves that they can benefit patients and in patients as compliant. The faithless physician is someone who injures patients, for example, with bad faith. Rather than a patient's agenda, for example, such a faithless physician allows his or her agenda to dictate practice often resulting in patient harm (Toon, 1993).

The cognate virtues of faith are also important for good medical practice, as the cognate vices of faithlessness are for bad medical practice. Belief in medical authorities is especially critical in training physicians and in their clinical practice. If physicians question every article of medical knowledge, then they would be incapacitated from or incapable of acting on the patient's behalf. Of course, doubt can be important when judging medical truth; however, it becomes a vice when used excessively or indiscriminately. Trust is an essential virtue for medical practice, especially for a healthful patient-physician relationship. Without the virtue of trust, the patient cannot benefit from the physician's skill and training. In fact, without this virtue the patient might as well not seek medical help. Mistrust, on either the patient's or the physician's part, simply destroys the patient-physician relationship and hinders any chance for healing. Moreover, the virtuous physician must be committed to the patient, to provide the best possible healthcare. A physician's allegiance must be to the patient and not to the physician, medical community, or even society. Without this commitment, the patient is vulnerable to medical abuses. Medical history is replete with such abuses, e.g. the Tuskegee syphilis study (Gray, 1998).

3.3.2 Hope and Hopelessness

Hope (ἐλπίς) is the second theological virtue, which Paul lists in his letter to the Corinthians. In his letter to the Romans, however, Paul provides a clear description of the virtue. For Paul, hope represents the patient expectation of God's ability to fulfill promises: "Now hope that is seen is not hope. For who hopes for what is seen? But if we hope for what we do not see, we wait for it with patience" (Romans 8:24–25, RSV). Based on Paul's letter to the Romans, especially 5:1–11 and 8:18–25, John Heil defines hope as "the act or the attitude of confident expectation for God's future salvific activity that arises from faith in what God has promised and/of already accomplished on our behalf" (1987, p. 6). Given this definition, Heil identifies four critical features of hope as a virtue. The first is the "objective basis" of hope. This basis is what God has achieved in the past, and it provides the root for future possible actions by God. The next feature is faith, which allows the Christian to engage hope; thus, hope is a derivative of faith. The third feature is what Heil calls the "subjective expression" of hope. It is attitude or act of expectation that God can keep promises. The final feature is the "future objective or goal" of hope. It is that end towards which one hopes; and, for the Christian it is the resurrected life in Christ.

Augustine also bases his discussion about the virtue of hope on Paul's letter to the Romans and on other New Testament sources, especially the letter to the Hebrews. And, he goes on to distinguish faith from hope in that the latter concerns "only what is good, only what is future, and only what affects the man who entertains the hope" (1961, p. 8). In other words, believers can have faith that the wicked are punished but they would certainly not hope for it. The issue is the good, since hope pertains to what is good or best for a person, i.e. one would hope for eternal reward rather than eternal punishment.[23] Moreover, hope involves only the future whereas faith includes also the present and past. Obviously, one cannot hope for that which is or has been. Hope, in other words, involves an element of uncertainty (Nunn, 2005). The future is what faith and hope share. For Augustine, therefore, "When, then, we believe that good is about to come, this is nothing else but to hope for it" (1961, p. 9). The relationship between faith and hope is derivative, with hope a consequence of faith. Finally, Augustine contrasts hope and fear, especially in terms of motivation to believe. Certainly, scripture informs believers that they ought to fear God, not simply because of possible punishment but also because of the hope in a glorious God who transcends creation.

For Thomas Aquinas (1966b), hope is a virtue since it is a disposition or habit, which causes human actions to be good or to attain a future good that is difficult but not impossible to achieve. For a Christian, that difficult but not impossible good is God and the eternal happiness obtained from knowing divine truth. As for the other theological virtues, God infuses hope into the believer through grace, which directs the human will to conform to the divine will. As for Augustine, Thomas also acknowledges that fear of God in terms of punishment is an important motivation for the virtue of hope. But, such punishment is not punitive but corrective. Following Aristotle's notion of virtue as a mean, Thomas situates hope between despair and presumption (Geach, 1977). Hope overcomes the vice of despair, which relinquishes any possibility of ever obtaining the good, whereas it does not devolve into presumption, which assumes that the obstructions towards achieving the good are trivial and easily surmounted without effort. Finally, Thomas argues that the relationship between hope and the divine law involves the good that God promises through one's obedience to that law, for "all the promises to be found in the Law tend to stir up hope" (1966b, p. 117). Without hope, the Christian cannot see clearly the future good that God intends.

"Philosophers," according to John Day, "seem to have abandoned Hope" (1969, p. 89).[24] With modern secularization comes no longer the need for religious hope.

---

[23] The assumption is that "all have sinned and fall short of the glory of God" (Romans 3:23, RSV). Consequently, believers hope for mercy and not vengeance.

[24] For contemporary expositions on the notion of hope, see J.P. Day (1991), Godfrey (1987); Shade (2001), and Sutherland (1989).

Rather than religious hope, science offers what Peter Geach calls an "impious" hope. "Some people," notes Geach, "... desperately look forward to a time when natural science will have progressed so far that we shall not need, in serious thinking, to talk of people's words, opinions, plans, and intentions, but only of physical and physiological states and events!" (1977, p. 49). Indeed, hope becomes devoid of either the moral or the good, under the critical gaze of contemporary philosophers (Nunn, 2005). Instead, these philosophers reduce hope to a statement and subject it to linguistic analysis. For example, Day (1969) distinguishes several locutions, including "*A* hopes that *P*", "*A* hopes for *P*", and "*A* hopes to *P*". Through such an analysis, he diminishes the meaning of hope to simply the fulfillment of two criteria: *A* wishes *P*, and *A* finds *P* probable. Robin Downie offers similar examples of hope statements. The first is "hope that" and refers to both desire of the object hoped for and belief in some probability of its occurrence. The next is "hopeful that" and includes the previous two criteria except the probability is positive not negative. Interestingly, Downie acknowledges this type of hope can be a religious virtue since it "implies a settled belief in certain propositions, and an accompanying spiritual serenity" (1963, p. 250). The final hope statement is "hope to do" and consists of the previous criteria but extends it to an intention to try to achieve the hoped for.

Other contemporary philosophers object to the wholesale reduction of hope, especially elimination of the moral or the good associated with hope. For example, Barbara Nunn (2005) agrees that hope includes wishing or desiring the object hoped for and the probability or possibility of its occurrence; however, she also incorporates the subject's judgment that the object hoped for is good in some respect. The subject's context or background often determines the goodness of the hoped-for object. Without this context, the subject may mistake hope for a virtue when it is not. Nunn gives the example of a babysitter who hopes to steal her employer's money. The babysitter definitely has the disposition of hope but not as a virtue. Certainly, the virtue of hope would preclude such a criminal act. Michael Quinn also argues for an additional criterion of hope, in which the subject hopes for that which is important or significant. In other words, a legitimate reason accounts for why a subject would hope for the object. Finally, hoping is linked morally to caring, according to Quinn: "if the occurrence of instances of caring is a necessary condition for living a life which is worthwhile then, chances are, hoping is an ingredient of a minimally worthwhile life" (1976, p. 63). Hope, then, as a modern virtue represents a means for transforming the world into a better place. To lose such hope would result in an impoverished and a pathetic world, devoid of compassion and empathy.

The cognate virtues for hope include expectation, aspiration, reassurance, cheerfulness, and gratefulness (Peterson and Seligman, 2004). Expectation as a virtue refers to looking forward to an object or event with anticipation. As with hope, no guarantee assures that an object is obtained or an event occurs—even through much effort on the part of a virtuous agent. The next cognate virtue is aspiration, in which the virtuous agent seeks an object or event with diligence and conscientiousness. As with hope, even though the virtuous agent desires a future object or event, circumstances surrounding them may be beyond that agent's control. Reassurance is the third cognate virtue and involves the positive assurance about a future object or event. Reassuring people are optimistic about the future, even though they are cognizant that such a future is only possible and not definite. The penultimate cognate virtue is cheerfulness. The cheerful person is full of happiness and joy about the future and its possibilities. Such a person does not succumb to the difficulties the future might hold but tackles them head on with a constructive and positive attitude. Gratefulness is the final cognate virtue and refers to a virtuous agent's gratitude and appreciation for future promises. In a sense, the grateful agent exhibits indebtedness for the realization of such promises.

Hopelessness is the vice that opposes the virtue of hope and as such it is, broadly speaking, lack of hope. Hopelessness is a vice, since it leads to a desperate and pitiable life in which a person is robbed of the future, especially a fulfilled and flourishing life. From a religious perspective, the hopeless person rejects God's promises for future life and happiness. Fundamentally, fear deprives a person of any hope. According to William Lynch (1974), the vice of hopelessness exhibits three key features. The first is a sense of impossibility, that nothing a person does has any real affect—"what a man must do he cannot; no matter what he does it leads to a sense of checkmate" (Lynch, 1974, p. 48). Such a state of affairs is due to a "constant cancellation" in which negative feelings curb any positive ones of success and leads to a sense of entrapment. The next feature is what Lynch calls "too-muchness." In other words, a person's life simply becomes unmanageable or out of control. "No matter what effort of the will," notes Lynch, "the project of managing such a world, of functioning within it, seems hopeless" (1974, p. 49). The final feature of hopelessness is futility in which the world simply becomes meaningless and devoid of any purpose. The world just ceases to make sense. The result is a person who is "deeply passive" and most unhappy. In all, the improbability of a hopeful future leads to a sense of helplessness in which the hopeless person lives a vicious and an impoverished life.

Despair, indolence, anxiety, despondency, and ungratefulness, are the cognate vices of hopelessness. Despair comes from the Latin word *desperare* and is a compound of the prefix *de* (reversing the action of a verb) and the verb *sperare* (to hope). Despair, then, means not to hope in the sense that a person is unable to look forward to obtaining an object or to an event happening. Despair often leads to the next vice, indolence, which refers to sloth or laziness. The indolent person is one who simply does not aspire to achieve and therefore fails to grasp hold of the future, since circumstances appear to be beyond his or her control. If the future is not attainable, why make an effort to reach for it? This vice, in turn, often results in the next vice—anxiety. The anxious person is one who is pessimistic about the future and is apprehensive about any chance for success. Such apprehension incapacitates a person and inhibits efforts to act. The next vice is despondency or joylessness, i.e. unable to experience joy, which generally ends in depression and melancholy or even suicide. Ungratefulness is the final cognate vice and involves a sense of ingratitude towards the future. The ungrateful person loathes the future, since that future is largely unrealizable.

Hope is a vital virtue for the practice of medicine (Coulehan, 2011). According to Pellegrino and Thomasma, "hope is essential to healing" (1996, p. 56). Without hope, a physician can offer no real effective treatment for patients (Bruhn, 1984). Rather, the hopeless physician instills a sense of dread and foreboding within patients. For those patients, such a physician, rather than engendering or nurturing hope for healing, destroys the possibility of healing. Not only has the disease robbed the patient of a future, but so has the hopelessness of the physician. The only thing the hopeless physician can offer a patient is despair. For, as Pellegrino and Thomasma note, "Raising too little hope induces despair [in patients]" (1996, p. 60). Such despair often leads patients to experience helplessness and anxiety and to forgo any expectation of healing (Lynch, 1974).

On the other hand, Pellegrino and Thomasma go on to note, "raising too much [hope] induces false expectations [in patients]" (1996, p. 60). This false or unrealistic hope does as much harm to the patient as no hope. Such hope does not take into consideration the limits of medicine. At some point, a patient is going to die. Ignoring this fact by physicians often cheats patients of the hope for a good or dignified death, with the false hope of wasted and ineffectual therapies (Miller, 2007). "The hopelessness of the sick comes largely," according to Lynch, "from an overextension of hope, an absolutizing of its range" (1974, p. 53). This hopelessness is the fruit of unrealistic expectations of medicine's effectiveness.

Hope and hopelessness are active areas of research within the healthcare professions, including medicine, nursing, and psychology. Researchers have proposed a number of models for hope, but Ezra Stotland (1969) put forward the most prevalent model or conceptualization, which has been used to quantify both the hope and hopelessness of patients, especially with respect to therapeutic outcomes.[25] According to Stotland's model, hope represents an anticipation of a particular goal that is of vital importance for a person, e.g. a medical cure. The importance of the goal motivates the person to obtain it, e.g. seeking a cure. However, if the person perceives that the goal is unobtainable then he or she may become anxious and may not obtain it. Hence, for Stotland a direct correlation exists between a person's perception of the goal's attainability and efforts to attain it. In other words, the more unattainable the goal the harder the person works to attain it until he or she realizes that circumstances are hopeless. "Hope," as Edith Raleigh summarizes Stotland's position, "is a component of adaptive action in a difficult situation, and hopelessness is a factor in maladaptive behavior" (2000, p. 435). Healthcare providers have an important obligation to instill hope or healthful behavior and not hopelessness or unhealthful behavior within patients to increase the likelihood of beneficial and not harmful outcomes.

3.3.3 Love and Lovelessness

Love is the final theological or transcendental virtue listed by Paul in his letter to the Corinthians, but it is not the least. "For now there are faith, hope, and love. But

[25] For an able review of efforts to quantify hope and hopelessness, see Farran et al. (1995).

of these three," writes Paul, "the greatest is love" (1 Corinthians 13:13, CEV). Of the three virtues, he considers it the greatest because someday faith will be realized and hope fulfilled but love will remain. Why? Because love characterizes not only the Christian but more importantly it characterizes God. In an epistle, the beloved disciple John writes, "God is love" (1 John 4:8, CEV). This does not mean that love is God; rather, the very nature of God is what makes love—love. "God is love because, and only because," claims Geach, "the Three Persons eternally love each other" (1977, p. 80). The same should also be true for Christians, since God makes them new in Christ's image and invites them into this relationship of divine love. Again, as John writes, "Love comes from God, and when we love each other, it shows that we have been given new life?" (1 John 4:7, CEV). Love, then, is the theological virtue infused through God's grace and reclaims or renews the *imago Dei* tarnished through original sin.

What then is this love? The ancient Greeks have several terms for it, including ἔρως, which is the fervor of intimate or driving, often sexual, love; φιλία, which is the enjoyment of friendship; and, στοργή, which is the affection found especially among family members (Comte-Sponville, 2001; Davidson, 2007; Lewis, 1960). The term most often used for love in the New Testament is αγάπη. Traditionally, the term means a welcoming or cherishing fondness, in contrast to the fervor of erotic love. However, the New Testament writers embrace agape love and define it as servant love and as altruistic or self-sacrificing love, especially in terms of Christ's death upon the cross to redeem the world. "God loved the people of this world so much," professes John in his gospel, "that he gave his only Son, so that everyone who has faith in him will have eternal life and never really die" (John 3:16, CEV). The virtue of love for the New Testament writers is a love that calls Christians to love not only God, themselves, and neighbors, but also their enemies (Matthew 5:34). In the Corinthian letter, Paul gives a comprehensive description of agape love: "Love is kind and patient, never jealous, boastful, proud, or rude. Love isn't selfish or quick tempered. It doesn't keep a record of wrongs that others do. Love rejoices in the truth, but not in evil. Love is always supportive, loyal, helpful, and trusting. Love never fails!" (1 Corinthians 13:4–8, CEV). Importantly, agape love is the foundation upon which the other theological, including the cardinal, virtues rest.

In a sermon on John's first epistle, Augustine makes a bold claim, "Love, and do what you will" (1984, p. 305). What makes this claim so bold is that not only is the moral life but also the whole life of a Christian grounded in agape love. Of course, God guarantees this love. Following scripture closely, he acknowledges that divine love makes us children of God and therefore each other's siblings. The basis of love for God and for one's neighbors (as well as enemies) must be a person's "intrinsic worth, not for some pleasure or advantage that we hope to derive from him [or her]" (Kent, 2001, p. 214). Love then is the ultimate source of the Christian life and animates every action. "If you keep silent, keep silent by love," admonishes Augustine, "if you speak, speak by love; if you correct, correct by love; if you pardon, pardon by love: let love be rooted in you, and from this root nothing but good can grow" (1984, p. 305). For Augustine, love is the greatest of the theological virtues since the person "who loves aright believes and hopes rightly" (1955, p. 409).

Without love, according to Augustine, a robust spiritual life is simply impossible. For without the love of God, only the baser human nature rules a person's actions.

For Thomas Aquinas (1975), love or charity is a virtue because it is a habit through which a person obtains the good or happiness, especially in terms of union with God. Following both scripture and Augustine, Thomas also maintains that love is the greatest of the theological virtues. Whereas faith and hope are important virtues or habits for attaining knowledge or understanding of the divine, they do not obtain God directly—only the virtue of love does that. Moreover, Thomas argues that love is "a foundation or a root because it supports and nourishes all the other virtues" (1975, p. 33). In other words, love is their efficient cause. Also, since love is the essence of Christianity, love of God, of one's neighbors and enemies, and of oneself, stems from the same act of the will and thereby perfects the person.[26] For, love is the virtue that incorporates people into Christ's salvific work on the cross. However, Thomas imposes an order upon love as to its objects in that God being the source of love must be loved prior to either one's neighbor or even oneself. Finally, Thomas notes that although love does not necessitate duty it does conform to precept, the precept to love God with one's whole heart and one's neighbor as oneself (Matthew 22:37–39).

Contemporary notions of love still resonate with traditional notions as a theological or transcendental virtue, as modern theologians attempt to provide an analytic definition of love. For example, Thomas Oord defines love theologically as an "act intentionally, in sympathetic/empathetic response to God and others, to promote overall well-being" (2010, p. 17). However, as for both faith and hope contemporary thinkers also promulgate secular versions of the virtue. Swanton (2003), for instance, defines love in a Kantian manner as a "coming close," not simply in terms of outcomes of loving acts but rather with respect to receptivity, i.e. seeing the world in a loving way, and to appreciation, i.e. approving of the world in a loving way. Her foundation is not religious or theological but psychological, whether individual or societal. Swanton divides love into self and universal types. Self-love is Nietzschean bonding with oneself, not strictly in power terms but psychologically with respect to self-worth, as a creative and vital human being. However, self-love is not the goal of the virtuous life but rather universal love is, which is particular and concrete (one cannot love ambiguously), as well as impartial (objective love) and unconditional (unrestricted love). As such, universal love is the basis of virtue ethics in which a person reaches out to oneself and others in an affectionate manner.

Frankena (1973) proposes an ethic of love (EL) in terms of virtue ethics. He identifies five features of such an ethic. The first is that EL is monistic in that love is the greatest of the virtues, to which the other virtues are subsidiary. Importantly, Frankena constrains love to function only as an ethical virtue in terms of neighbor-love and discards God-love simply as religious love reducible to ethical

---

[26] Interestingly, Thomas excludes demons from the list of persons that a Christian must love since they are damned and loving them would make a mockery of God's just judgment of them (1975, pp. 111–115).

love.[27] The next feature is that EL must be aretaic, as opposed to deontic. As aretaic, EL focuses on the agent's character rather than on rules and regulations. The third feature specifies the second feature in terms of the agent as a person who is, rather than as a person who does. Motivation, character, and traits of a virtuous agent are what is important for EL. The next feature extends his analysis to consider the declaration, "Be loving!", not as a command or rule to be followed mindlessly but as a charge to act towards the good of oneself or another. The final feature caps the analysis by taking love not as an ethical or a normative term but as a psychological term that under girds a disposition for how one should be as one acts. Although Frankena is sympathetic to EL, he argues that love must be supplemented by another virtue, for him justice, so as to avoid problems associated with relativism or situationism associated with the bold declaration of "Be loving!"

The cognate virtues of love include compassion or sympathy, empathy, humility, forgiveness, and loyalty (Peterson and Seligman, 2004; Toledo-Pereyra, 2006). The first two, compassion or sympathy and empathy form a specific set of cognate virtues (Chismar, 1988; Comte-Sponville, 2001; Darwall, 1998; Davis, 1996; Eisenberg, 2000; Snow, 2000; Taylor, 2002; White, 2008). They are often used interchangeably, since the person expressing the virtue of compassion, sympathy, or empathy, generally desires to alleviate the pain and suffering of another in a benevolent or magnanimous manner. However, they are different from one another as virtues. Compassion and sympathy are terms from Latin and Greek, respectively, which translate "to suffer with," while empathy "to suffer in." Compassion and sympathy refer to the virtue of being able to pull alongside another person and to share in that person's pain and suffering, in order to help. Empathy, on the other hand, refers to the virtue of being able to project oneself into the other person's pain or suffering and to suffer what the other suffers, if only momentarily. Whereas compassion and sympathy maintain a separation between the sufferer and the sympathetic person with respect to pain and suffering, which remain the experience of the sufferer although the person sympathizing is cognizant of the sufferer's experience, for empathy both the sufferer and the person empathizing share in some respect the experience.

The next three cognate virtues of love also form a loose set in terms of being dispositions of a loving or caring agent. Humility is a virtue in terms of not simply having a low opinion of oneself but rather with respect to having a reasonable or realistic perspective of one's abilities and accomplishments (Richards, 1988). Humility, as an ability to asses one's importance modestly or soberly, is a cognate virtue of love because "humility leads to love. . .without humility, the self comes to occupy all the available space and sees the other person as an object—not of love but of concupiscence!—or as an enemy" (Comte-Sponville, 2001, p. 147). The next cognate virtue of love, forgiveness, refers to not being unduly susceptible to injury or not to over react to harm but a capacity to pardon it (Griswold, 2007; McGary,

[27] Interestingly, Frankena supports this reduction by quoting Galatians 5:14, which states that the Jewish law may be summarized in terms of loving one's neighbor.

1989). For, "undue sensitivity to injury coupled with unreadiness to forgive creates a much worse situation" (Downie, 1965, p. 134). Often associated with forgiveness is mercy, which, as a virtue, imputes leniency or clemency upon a perpetrator of harm or injury. It is a supererogatory virtue in the sense that the wronged person does not demand the full extent of justice but rather extends charity to the one responsible for the wrong (Tasioulas, 2003). Finally, loyalty is a cognate virtue in which a virtuous agent is devoted to another person or institution in terms of promoting that person's or institution's good (Fletcher, 1993; Royce, 1995). Contra R.E. Ewin, one need not suspend good judgment in order to be loyal; rather, good judgment, as for other important virtues, is imperative for the loyal agent.[28]

Lovelessness is the vice counterpoised to the virtue of love and in a general sense refers to the habit or disposition of being unloving or without love. The vice has three versions, according to Thomas Hurka, "hatred of good, love of evil, or indifference to good or evil" (2001, p. 177). The first version is a cold vice, in which the loveless agent abhors or detests the good and avoids doing it at all costs. The second version, on the other hand, is a cruel vice, in which the loveless agent seeks out and enjoys doing what is morally bad or corrupt. The two versions, although similar in terms of being loveless, do differ. Although a loveless agent who hates good, for example, would not hurt necessarily or intentionally another, an agent who loves evil would. In a sense, the loveless agent who hates good may not do evil intentionally, whereas the agent who loves evil cannot help but do it intentionally. Part of the reason for this difference is that good is real while evil is not, so that the agent who hates good hates something while the agent who loves evil loves nothing and becomes consumed by it.[29] The third version is a pathetic and insidious vice, in which the loveless agent simply does not care about either good or evil. This indifference is a contemptible vice, since such an agent is unwilling to contemplate whether a particular action is good or bad. Rather, the indifferent loveless agent stands by without any concern for whether the agent's or even someone else's actions are morally justifiable or not.

Finally, the vice of lovelessness undermines the other virtues, especially faith and hope. As Augustine notes, "he who does not love believes in vain, even if what he believes is true; he hopes in vain, even if what he hopes for is generally agreed to pertain to true happiness, unless he believes and hopes for this: that he may through prayer obtain the gift of love. For," as he continues, "although it is true that he cannot hope without love, it may be that there is something without which, if he does not love it, he cannot realize the object of his hopes" (1955, p. 409). The loveless agent,

[28] Ewin (1992) defines loyalty as sticking with something or someone, which may at times require suspension of good judgment. Although Erwin acknowledges that loyalty may be connected with other virtues, he fails to realize that good judgment is an important virtue often associated with other virtues, without which the virtues could be vices.

[29] This distinction between hatred of good and love of evil is strictly analytic in the sense that hatred can also consume a loveless agent so that in the end the result of one who hates good and another who loves evil is indistinguishable. Thus, both types of vicious agents can hurt another innocent person.

from the religious or transcendent perspective, is incapable of any true virtue that results in lasting significance, i.e. of loving God and neighbor—or even oneself. In fact, as Thomas Aquinas (1975) argues, hatred or lovelessness is the capital vice, just as love is the capital virtue. Rather than leading to a fulfilled life, ultimately in union with God, it robs the loveless agent even of the agent's human nature, reducing the agent to a pitiable existence devoid of the fruits of true happiness or fulfillment.

The cognate vices of lovelessness are incompassion or unsympathy and apathy, pride, unforgiveness, and disloyalty. The vices of incompassion or unsympathy and apathy comprise a set, since an agent exhibiting these vices is often malevolent, petty, or cruel and seeks harming or avoids helping another, or is simply indifferent to the suffering of another. However, they do differ in that the former pertains to an inability to understand while the latter to feel the suffering of another person.[30] The next three vices form a loose set that are generally associated with the loveless agent. The prideful agent is arrogant and often disdains others, whom the agent feels is of less worth than the agent. Next, the unforgiving agent is often overly sensitive to personal injury by another person and seeks not justice only but also revenge on the person causing the agent injury. Finally, the disloyal agent is incapable of being true to another person since the agent is devoted to obtaining and promoting only the agent's own good, even at the expense of another person's good.

For some, love or charity is at the heart of medicine in terms not only of its practice but also its ethics and morality (Bryan, 2007a; Pellegrino and Thomasma, 1996; Underwood, 2008). As noted earlier, medicine has a moral obligation to the patient since it seeks the patient's good. Love, as being the chief virtue, is the means by which to discharge medicine's moral duty to benefit a patient. For Pellegrino and Thomasma (1996), love, or their preferred term charity, is the ordering principle of both medicine's practice and ethics but not in an absolute sense. The role of charity in medicine can range from a minimalist position of "not harming the patient" (*primum non nocere*), to an intermediate position of not simply avoiding harm but benefiting the patient, or ultimately to the extreme position of "heroic sacrifices" in which the physician treats patients with no regard to personal agenda. Bryan (2007a) makes a similar claim for the role of love in medical practice in terms of the Samaritan contract, with the minimally decent, good, and splendid Samaritan.[31] However, these commentators are quick to point out at the latter position represents an ideal that few healthcare providers achieve—"nevertheless an ideal that defines us physicians" (Bryan, 2007a, p. 75). Although such an ideal is hard to achieve at a global level because of other demands on a physician, including personal health,

---

[30] The apathetic, as opposed to the unsympathetic, agent may be unable to feel the other person's suffering, but that agent may be able to feel delight in the person's suffering.

[31] See Chalmers Clark (2005) for additional discussion of the notion of the Samaritan contract for medical practice.

family obligations, and social responsibilities, it is certainly achievable at a local or constrained level as long as love motivates a physician's actions.[32]

The cognate virtues of love are also critical for a robust medical practice. Commentators recognize in particular the essential nature of compassion or sympathy and empathy for the effective practice of medicine (Elliott et al., 2011; Graber and Mitcham, 2004). In remarks on the role of compassion, for example, Pellegrino and Thomasma note, "the good physician cosuffers with the patient" (1996, p. 79). They divide compassion into moral and intellectual dimensions. The former refers to a physician's ability to feel what an individual patient is experiencing vis-à-vis illness, while the latter "to comprehend, assess, and weigh the uniqueness of this patient's predicament of illness" (1996, p. 80). Both dimensions of compassion are critical for a physician to connect with a patient emotionally in order to address a patient's existential fears concerning illness. Empathy, as distinguished from compassion or sympathy, is for some the "sine qua non to improve clinical understanding, to reduce misunderstandings, and to identify, analyse, and handle moral challenges in medicine" (Pedersen, 2008, p. 326).[33] This virtue allows a physician not only to show concern but also to identify on a deeper existential level with the patient's fears and suffering (Zinn, 1993). In fact, empirical studies indicate that empathetic physicians are "more effective" clinically than those who are not empathetic (Larson and Yao, 2005; Neuman et al., 2009).[34]

The other three cognate virtues of love are also important in medical practice. The virtue of humility, in particular, plays a significant role in a healthful or therapeutic patient-physician relationship. Karen Lebacqz (1992), for example, contends that humility is critical for delivering the medical "goods." She identifies two dimensions to the role of humility in medicine. The first is a relational dimension in which a physician acknowledges and utilizes the patient's role in forging a robust patient-physician relationship. The second is an epistemological dimension in which the physician recognizes the limits of medical knowledge and the need for the patient's input to revise or even to generate new paradigms for effectively treating patients. The next cognate virtue, forgiveness, goes to the very heart of what it means to be a physician qua human being. Patients often fail in following a physician's prescribed treatment protocol, often being non-compliant for one reason or another. Physicians must excuse, within limits that do not result in patient harm, such patient behavior

---

[32] Medicine is a vocation or calling and physicians are called to love their patients. But, as long as the love is genuine then physicians cannot be faulted because they did not act heroically. After all, in the parable of the talents, Christ demarcates between those given more and those less. The parable's point is that each person uses what is given to him or her to the fullest.

[33] To that end, Reidar Pedersen proposes a definition of empathy based on Gadamer's hermeneutics that incorporates the patients historical and community contexts. Pederesen then defines empathy as an "appropriate understanding of another human being" (2008, p. 332). Appropriate refers to the specific context, and understanding to the entire constellation of human activities not simply cognitive processes that compose comprehension.

[34] For a current review of these studies, see Pedersen (2009).

and not abandon their patients but extend grace to them.[35] Finally, the cognate virtue of loyalty plays a major role in medical practice. Traditionally, a physician's loyalty is to the patient but other elements often vie for that loyalty (Murray, 1986). In an insightful analysis relating Royce's philosophy of loyalty to the clinic, Griffin Trotter (1997) argues that medicine must reshape its loyalty or what medicine is committed to, in order to redefine its identity vis-à-vis its tradition and to develop a vibrant ethics for the future.

Just as love and its cognate virtues are at the center of a vibrant medical practice that seeks the patient's good, so lovelessess and its cognate vices are at the heart of a medicine that inflicts harm or even death onto vulnerable patients. "Rudeness, inaccessibility, abruptness, refusal to treat for economic reasons, discrimination because of social class, ethnicity, etc.," which Pellegrino and Thomasma recognize as some of the fruits of this vice and its cognate vices, "are not reconcilable with a charity-based ethic of medicine" (1996, p. 77). Three types of loveless physicians surface, when applying Hurka's versions of this vice to medicine. The first is a physician who detests or resents patients, who often frustrate the physician's efforts to cure them. Such a physician may generally refer to patients in pejorative terms, such as gomers (Leiderman and Grisso, 1985).[36] The second type of loveless physician is one who loves evil and intentionally inflicts pain and suffering onto patients. The last and probably most prevalent type are loveless physicians who are simply indifferent to the patient's pain and suffering. These physicians are most likely suffering from burnout resulting from the demands and abuses of their own practice.

As for the cognate vices of lovelessness, George Dunea recounts the clinical case of a young surgeon who, upon returning from the tropics, experienced such severe chills and diarrhea that she rushed to an emergency room of a prominent hospital. Unfortunately, she encountered a standard of care that Dunea could only describe as uncompassionate—busy and uncommunicative interns and residents, hard to reach and often absent physicians, ineffective antibiotics, ward transfer and repetitive medical histories, and questionable medical procedures. As Dunea concludes, "nobody's reputation would have been harmed if the doctors of that prestigious institution had spared a few minutes of their valuable time to talk to the patient and her family, listen, explain, and assuage their anxieties" (2005, p. 243). Dunea's story also illustrates other cognate vices of lovelessness. For example, the surgeon-patient thought that she should undergo surgery immediately, based on a particular test result, but surgery was not performed for another several days and only then as an emergency. Did pride on the part of the hospital's medical staff interfere with its quality of care? Of course, this story raises other questions concerning the loyalties

[35] Moreover, every physician fails in some capacity and is need of forgiveness, either from patients or peers. Such forgiveness allows physicians to practice medicine without fear of reprisal (Brody, 1998; Sulmasy, 2001).

[36] Although such terms as gomer and grok, among others, may serve a linguistic function (Gordon, 1983), they still belie a deep-seated hatred for the patient that if uncheck during residency often leads not only to patient harm but also to physician burnout.

of the hospital staff and the role forgiveness plays in the clinical lives of both health-care providers and patients. Unfortunately, such stories of loveless medical practice are simply too common and range from the seemingly trivial to the egregious and potentially criminal.

3.4 Summary

The virtuous physician is someone who dons the traditional cardinal (intellectual and ethical) and theological or transcendental (whether in their religious or secular form) virtues and the associated cognate virtues to benefit patients and to meet not only their physical needs but also their emotional, existential, and—if need be—spiritual needs. However, the virtuous physician is also someone who understands how best to order and prioritize these virtues so to live the eudemonic life and to practice good or virtuous medicine so as to benefit patients. In contrast, the unvirtuous physician dons the vices and the cognate vices, and in doing so often causes the patient grievous harm or even death. I must note, nevertheless, that, seldom, if ever, are physicians either completely virtuous or unvirtuous. Rather, physicians are often virtuous in some respects and unvirtuous in others. For, many physicians are at a particular stage in either their virtuous or unvirtuous development. Some are gaining in virtue, while others in vice. Moreover, physicians generally value one virtue over another. Some may find that one virtue, such as courage or loyalty, more pertinent to medical practice, while others may find many or even all virtues relevant. In addition, the pertinent virtue may change with the circumstances or with maturity—so that courage may be necessary under one circumstance but not under another, or as a physician develops a virtuous practice the virtue of justice may take precedence over an earlier one such as courage. Finally, to presage my own position developed later, I believe that the chief virtue of the virtuous physician is the compound virtue of prudent love (and of the unvirtuous physician, loveless imprudence). But, before arriving at that position, I must first discuss the chief ontological virtue of caring and its two ontic virtues of care and competence, along with the chief ontological vice of uncaring and its two ontic vices of carelessness and incompetence, which I take up in the next chapter.

References

Adams, R.A. 2006. *A theory of virtue: Excellence in being for the good.* Oxford, UK: Claredon Press.

Alexander, T. 1987. Arche, Dike, Phusis: The origins of natural justice in Anaximander. *Southwest Philosophical Studies* 10:11–20.

Aquinas, Thomas 1966a. *Summa theologiæ, volume 42, Courage (2a2æ. 123–40)*, (trans.) A. Ross and P.G. Walsh. New York, NY: McGraw-Hill.

Aquinas, Thomas 1966b. *Summa theologiæ, volume 33, Hope (2a2æ. 17–22)*, (trans.) W.J. Hill. New York, NY: McGraw-Hill.

Aquinas, Thomas 1974a. *Summa theologiæ, volume 36, Prudence (2a2æ. 47–56)*, (trans.) T. Gilby. New York, NY: McGraw-Hill.

Aquinas, Thomas 1974b. *Summa theologiæ, volume 31, Faith (2a2æ. 1–7)*, (trans.) T.C. O'Brien. New York, NY: McGraw-Hill.

Aquinas, Thomas 1975. *Summa theologiæ, volume 34, Charity (2a2æ. 23–33)*, (trans.) R.J. Batten. New York, NY: McGraw-Hill.

Aristotle . 1998 *The Nicomachean ethics*, (trans.) W.D. Ross. New York, NY: Oxford University Press.

Audi, R. 1991. Faith, belief, and rationality. *Philosophical Perspectives* 5:213–239.

Augustine, A. 1955. *Confessions and Enchiridion*, (trans.) A.C. Outler. Philadelphia, PA: Westminster.

Augustine, A. 1961. *The enchiridion of faith, hope, and love*. Washington DC: Regnery Publishing.

Augustine, A. 1984. *Selected writings*, (trans.) M.T. Clark. Mahwah, NJ: Paulist Press.

Barry, B. 1989. *Theories of justice: A treatise on social justice*, volume 1. Berkeley, CA: University of California Press.

Beauchamp, T.L., and J.F. Childress. 2001. *Principles of biomedical ethics*, 5th edition. New York, NY: Oxford University Press.

Begley, A.M. 2008. Truth-telling, honesty and compassion: A virtue-based exploration of a dilemma in practice. *International Journal of Nursing Practice* 14:336–341.

Brody, H. 1998. The family physician: What sort of person? *Family Medicine* 30:589–593.

Brown, S.F. 2002. The theological virtue of faith: An invitation to an ecclesial life of truth (IIa IIae, qq. 1–16). In *The ethics of Aquinas*, ed. S.J. Pope, 221–231. Washington, DC: Georgetown University Press.

Bruhn, J.G. 1984. Therapeutic value of hope. *Southern Medical Journal* 77:215–219.

Bryan, C.S. 2005. The seven basic virtues in medicine. II. Justice. *Journal of the South Carolina Medical Association* 101:388–390.

Bryan, C.S. 2006. The seven basic virtues in medicine. V. Faith. *Journal of the South Carolina Medical Association* 102:348–349.

Bryan, C.S. 2007a. The seven basic virtues in medicine. VII. Love. *Journal of the South Carolina Medical Association* 103:74–75.

Callan, E. 1994. Impartiality and virtue. *Journal of Value Inquiry* 28:401–414.

Carr, M.F. 2001. *Passionate deliberation: Emotion, temperance, and the care ethic in clinical medicine*. Dordrecht: Kluwer.

Carson, S. 2006. Phronêsis. In *Encyclopedia of Philosophy*, 2nd edition, volume 7, ed. D.M. Borchert, 27–28. Detroit, MI: Macmillan.

Carter, S.L. 1996. The insufficiency of honesty. *Atlantic Monthly* 277:74–76.

Chappell, T. 1996. Why is faith a virtue? *Religious Studies* 32:27–36.

Chismar, D. 1988. Empathy and sympathy: The importance difference. *Journal of Value Inquiry* 22:257–266.

Cicero, M.T. 1949. *De invention*, (trans.) H.M. Hubbell. Cambridge, MA: Harvard University Press.

Clark, C.A. 2005. In harm's way: AMA physicians and the duty to treat. *Journal of Medicine and Philosophy* 30:65–87.

Comte-Sponville, A. 2001. *A small treatise on the great virtues*, (trans.) C. Temerson. New York, NY: Metropolitan Books.

Coulehan, J. 2011. Deep hope: A song without words. *Theoretical Medicine and Bioethics* 32:143–160.

Curzer, H.J. 1995. Aristotle's account of the virtue of justice. *Apeiron* 28:207–238.

Cushman, R.E. 1950. Faith and reason in the thought of St. Augustine. *Church History* 19:271–294.

Daniels, N. (ed.) 1989. *Reading Rawls: Critical studies on Rawls' 'A theory of justice'*. Stanford, CA: Stanford University Press.

Darwall, S. 1998. Empathy, sympathy, care. *Philosophical Studies* 89:261–282.

Da Silva, C.H.M., R.L.G. Cunha, R.B. Tonaco, T.M. Cunha, A.C.B. Diniz, G.G. Domingos, J.D. Silva, M.V.V. Santos, M.G. Antoun, and R.L. De Paula. 2003. Not telling the truth in the patient-physician relationship. *Bioethics* 17:417–424.
Davidson, J. 2007. *The Greeks and Greek love*. New York, NY: Random House.
Davis, M.H. 1996. *Empathy: A social psychological approach*. Boulder, CO: Westview Press.
Day, J.P. 1969. Hope. *American Philosophical Quarterly* 6:89–101.
Day, J.P. 1991. *Hope, a philosophical inquiry*. Helsinki: Societas Philosophica Fennica.
Dekkers, W., and B. Gordijn. 2007. Practical wisdom in medicine and health care. *Medicine, Health Care and Philosophy* 10:231–232.
Downie, R.S. 1963. Hope. *Philosophical Quarterly* 15:128–134.
Downie, R.S. 1965. Forgiveness. *Philosophy and Phenomenological Research* 24:248–251.
Drane, J.F. 1995. *Becoming a good doctor: The place of virtue and character in medical ethics*, 2nd edition. Kansas City, MO: Sheed & Ward.
Dunea, G. 2005. Uncompassionate care. *British Medical Journal* 331:243.
Eisenberg, N. 2000. Empathy and sympathy. In *Handbook of emotions*, 2nd edition, eds. M. Lewis and J.M. Haviland-Jones, 677–691. New York, NY: Guilford Press.
Elliott, R., A.C. Bohart, J.C. Watson, and L.S. Greenberg. 2011. Empathy. *Psychotherapy* 48:43–49.
Ewin, R.E. 1992. Loyalty and virtues. *Philosophical Quarterly* 42:403–419.
Farran, C.J., K.A. Heath, and J.M. Popovich. 1995. *Hope and hopelessness*. Thousand Oaks, CA: Sage.
Fleischacker, S. 2004. *A short history of distributive justice*. Cambridge, MA: Harvard University Press.
Fletcher, G.P. 1993. *Loyalty: An essay on the morality of relationships*. New York, NY: Oxford University Press.
Frankena, W.K. 1973. The ethics of love conceived as an ethics of virtue. *Journal of Religious Ethics* 1:21–36.
Fricker, M. 2007. *Epistemic injustice: Power and the ethics of knowing*. New York, NY: Oxford University Press.
Gagarin, M. 1974. DIKĒ in archaic Greek thought. *Classical Philology* 69:186–197.
Garrigou-Langrange, R. 1965. *The theological virtues, volume one, on faith*, (trans.) T.K. O'Reilly . St. Louis, MO: B. Herder Book Co.
Geach, P.T. 1977 *The virtues*. Cambridge, UK: Cambridge University Press.
Gerhard, W.A. 1945. The intellectual virtue of prudence. *Thomist* 8:413–456.
Gert, B. 1995. Moral impartiality. *Midwest Studies in Philosophy* 20:120–127.
Gert, B. 2005. *Morality: Its nature and justification*, revised edition. New York, NY: Oxford University Press.
Godfrey, J.J. 1987. *A philosophy of human hope*. Dordrecht: Martinus Nijhoff.
Gordon, D.P. 1983. Hospital slang for patients: Crocks, gomers, groks, and others. *Language in Society* 12:173–185.
Graber, D.R., and M.D. Mitcham. 2004. Compassionate clinicians take patient care beyond the ordinary. *Holistic Nursing Practice* 18:87–94.
Gray, F.D. 1998. *The Tuskegee syphilis study: The real story and beyond*. Montgomery, AL: New South Books.
Griset, P.L. 1991. *Determinate sentencing: The promise and the reality of retributive justice*. Albany, NY: State University of New York Press.
Griswold, C.L. 2007. *Forgiveness: A philosophical exploration*. New York, NY: Cambridge University Press.
Hardin, R. 2002. *Trust and trustworthiness*. New York, NY: Russell Sage Foundation.
Hariman, R. 2003. Theory without modernity. In *Prudence: Classical virtue, postmodern practice*, ed. R. Hariman, 1–32. University Park, PA: Pennsylvania State University Press.
Havelock, E.A. 1978. *The Greek conception of justice: From its shadow in Homer to its substance in Plato*. Cambridge, MA: Harvard University Press.

Haynes-Curtis, C. 1988. The 'faith' of bad faith. *Philosophy* 63:269–275.
Heil, J.P. 1987. *Romans—Paul's letter of hope*. Rome: Biblical Institute Press.
Hofman, B. 2002. Medicine as practical wisdom (phronesis). *Poiesis & Praxis* 1:135–149.
Huang, X. 2007. Justice as a virtue: An analysis of Aristotle's virtue of justice. *Frontiers of Philosophy in China* 2:265–279.
Hurka, T. 2001. *Virtue, vice, and value*. New York, NY: Oxford University Press.
Jenkins, J.I. 1997. *Knowledge and faith in Thomas Aquinas*. New York, NY: Cambridge University Press.
Johnstone, G. 2002. *Restorative justice: Ideas, values, debates*. Portland, OR: Willan Publishing.
Kahn, C.H. 1988. Plato's Charmides and the proleptic reading of Socratic dialogues. *Journal of Philosophy* 85:541–549.
Keenan, J.F. 1995. Proposing cardinal virtues. *Theological Studies* 56:709–729.
Kent, B. 2001. Augustine's ethics. In *The Cambridge companion to Augustine*, eds. E. Stump and N. Kretzman, 205–233. New York, NY: Cambridge University Press.
Kreeft, P. 1992. *Back to virtue: Traditional moral wisdom for modern moral confusion*. San Francisco, CA: Ignatius Press.
Kukathas, C., and P. Pettit. 1990. *Rawls: 'A theory of justice' and its critics*. Stanford, CA: Stanford University Press.
Lamont, J.R.T. 2004. *Divine faith*. Burlington, VT: Ashgate.
Larson, C.W.R. 1951. The Platonic Synonyms, ΔΚΑΟΣΥΝΗ and ΣΩΦΡΟΣΥΝΗ. *American Journal of Philology* 72:395–414.
Larson, E.B., and X. Yao. 2005. Clinical empathy as emotional labor in the patient-physician relationship. *Journal of American Medical Association* 293:1100–1106.
Lebacqz, K. 1992. Humility in health care. *Journal of Medicine and Philosophy* 17:291–307.
Leiderman, D.B., and J.-A. Grisso. 1985. The gomer phenomenon. *Journal of Health and Social Behavior* 26:222–232.
Lewis, C.S. 1960. *The four loves*. New York, NY: Harcourt, Brace.
Lonergan, B.J.F. 1978. *Insight: A study of human understanding*. New York, NY: Harper & Row.
Lucas, J.R. 1972. Justice. *Philosophy* 47:229–248.
Lynch, W.F. 1974. *Images of hope: Imagination as healer of the hopeless*. Notre Dame, IN: University of Notre Dame Press.
MacIntyre, A. 1988. *Whose justice? Which rationality?* Notre Dame, IN: Notre Dame University Press.
Mandle, J. 2009. *Rawls's 'A theory of justice': An introduction*. Cambridge, MA: Harvard University Press.
McFall, L. 1987. Integrity. *Ethics* 98:5–20.
McGary, H. 1989. Forgiveness. *American Philosophical Quarterly* 26:343–351.
McKinnon, C. 1999. *Character, virtue theories, and the vices*. Orchard Park, NY: Broadview Press.
Miller, J.F. 2007. Hope: A construct central to nursing. *Nursing Forum* 42:12–19.
Murray, T.H. 1986. Divided loyalties for physicians: Social context and moral problems. *Social Science & Medicine* 23:827–832.
Neidhardt, W.J. 1984. Realistic faith seeking understanding: A structured model of human knowing. *Journal of American Scientific Affiliation* 36:42–45.
Neuman, M., J. Bensing, S. Mercer, N. Ernstmann, O. Ommen, and H. Pfaff. 2009. Analyzing the "nature" and "specific effectiveness" of clinical empathy: A theoretical overview and contribution towards a theory-based research agenda. *Patient Education and Counseling* 74:339–346.
North, H. 1966. *Sophrosyne: Self-knowledge and self-restraint in Greek literature*. Ithaca, NY: Cornell University Press.
Nozick, R. 1974. *Anarchy, state, and utopia*. New York, NY: Basic Books.
Nunn, B.V. 2005. Getting clear what hope is. In *Interdisciplinary perspectives on hope*, ed. J.A. Eliott, 63–77. Hauppauge, NY: Nova Science Publishers.
Oord, T.J. 2010. *The nature of love: A theology*. St. Louis, MO: Chalice Press.

Pappas, N. 2003. *Plato and the Republic*, 2nd edition. London: Routledge.
Pedersen, R. 2008. Empathy: A wolf in sheep's clothing? *Medicine, Health Care and Philosophy* 11:325–335.
Pedersen, R. 2009. Empirical research on empathy in medicine—a critical review. *Patient Education and Counseling* 76:307–322.
Pellegrino, E.D., and D.C. Thomasma. 1981. *A philosophical basis of medical practice: Toward a philosophy and ethic of the healing professions*. New York, NY: Oxford University Press.
Pellegrino, E.D., and D.C. Thomasma. 1993. *The virtues in medical practice*. New York, NY: Oxford University Press.
Pellegrino, E.D., and D.C. Thomasma. 1996. *The Christian virtues in medical practice*. Washington, DC: Georgetown University Press.
Penelhum, T. 1977. The analysis of faith in St. Thomas Aquinas. *Religious Studies* 13:133–154.
Perkins, J. 2010. Faith is not a virtue. *Australian Humanist* 97:17–18.
Peterson, C., and M.E.P. Seligman. 2004. *Character strengths and virtues: A handbook and classification*. New York, NY: Oxford University Press.
Pieper, J. 1986. *Faith, hope, love*, (trans.) M.F. McCarthy. San Francisco, CA: Ignatius Press.
Plato . 1996 The republic, (trans.) R.W. Sterling and W.C Scott. New York, NY: Norton.
Polanyi, M. 1946. *Science, faith and society*. London: Oxford University Press.
Prior, W.J. 1991. *Virtue and knowledge: Introduction to ancient Greek ethics*. London: Routledge.
Quinn, M.S. 1976. Hoping. *Southwestern Journal of Philosophy* 7:53–65.
Raleigh, E.D.H. 2000. Hope and hopelessness. In *Handbook of stress, coping, and health: Implications for nursing research, theory, and practice*, ed. V.H. Rise, 437–459. Thousand Oaks, CA: Sage.
Rawls, J. 2005. *A theory of justice: Original edition*. Cambridge, MA: Harvard University Press.
Rhodes, R. 2005. Justice in medicine and public health. *Cambridge Quarterly in Healthcare Ethics* 14:13–26.
Richards, N. 1988. Is humility a virtue? *American Philosophical Quarterly* 25:253–259.
Roberts, R.C., and W.J. Wood. 2007. *Intellectual virtues: An essay in regulative epistemology*. New York, NY: Oxford University Press.
Robinson, E.A. 2004. *These three: The theological virtues of faith, hope, and love*. Cleveland, OH: Pilgrim Press.
Roemer, J.E. 1996. *Theories of distributive justice*. Cambridge, MA: Harvard University Press.
Royce, J. 1995. *The philosophy of loyalty*. Nashville, TN: Vanderbilt University Press.
Salmon, N. 1995. Being of two minds: Belief with doubt. *Noûs* 29:1–20.
Santoni, R.E. 1995. *Bad faith, good faith, and authenticity in Sartre's early philosophy*. Philadelphia, PA: Temple University Press.
Sartre, J.P. 1956. *Being and nothingness: An essay on phenomenological ontology*, (trans.) H.E. Barnes. New York, NY: Philosophical Library.
Sen, A. 2009. *The idea of justice*. Cambridge, MA: Harvard University Press.
Shade, P. 2001. *Habits of hope: A pragmatic theory*. Nashville, TN: Vanderbilt University Press.
Shelp, E.E. 1984. Courage: A neglected virtue in the patient-physician relationship. *Social Science & Medicine* 18:350–361.
Shklar, J. 1989. Giving injustice its due. *Yale Law Journal* 98:1135–1150.
Shklar, J.N. 1990. *The faces of injustice*. New Haven, CT: Yale University Press.
Smith, T. 2003. The metaphysical case for honesty. *Journal of Value Inquiry* 37:517–531.
Snow, N.E. 2000. Empathy. *American Philosophical Quarterly* 37:65–78.
Stotland, E. 1969. *The psychology of hope*. San Francisco, CA: Jossey-Bass.
Stubenberg, L. 1988. The principle of disbelief. *Analysis* 48:184–190.
Sulmasy, D.P. 2000. Should medical schools be schools for virtue? *Journal of General Internal Medicine* 15:514–516.
Sulmasy, D.P. 2001. At Wit's end. *Journal of General Internal Medicine* 16:335–338.
Sutherland, S. 1989. Hope. In *The philosophy in Christianity*, ed. G. Vesey, 193–206. Cambridge, UK: Cambridge University Press.

Swanton, C. 2003. *Virtue ethics: A pluralistic view*. New York, NY: Oxford University Press.
Tasioulas, J. 2003. Mercy. *Proceedings of the Aristotelian Society* 103:101–132.
Taylor, C. 2002. *Sympathy: A philosophical analysis*. New York, NY: Palgrave Macmillan.
Telfer, E. 1990. Temperance. *Journal of Medical Ethics* 16:157–159.
Toledo-Pereyra, L.H. 2006. Loyalty. *Journal of Investigative Surgery* 19:275–277.
Toon, P.D. 1993. After bioethics and towards virtue? *Journal of Medical Ethics* 19:17–18.
Trotter, G. 1997. *The loyal physician: Roycean ethics and the practice of medicine*. Nashville, TN: Vanderbilt University Press.
Underwood, L.G. 2008. Compassionate love: a framework for research. In *The science of compassionate love: theory, research, and applications*, eds. B. Fehr, S. Sprecher, and L.G. Underwood, 3–25. Malden, MA: Wiley-Blackwell.
Vlastos, G. 1947. Equality and justice in early Greek cosmologies. *Classical Philology* 42:156–178.
Wallace, J.D. 1978. *Virtues and vices*. Ithaca, NY: Cornell University Press.
Walton, D.N. 1986. *Courage: a philosophical investigation*. Berkeley, CA: University of California Press.
Weiler, J.M.P. 1978. Why do we punish? The case for retributive justice. *University of British Columbia Law Report* 12:295–319.
Wetzel, J. 1992. *Augustine and the limits of virtue*. New York, NY: Cambridge University Press.
White, R. 2008. *Radical virtues: Moral wisdom and the ethics of contemporary life*. Lanham, MD: Rowman & Littlefield.
Widdershoven-Heerding, I. 1987. Medicine as a form of practical understanding. *Theoretical Medicine* 8:179–185.
Young, C.M. 1988. Aristotle on temperance. *Philosophical Review* 97:521–542.
Zagzebski, L.T. 1996. *Virtues of the mind: An inquiry into the nature of virtue and the ethical foundations of knowledge*. New York, NY: Cambridge University Press.
Zehr, H. 1997. Restorative justice: The concept. *Corrections Today* 7:68–70.
Zinn, W. 1993. The empathic physician. *Archives of Internal Medicine* 153:306–312.

Chapter 4
On Caring and Uncaring

I explicate the notion of virtuous physician in terms of caring not just as a virtue but as the chief virtue of medical practice, which is prior ontologically to the other virtues discussed in the previous chapter. Many medical or healthcare professionals place caring at a strategic or central location in healthcare practice (Barbour, 1995; Barker, 2000; Bevis, 1988; Brody, 1988; Duffy, 2009; Fry, 1991; Galvin, 2010; Jecker and Reich, 2004; Knowlden, 1990; Ledesma, 2011; Leininger, 1988; Parse, 1988; van Hooft, 2006; Watson, 2008).[1] In general, caring is an authentic and a deeply felt compassion or empathy that results in genuinely helping and not harming the sick and dying. It is a virtue because caring represents a character trait or disposition that allows its possessor to achieve or deliver not only the epistemological and ethical goods; but it also contributes to the overall goodness of the possessor, as well as others. Importantly, then, caring is the chief ontological virtue for practicing medicine, because it makes possible the virtues discussed in Chapter 3. In other words caring undergirds, for example, a physician's authentic understanding of a patient as an individual—from an epistemological perspective—and buttresses right or good action on the physician's part for that patient—from an ethical or moral perspective.

For medical or clinical practice, the ontological virtue of caring manifests itself chiefly in terms of two ontic virtues, care and competence, including the other virtues discussed in the previous chapter.[2] Initially, I define the two virtues in general terms; and then I explore the relationship between them, with respect to the

[1] Interestingly, the majority of these professionals are nurses, even though some nurses question caring or care as the basis for nursing practice or ethics (Warelow, 1996). Pamela Salsberry (1992), for example, argues that virtue theories based on caring do not provide an adequate replacement for duty-based theories of nursing practice.

[2] Traditionally, the ontological refers to fundamental or basic existence, i.e. what is prior to or makes existence actual, while the ontic refers to actual existence or something's facticity. Thus, caring as an ontological virtue is prior to or makes possible the existence of other virtues like care and competence, which are instantiations of caring. Although the use of caring and care might appear strained at times in subsequent discussion, still the distinction is important in terms of explicating precisely the philosophical foundation of the notion of virtuous physician and the specific roles the ontological virtue of caring and the ontic virtue of care play in that notion.

J.A. Marcum, *The Virtuous Physician*, Philosophy and Medicine 114,
DOI 10.1007/978-94-007-2706-9_4, © Springer Science+Business Media B.V. 2012

notion of virtuous physician. As a virtue, care represents not only a natural concern motivating a physician to help, i.e. to care about, but also an ability to act on the part of a patient, i.e. to take care of. I call the first type of care, motivational care or $care_1$. However, this care cannot guarantee the ability of a virtuous physician to take care of a patient. To that end, such a physician must be competent in the practice of medicine both as patient–centered art (ethical competence) and as evidence–based science (technical competence). Only then can a virtuous physician take care of the individual patient's bodily and existential needs. I call the second type of care behavioral care or $care_2$. Importantly, although motivational care ($care_1$) precedes competence, competence—both technical and ethical—is required for behavioral care ($care_2$) of the patient by the virtuous physician. In turn, $care_2$ reinforces $care_1$ thereby enhancing the overall quality of care and establishing a caring patient-physician relationship.

The discussion next turns to the ontological vice of the unvirtuous physician—uncaring. As an ontological vice, uncaring makes possible a physician's partial or even distorted understanding of an individual patient and wrong or unethical actions on the part of the physician for that patient, which often leads to patient harm and sometimes death. Uncaring makes possible two ontic vices—carelessness and incompetence. Carelessness represents indifference for the welfare of others or an unwillingness to help others, while incompetence an inability to help others. In like manner, an unvirtuous physician exhibits the two ontic vices of carelessness and incompetence in a relationship similar to that for the ontic virtues of care and competence. Such a physician is $careless_1$ in that he or she lacks proper concern or motivation to help a patient and often disregards the patient's bodily and/or existential needs or subverts them to other concerns not focused on the patient's wellbeing. For example, a $careless_1$ physician might worry about what tests to order to prevent a legal suite rather than to gain the necessary evidence to diagnose a patient's illness accurately and economically. This behavior on the part of the $careless_1$ physician often leads to both technical and ethical incompetence, such that the physician is $careless_2$ with respect to—and thereby does not meet or satisfy—a patient's bodily and/or existential needs. In other words, disingenuous concerns subvert a genuine concern to help the patient in the healing process and may often lead to patient harm or even death.

4.1 Caring

What is caring? Although the notion of caring has a long history and has defied a precise or consensus definition, caring generally represents a disposition or an attitude in which a person exhibits a deeply felt concern or empathy either for others or even for oneself (Held, 2006; Reich, 2004; Slote, 2007; van Hooft, 1995). For example, Madeleine Leininger defines caring as "the direct (or indirect) nurturant and skillful activities, processes, and decisions related to assisting people in such a manner that reflects behavioral attributes which are empathetic, supportive, compassionate, protective, succorant, educational, and others dependent upon the needs, problems, values, and goals of the individual or group being assisted" (1988, p. 4). Others define caring based on helping another to achieve a particular end-point or goal, whether health, happiness, or maturity. Rosemarie Parse, for instance, defines caring in terms of joyfulness: "caring is risking being with someone towards a moment of joy" (1988, p. 130). Moreover, others, especially feminists, define caring in relational terms. The feminist Nel Noddings (1984), for example, defines caring with respect to the bond that forges when the "one-caring" steps into the world of the "one cared-for" to assist in meeting that person's needs.

4.1.1 Mayeroff's Notion of Caring

Milton Mayeroff provides probably the best-known and most widely recognized definition of caring. According to Mayeroff, caring on the part of a person for another person fundamentally is to assist that person cared for to achieve his or her innate or natural potential. "To care for another person, in the most significant sense," writes Mayeroff, "is to help him grow and actualize himself" (1971, p. 1).[3] In other words, caring on the part of a person for another person fundamentally is to assist the cared-for person in achieving or realizing his or her innate or natural potential for a fulfilled and meaningful life. As a person takes care of or cares for another, the cared-for person becomes what he or she is meant or intended to be.

However, the trajectory of growth and fulfillment associated with caring is not simply one-way; rather, the person who does the caring grows and is self-actualized through his or her caring for another person. Although caring appears asymmetrical, it is reciprocal in that the caring person benefits from the care given to the other, especially in terms of finding meaning and fulfillment for one's life—to take care of others and to be cared for by others. Importantly for Mayeroff, taking care of another is a process that unfolds over time, as both persons bring into being through their cared-for lives what was originally or initially present or possible rudimentarily. Caring is developmental in nature in that the relationship between two persons matures as both realize their potential and fulfill their lives.[4]

According to Mayeroff, authentic caring exhibits a particular pattern in terms of helping someone or something to grow and to be actualized. The components of this pattern are:

> I experience the other as an extension of myself and also as independent and with the need to grow; I experience the other's development as bound up with my own sense of well-being; and I feel needed by it for that growing. I respond affirmatively and with devotion to the other's need, guided by the direction of its growth (1971, p. 6).

---

[3] Mayeroff's notion of caring operates differently depending on context. For example, in terms of parenting caring involves an intimacy and affection not found in professional relationships such as the patient-physician relationship.

[4] Mayeroff adds that one can take care of not only persons but also things, such as ideas or physical objects.

The first component represents the close bond that exists between the person caring and the person cared for. The caring relationship, however, is not dysfunctional dependency or parasitism, since the cared-for person or object is independent or free of the caring person. The person caring does not try to dominate the person cared for but permits that person freedom to develop as needed. The basis of the caring relationship is two individuals who willingly participate in a relationship that results in growth and actualization of the cared-for person. But, as noted above, the growth and actualization of this caring relationship is not simply one-way. The growth and actualization of the caring person is also tied up with the cared-for person. The next component is an affirmation of the person caring by the person cared for. The caring relationship is not only reciprocal, according to the second component, but also reflexive in that the person cared for supports the relationship through his or her growth and actualization, as well as his or her eventual ability to care for others.

According to Mayeroff, the final component of the caring pattern involves three features. The first is the positive response of the person caring for the needs of the person cared for to grow and to be actualized. The caring person encourages and supports the cared-for person. The next is the devotion or commitment on the part of the caring person for the needs of the cared-for person. Commitment and caring go hand-in-hand, claims Mayeroff, such that when commitment ceases so does caring. The final feature is the control of the direction for the growth and actualization of the person cared for. The impetus for that control is not the caring person but rather the cared-for person. “In helping the other grow,” observes Mayeroff, “I do not impose my own direction; rather, I allow the direction of the other’s growth to guide what I do, to help determine how I am to respond and what is relevant to such a response” (1971, p. 5). However, given the reciprocal and reflexive nature of the caring relationship the benefit is not simply for the person cared for. Rather, the person caring also grows and is actualized in ways unanticipated either by the caring person or by the cared-for person.

Besides the pattern of the caring relationship, Mayeroff identifies over a half-dozen features of caring. Knowing, although not the first to come to mind, is a crucial feature of a caring person. A person caring for another person must know that person’s needs for growth and actualization. Another important feature of caring is the ability to learn from the past. The caring person must assess whether his or her actions have truly benefited the person in need of care. The next is patience in which the caring person, according to Mayeroff, permits “the other to grow in its own time and in its own way” (1971, p. 12). Associated with patience is tolerance. Honesty is another important feature of caring. The caring person for another must be truly open and communicate with integrity. The person caring must also be able to let go and trust that the person cared for has the capability to grow in his or her own time and way. Moreover, the caring person must trust in his or her abilities to care and to be trustworthy as a caregiver. Another feature of caring is humility, in which the caring person realizes his or her limitations and is willing to learn from others—particularly the person needing care. Hope is also an important feature of caring, for the caring person genuinely believes that he or she can help the person in need of care to grow. Finally, courage as a feature of caring is the caring person's ability to face the unknown regardless of the outcome.

According to Mayeroff, caring for people involves a structure of "being with" and "being for" them. Being with the person who needs caring is necessary in order to know the needs of the person who seeks help to grow. In other words, to enter the world of the person needing help the caring person must pull alongside that person. "Instead of merely looking at him in a detached way from the outside, as if he were a specimen, I must be able," contends Mayeroff, "to be *with* him in this world, 'going' into his world in order to sense from the inside what life is like for him, what he is striving to be, and what he requires to grow" (1971, p. 30). By entering the other person's world, the caring person comes to appreciate and understand truly what another person needs to grow and to connect with that person at a deep level. However, being with the person who needs care does not mean that the caring person loses himself or herself in the other. Rather, the integrity of the caring person is critical for helping another person, i.e. of being for that person. "In caring," observes Mayeroff, "my being *with* the other person is bound up with being *for* him as well" (1971, p. 31). If the person caring for another loses his or her integrity, then that person is unable to make an objective evaluation of what needs to be accomplished to help the other. According to Mayeroff, the same structure of caring is also required for caring for oneself. In fact, taking care of oneself takes precedence over taking care of another person in the sense that only a person who knows how to take care of oneself "can properly understand and appreciate growth in another" (Mayeroff, 1971, p. 35).

Maurice Hamington also provides a definition for caring similar to Mayeroff's. "Care is committed to the flourishing and growth of individuals," according to Hamington, "yet acknowledges our interconnectedness and interdependence" (2004, p. 3). However, he adds an important dimension to the notion of caring—embodiment. Critical then to the flourishing and growth of the cared-for person by a caring person is the bodily dimension. "Embodied care centers," stresses Hamington, "not on theoretical or abstract understandings of right and wrong but on affective, embodied, and connected notions of morality" (2004, p. 32). He identifies three key components to what he calls embodied caring. The first is caring knowledge, which involves knowledge the body attains through implicit or tacit means. Such knowledge represents an extension of traditionally attained knowledge through explicit communication. Caring knowledge is important for the ability to care for others, especially when the person needing care cannot articulate his or her needs. This knowledge also leads to various habits, chief of which are the caring habits. These habits represent not rote behaviors but the body's physical and dynamic practices that result in a flourishing or fulfilled life, not only for the person cared for but also for the person caring. They are skills that the body learns in order to help others in need of care, often without much forethought. Moreover, these habits have a profound impact upon the types of inhabited bodies. Caring habits lead to healthful bodies, uncaring habits to unhealthful bodies. Finally, caring imagination provides the means for transcending physical limitations that hinder the communication of caring knowledge, by utilizing bodily experiences. This imaginative component of embodied care empowers the person caring for another to transmit what is essential for a person's flourishing and growth, especially in terms of empathizing with another person's plight.

Also utilizing Mayeroff's notion of caring, Stan van Hooft (1995, 1996) identifies two dimensions of caring: motivational and behavioral. "As motivation," notes van Hooft, "the word [caring] can refer to being fond of someone, feeling sympathy or empathy for them, being concerned for their well-being, or having a professional commitment to seeing to their needs" (1996, p. 83). At root, caring represents the psychological drive or stimulus that compels people to be concerned about either themselves or others and to commit themselves to the welfare of others. Caring provides the motive to act on the behalf of self and others. It provides the impetus for behaving in a solicitous fashion. Caring, however, is also behavioral. "As behavior," writes van Hooft, "the word [caring] often refers to looking after people and seeing to their needs, whether in the context of the healthcare professions, social work, teaching, parenting and other familial relationships, and so forth" (1996, p. 83). The behavioral dimension of caring involves not just the psychological but importantly the physical ability or capacity to care for self or others. It represents the actual doing of caring and involves techniques for carrying out or performing caring actions.

Finally, Simone Roach (1992), who employs Mayroff's definition of caring as well, identifies five attributes—what she calls the five C's—of caring. The first is compassion, which Roach defines "as a way of living born out of an awareness of one's relationship to all living creatures; engendering a response of participation in the experience of another; a sensitivity to the pain and brokenness of the other; a quality of presence which allows one to share with and make room for the other" (1992, p. 58). Compassion is a way of life, a way of being in the world with others that is open to life's joys and sorrows. Competence is the next C-attribute, which Roach defines "as the state of having the knowledge, judgment, skills, energy, experience and motivation required to respond adequately to the demands of one's professional responsibilities" (1992, p. 61). Competence and caring go hand-in-hand, since caring without competence often leads to harm, while competence without caring is often inhumane. The next attribute is confidence, which Roach defines "as the quality which fosters trusting relationships" (1992, p. 62). Without confidence in the caregiver, the person receiving care cannot trust the caregiver and the caregiver cannot provide care. The fourth C-attribute is conscience, which Roach defines "as a state of moral awareness; a compass directing one's behavior according to the moral fitness of things" (1992, p. 63). Conscience is the moral ground, which allows the caregiver to value correctly and to provide care adequately to one in need. The final attribute is commitment, which Roach defines "as a complex affective response characterized by a convergence between one's desires and one's obligations, and by a deliberate choice to act in accordance with them" (1992, p. 65). Commitment reflects the investment a person makes to care for another person.[5]

[5] Roach also identifies other C-attributes, such as comportment and communication, in caring.

4.1.2 Models of Caring

A variety of theoretical models for understanding caring exists, including cultural, feministic, humanistic, narrative, and scientific (Barbour, 1995; Fry, 1990; Marcum, 2008). For example, the cultural model explains caring in terms of civil, ethnic, and social beliefs, contexts, practices, and values. In addition, this model views caring as relative to a particular historical period so that what was once caring in one period, for instance, is no longer caring in another. Sara Fry (1991) combines the various traditional and contemporary models of caring into two categories of models: obligation-oriented and covenant-oriented models. Obligation-oriented models focus on duties towards the patient's good. For example, Fry cites the example of Pellegrino's model of caring in which healthcare professionals have an obligation to meet a patient's needs by promoting his or her good vis-à-vis healthcare. According to Fry, these models are problematic in terms of defining what constitutes the patient's good. Covenant-oriented models focus not on obligation but on the fidelity of the relationship between healthcare providers and a patient, especially with respect to the patient's dignity and autonomy. These models are problematic in terms of technical competence, although Fry notes that caring and competence should go hand-in-hand in good healthcare practice.

In an effort to address the problematic areas of both obligation-oriented and covenant-oriented models, Fry (1991) proposes a pluralistic model of caring in which both models converge. Beginning with the compassion that animates both models, she acknowledges, "the parties to a covenantal relationship are, in certain respects, under obligation to behave in certain ways within the relationship" (Fry, 1991, p. 167). The obligation in the pluralistic model, however, is not legalistic in terms of duties that a healthcare professional must provide but rather volitional in terms of duties that such a professional wants to provide, especially in a compassionate and caring manner. Importantly, the covenantal dimension of the pluralistic model maintains the independence in the healthcare provider-patient relationship through respect for the patient's autonomy and maintenance of the patient's dignity. For Fry, the fidelity of the relationship acts as a lens through which "caring becomes the possibility of giving help and receiving help" (1991, p. 167).

4.1.3 Is Caring a Virtue?

Now that the notion of caring is better clarified in terms of definitions and models, the critical question of whether caring is a virtue requires addressing. According to Peter Allmark, caring is not a virtue but rather a value. "Caring is not a virtue," insists Allmark, "but having a virtue involves caring about the right things in the right way" (1998, p. 467). In other words, caring is a kind of valuing or a quality that allows a person to see the worth of something rather than a virtue or quality "possessed by something which helps it fulfill its function well" in order to flourish (Allmark, 1998, p. 467). He illustrates virtues by such qualities like courage, which allows one to live an eudaimonic or a fulfilled life in the face of fear. The problem with Allmark's analysis, however, is that he defines caring too narrowly in terms of existential burdens, e.g. the cares of the world. Certainly caring represents a person's existential concern about life and its manifold burdens, but it is also a trait not only of a life well lived but also of the person who lives that life well and to the fullest. Caring, then, as I utilize it, is an expansive or unrestrained notion that not only encompasses caring as value but also, and more importantly, caring as virtue.

Consequently, caring—as noted in the above definitions—is more than simply a quality valuing the good; rather, it is also a quality or disposition equipping its possessor to flourish and thrive in the face of life's many difficulties and hardships. Caring is a virtue and conforms to many if not all of the definitions for virtue given in Chapter 2. For example, caring certainly represents an Aristotelian virtue or mean between the vices of an excess and a deficiency of caring. An excess of caring is when a person cared for is controlled and manipulated by a person caring, while deficiency is when a person has little or no care or concern for a person in need—whether another person or even oneself. Caring, as an Aristotelian virtue, represents a character state or trait that a caring person chooses based on practical deliberation for another's need. In addition, caring also conforms to contemporary virtue theory criteria. For example, caring is a McKinnonian virtue in that it is a deliberate character quality of someone who wants to be with another person or oneself rationally, emotionally, and ethically in ways benefiting not only the other person in need but also oneself. Moreover, the reasons for caring represent those that people ought to have for treating others. In other words, caring, like all virtues, is required for living a well-rounded and flourishing life.

Caring, however, is not only a virtue—it is a rather unique virtue, unique in the sense that it is distinctively connected to a variety of virtues. For example, Robert Veatch, expounding on the nature of care within ethical theories, suggests that caring represents not just a single virtue but rather a cluster of virtues. Discussing the relevance of caring for the healthcare professions, Veatch notes, "*care* is the name not of a single virtue or a synonym for virtue theory in general, but it is the name for a particular cluster of virtues, so that *care theory* is a species of virtue theory that identifies a package of virtues as the appropriate ones for certain relations, especially those in health care" (1998, p. 212). He goes on to identify several traditional virtues closely associated with care, including benevolence, compassion, faithfulness, and humaneness. Such a package of virtues, centered in care, helps to define the nature of virtue theory. The issue that Veatch leaves unsettled, however, is the exact relationship between the traditional virtues and caring per se. What I propose is that caring is the fundamentally ontological virtue, in a Heideggerian sense, which makes possible the existence and expression not only of a discrete set of virtues but also of the traditional virtues.

In *Being and Time*, Martin Heidegger identifies care as the very essence or being of human nature. Beginning with the human condition founded in the phenomenon of *Angst*, Heidegger grounds the wholeness of human existence in care: "Its being reveals itself as *care*" (1996, p. 171). Heideggerian care, or my term caring, represents the ontological, as well as the existential, foundation for human being-in-the-world, in terms of its relationship to both inanimate and animate objects. For Heidegger, the notion of care helps to address the meaning of being. It does so in the sense that humans are solicitous not only about their present existence but also about their future. The nature of care exhibits itself with respect to a fundamental ontological structure. That structure, according to Heidegger, consists of three elements: "being-ahead-of-itself" in that humans project themselves into the future in terms of their possibilities and potentials, "already-being-in-the-world" in that humans find themselves in a world occupied by objects and persons, and "being-together-with" in that humans find themselves fundamentally related to that world of objects and persons. Importantly for Heidegger, authentic care represents an ability to leap ahead of another person and to anticipate that person's needs, as opposed to inauthentic care, which represents a person leaping into another's needs and taking over in order to control and manipulate that person for ends other than the person's needs.

Employing Heidegger's notion of care and extending it, van Hooft (1996) proposes a notion of deep caring. Deep caring involves engagement of both life's temporality and its integrity, which is the *telos* of a person's life. Temporality, in accord with Heidegger, comprises a person's past, present, and future, with a person acting from out of the past through the present towards the future. It captures deep caring's chronological dynamism in which a caring person reaches out in the present from past experiences in a solicitous manner not only to a world of objects but also to a world of others and self to meet future needs and concerns. Those future needs and concerns converge on the person's integrity at four levels, according to van Hooft. The first is the biological level and pertains to a person's basic physical requirements for sustaining life. The next level is perceptual-reactive and involves a person's involuntary but conscious behavior, while the next level is evaluative-proactive and concerns a person's voluntary and purposeful behavior. The final level is spiritual, which is unique to humans, and includes a person's cognitive efforts to make sense of or to give meaning to life. Deep caring is the response of a caring person to maintain the integrity of another or oneself at these various levels during the other's or one's lifetime. "The function of deep caring," concludes van Hooft, "is to integrate our living and to give it world-relational and intersubjective meaning" (1996, p. 85).

Although van Hooft's deep caring captures what caring means or signifies for humans to act in a solicitous fashion for another person or oneself, it fails to provide the means or agency for acting in a caring way. I propose that caring is the chief ontological virtue that makes feasible such solicitous action by making possible existentially the cardinal and theological virtues of a virtuous person or agent. Caring makes virtues possible existentially in both their motivational and behavioral dimensions. Motivationally, caring makes possible the intellectual virtues in that the epistemically virtuous person cares about knowledge or justified true belief, the ethical virtues in that the morally virtuous person cares about good actions, and the theological virtues in that the theologically virtuous person cares about the transcendent meaning or significance of life. Behaviorally, caring makes possible the epistemic virtues in that the epistemically virtuous person cares for or takes care of knowledge or justified true belief by discovering it, the ethical virtues in that the ethically virtuous person cares for or takes care of right actions by doing them, and the theological virtues in that the theologically virtuous person cares for or takes care of the transcendent by believing it. Finally, caring manifests itself in the life of a virtuous person, especially a virtuous physician, in terms of the ontic virtues of care and competence—to which I now turn.

4.1.4 Care

Whereas caring is an ontological virtue, which makes possible the traditional virtues that compose the virtuous life, care is an ontic virtue that instantiates, in part, caring (competence is the other ontic virtue that completes caring's instantiation). As such, care reflects the various definitions for caring. For example, Pettersen (2008) defines care as a virtue in Aristotelian terms, i.e. as a mean between the deficiency of selfishness and the excess of self-sacrifice. Noddings (1984) identifies three uses or meanings of the term care. The first is a burden or anxiety that one feels for another person or oneself. Care represents a mental state of suffering over something or someone. Next, it is a consideration or tendency towards helping or to care about another person. It is the basic feeling or instinct that a caring person may have about another. Finally, according to Noddings, "care may mean to be charged with the protection, welfare, or maintenance of something or someone" (1984, p. 9). Such care involves not only a charge to care for but the actual caring itself. In other words, care includes the ability or capacity of a person to care for another but to do what a caring person does in order to meet the needs of another person or oneself.

As an ontic virtue, care is a derivative of caring; and, it reflects how a person actually wants to be or exist. Care addresses ontically the question of what a person is doing rather than ontologically what a person is being, when he or she is caring. In other words, an agent who cares is someone who actually cares for others. However, care is not reducible to one of the cardinal or theological virtues; rather, it is an intermediate virtue or a proximate virtue to caring that mediates between caring and the traditional virtues, especially the theological or transcendent virtue of love. As such, care as doing, like caring as being, exhibits both motivational and behavioral dimensions. As motivational, care represents the genuine desire to care about or to act in an authentic caring manner towards another person or oneself. I call this type of care, motivational care or $care_1$. According to Noddings, to care about reflects "an internal state of readiness to try to care for whoever crosses our path" (1984, p. 18). As behavioral, care is the tangible capacity or ability to care for, to take care of, or to provide the care that another person or oneself actually needs. I call this type of care, behavioral care or $care_2$. Whereas $care_1$ represents the intention, readiness, or motivation to care for another person or oneself, $care_2$ is the ability to do so.

4.1.4.1 Care Ethics

In the 1980s and 1990s, the notion of care gives rise to an ethic of care or a care ethic, especially among feminists (Clement, 1996; Held, 1995, 2006; Larrabee, 1993;

Noddings, 1984). Care ethicists cite Carol Gilligan's 1982 classic, *In a different voice*, as the foundation for this ethic. In that book, Gilligan proposes an approach to moral development different from that proposed earlier by Lawrence Kohlberg. Kohlberg (1981) concludes from empirical studies that moral development occurs in six stages, beginning with an immature individual who avoids punishment and culminating in a mature individual who acts according to (Kantian) universal ethical principles. According to Kohlberg, justice is the orienting principle of moral development. Interestingly, he uses only males as subjects in the studies. Gilligan (1982), a student of Kohlberg, conducts studies using females. She insists that the orientation of moral development is not simply principles, such as justice, but it is also relational, in terms of caring for one another. "The ideal of care is thus an activity of relationship," concludes Gilligan, "of seeing and responding to need, taking care of the world by sustaining the web of connection so that no one is left alone" (1982, p. 62). For her, this ideal is the basis for an ethic of care as compared to Kohlberg's ethic of justice.

The ethics of care is now a cottage industry, eclipsing its feminist origins and perspectives. For example, Tove Pettersen (2008), based on Gilligan's work, identifies three notions of care corresponding to three levels associated with an agent's moral development. The first is selfish care, in which the caring agent focuses on the self, especially to guard oneself from abuse and harm. Altruistic care is the next notion and involves the caring agent's ability to reach beyond oneself to another person in need and to meet that need, even at a possible cost to the agent. The final notion is mature care, the culmination of the developmental process, in which the relationship between the caring agent as self and the other in need becomes the focus for ethical deliberation, especially in terms of distributing mature care. What drives mature care, as well as altruistic care, is the notion of co-feeling as distinct from empathy. "Co-feeling," writes Pettersen, "is the ability to participate in the feelings of others, through the act of 'affective imagination', without (con)fusing self with others on the one hand, or, on the other hand, merely observing the other's feelings from a distance" (2008, p. 57). Although he acknowledges problems with the distinction between co-feeling and empathy, still Pettersen is sympathetic and committed to co-feeling.[6]

Given limited and strained resources, the distribution of mature care is often problematic for morally developed caring agents. To address this problem, Pettersen divides mature care into thin and thick varieties. The former pertains to "what the carer does for people she doesn't know well, when her information about them is general and impersonal," while the latter to "a personal relationship between the carer and the cared-for, where one's knowledge of the other is detailed and discriminating" (Pettersen, 2008, p. 114). The mature caring agent is able to deliberate

[6] The distinction between co-feeling and empathy is Gilligan's, in which she takes empathy to be identification with another's feelings in contrast to co-feeling which maintains the autonomy or integrity of the caring agent, as well as the person cared for. Michael Slote (2007), among others, criticizes this distinction claiming that empathy, based on current psychological research, is a multifaceted notion and that co-feeling is simply one species of empathy.

in such a way that minimizes harm, maximizes good, and distributes care goods fairly, through thin and thick caring. A mature caregiver may engage, for instance, in thin caring if resources are restricted and the person in need of care is unfamiliar. However, thin and thick care does not always dictate the preference of care. For example, a caring agent may give care preferentially to those whom the agent does not known personally but whom the agent feels have a greater need in contrast to a good friend whose needs are minimal. In this example, thin care takes precedence over thick care and represents mature care in action with respect to a mature caregiver deliberating over the distribution of care goods.

The notion of mature care, according to Pettersen, has important implications for professions, such as medicine. To that end, he distinguishes between symmetric and asymmetric care. Two people in a symmetric care relationship are equals, and they distribute the care goods reciprocally. In an asymmetric relationship, the two people are unequal and the care goods are not distributed equally. He gives an example of the nurse-patient relationship. In this relationship, the flow of care is in the patient's direction, although Pettersen acknowledges that the nurse might receive the patient's gratitude for the care given. Indeed, he goes on to admit that caring for the patient may be reward enough for the nurse, especially in terms of thin care when the patient may be unconscious. The goal of the nurse or healthcare provider is to move beyond thin to thick care, as the provider obtains more information about the patient permitting a deeper relationship to develop with the patient. Finally, he notes that mature care is not predicated on a paternalistic relationship, as was once the norm in healthcare.

Another important example of the global appeal of care ethics is Michael Slote (2007). In contrast to Gilligan and Pettersen, Slote proposes a care ethic grounded in a robust notion of empathy rather than Gilligan's or Pettersen's notion of co-feeling. He makes empathy the *sine qua non* of morality, in that empathy is the basis for moral deliberation and action. To that end, he enlists recent psychological research on empathy. Specifically, he utilizes Daniel Batson's *The altruism question* and Martin Hoffman's *Empathy and moral development*. From the former, he uses the "empathy-altruism" hypothesis, which states that people who identify with the need of another person often act to meet that need, generally in an altruistic manner. From the latter, he employs the notion of "empathic identification," which refers to the appropriateness of empathy vis-à-vis the situation of the person requiring empathy than the person showing empathy. From this literature, Slote concludes, "differences in strength of force of empathy make a difference to how much we care about the fate of others in various different situations" (2007, p. 15). The more empathic a caring agent is, then, the more moral the agent's behavior and actions. Why should this be? Although Slote admits he does not have a convincing answer to this question, he does have an intuition that the more a caring agent understands and identifies with the needs of others, the more likely the agent acts rightly in accord with those needs.

Slote also raises the issue over the nature of the relationship between care and virtue ethics. Although he is neutral to the relationship, others are not. For example, Virginia Held claims that care ethics is not a species of virtue ethics. "In my view," writes Held, "although there are similarities between them and although to be caring is no doubt a virtue, the ethics of care is not simply a kind of virtue ethics. Virtue ethics," she explains, "focuses especially on the states of character of individuals, whereas the ethics of care concerns itself especially with caring relations. Caring relations," she concludes, "have primary value" (2006, p. 19).[7] Moreover, she goes on to acknowledge that care and virtue ethics may be intimately related to one another, if the latter ethics is broadened to include more than simply dispositions or motives for acting. However, she still believes that care ethics is distinct from virtue ethics, since the former focuses on social relations and the values and practices that support these relations while the former simply on dispositions. But, others view care ethics as a species of virtue ethics. For example, Raja Halwani (2003) argues that virtue ethics can subsume care ethics, since the former ethics extends to "care an important status among the virtues" (2003, p. 169). Care qua virtue has this status because virtue ethics emphasizes the social and relational nature of human interactions, which is central to care and its ethics. Certainly, the relationship of care and virtue ethics depends on how one defines the relevant terms and their associations.[8]

4.1.4.2 Peabody's Notion of Patient Care

The virtue of care and the ethics of care are instrumental in the practice of contemporary healthcare (Benner, 1997; Cates and Lauritzen, 2001; Tong, 1998). Francis Peabody, an early twentieth century Boston City Hospital clinician and Harvard Medical School professor, best illustrates this ontic virtue's relevance for healthcare.[9] In a lecture entitled, "The care of the patient," delivered at Harvard Medical School, Peabody warns his audience that both the science of medicine and its art are not antagonistic enterprises but rather supplementary or complementary to each other. In other words, physicians must not only know the disease mechanism of an illness but they must also understand what that disease means in terms of the patient's illness experience and life. As Peabody so famously articulates his position at the conclusion of the paper, "One of the essential qualities of the clinician is interest in humanity, for the secret of the care of the patient is in caring for the patient" (1927, p. 882). The underlying quality or virtue of a good clinician is a general interest in humanity at large and in patients specifically. Without that broad interest in the human condition in which people share, according to Peabody, clinicians cannot adequately help patients because they cannot connect with them to form a genuinely caring human relationship. Thus, clinicians must first be motivated to help or care about ($care_1$) patients before they can help or take care of ($care_2$) them. In other words, clinicians must connect with their patients at a rudimentary level emotionally or existentially to care for them effectively.

---

7 Interestingly, Held considers Slote a prime exponent of care ethics as a species of virtue ethics.

8 Personally, I envision care ethics as a species of virtue ethics since care is a central virtue to the larger enterprise of virtue ethics.

9 For biographical material on Peabody, see Paul (1991) and F.G. Peabody (1933).

To take $care_2$ of the patient, according to Peabody, requires a rather comprehensive "clinical picture" of the patient. He claims that this picture is "not just a photograph of a man sick in bed; it is an impressionistic painting of the patient surrounded by his home, his work, his relations, his friends, his joys, sorrows, hopes, and fears" (1927, p. 878). Given this picture's complexity, the forces that have an impact on the patient's illness experience are simply more than just the organic, rather they may also include the psychological, the emotional, or even the spiritual. The physician must be motivated ($care_1$) and equipped ($care_2$) to deal with or care for all the forces influencing the patient's illness experience and medical outcome. Using another analogy, Peabody encourages clinicians to use low-power magnification initially to understand patients and their world, especially their world of illness, before switching to high-power to determine the disease's etiology and treatment. Moreover, Peabody realizes that physicians cannot single-handedly treat patients in a comprehensive fashion. Other members of a healthcare team must also play their part. For example, he notes that after a physician listens to the economic concerns of a patient with heart disease the clinician must then refer the patient to a social worker who can then help, for instance, the patient find appropriate employment. Thus, to care about ($care_1$) and to take care of ($care_2$) the patient involves not only a desire to help the patient but also an ability to do so. A physician must care about patients (personally or existentionally) in order to take care of them (physically or technically).[10]

4.1.5 Competence

A gap exists, however, between a person who cares about ($care_1$) another person or oneself and that caring person's ability to take care of ($care_2$) another or oneself. To fill that gap requires competence on the part of the caring person. Peabody fully appreciates the gap that exists between the general desire to care about patients and the capacity to take care of them. Although a person may authentically $care_1$ about another person, a caring agent cannot genuinely take $care_2$ of another unless the caring agent is competent to do so. Competence in medical practice, especially in America, is a major concern of both patients and professional healthcare providers (Good, 1998). Although definitions of competence vary, competence in general or broadly construed refers to an "ability to perform a task" (Carraccio et al., 2004, p. 252). The task's performance must result in a quality outcome according to specific standards or criteria, often defined by a community of professional practitioners. Minimally, the performance depends upon acquisition of specialized knowledge and the application of that knowledge through specific skills.

[10] Peabody not only had a tremendous impact on his generation of clinicians and the next generation following him, but his impact also continues today (Hollingsworth, 1995; Tishler, 1992; 2010).

Ronald Epstein and Edward Hundert offer a comprehensive definition of competence, especially for clinical practice: "the habitual and judicious use of communication, knowledge, technical skills, clinical reasoning, emotions, values, and reflection in daily practice for the benefit of individual and community being served" (2002, p. 226). Competence, then, is a multifaceted virtue that requires certain character traits of its possessor. For the purposes of the present discussion, Epstein and Hundert's definition of competence suffices, although I divide competence broadly into a technical and an ethical performing or doing. Technical competence refers to an ability to carry out practical and mechanical activities and procedures safely and correctly, as well as efficiently. This competence is generally discipline specific in healthcare. For example, surgeons often define how best to conduct a particular surgical procedure (Larkin et al., 2005).

Ethical competence represents a capacity to conduct oneself morally in an upright and a principled manner and to demarcate between right and wrong or good and bad. Ethical competence, especially within healthcare, involves keen moral perception, insightful moral judgment, and appropriate and right behavior (Jormsri et al., 2005). Generally, professional communities define ethical competence in terms of codes of conduct. However, a code may not be able to cover every ethical or moral ambiguity that arises in the practice of a trade, especially like medicine. Virtues certainly assist in providing guidance, when codes fail to offer any guidance. However, practitioners cannot invoke the virtues in an unconscious fashion; rather, ethically virtuous practitioners must act in a reflective manner (Eriksson et al., 2007). In other words, such virtuous practitioners can be truly ethical or act ethically only when they have adequately deliberated over the ethical conundrums to arrive at just and caring actions.

Competence is generally composed of particular competencies, where a competency is "a complex set of behaviors built on the components of knowledge, skills, and attitudes" (Carraccio et al., 2004, p. 252). The Accreditation Council for Graduate Medical Education (ACGME, 2007) identifies six essential competencies for the practice of medicine (Carraccio et al., 2004; Larkin et al., 2005). The first is patient care, which refers to the ability of healthcare professionals to provide appropriate and effective healthcare in a compassionate manner. The next competency is medical knowledge, which includes not only the acquisition of established and current medical knowledge but also the capacity to apply that knowledge effectively to patient care. The third competency is practice-based learning and improvement. According to this competency, healthcare providers are able "to investigate and evaluate their patients, to appraise and assimilate scientific evidence, and to continuously improve patient care based on constant self-evaluation and life-long learning" (ACGME, 2007). Interpersonal and communication skills, which facilitate and ensure the efficient exchange of accurate information between healthcare providers and patients, represent the next competency. The fifth competency is professionalism, which involves performing duties and commitments in a responsible fashion—often with respect to a code of ethical conduct. Systems-based practice is the final competency and includes the ability to work within the larger healthcare system to provide total care for the patient.

Epstein and Hundert (2002) identify seven dimensions of competence, especially for medical and healthcare professionals. The first is cognitive and includes the basic knowledge associated with the profession, along with abstract problem-solving and communication skills, information management, and learning from experience particularly in terms of its tacit or personal aspects. The technical dimension refers to practical skills associated with medical and surgical procedures and practices. Next is the integrative dimension, which involves the ability to incorporate clinical, scientific, and humanistic information to make judicious medical judgments and decisions. A competent clinician is not only able "to solve problems with clear-cut solutions. . .[but is also able] to manage ambiguous problems, tolerate uncertainty, and make decisions with limited information" (Epstein and Hundert, 2002, p. 227). The context dimension of competence involves the "ecology" of a clinical setting and the capacity of a professional to play an appropriate "role" in that setting vis-à-vis a patient's healthcare needs. The fifth dimension of competence is the ability of a physician to build a strong therapeutic relationship with patients, which includes a team of healthcare professionals. The affective and moral represent the next to last dimensions of competence and refer, in part, to a physician's emotional intelligence to connect at a fundamental level with patients and their illness experiences. The final dimension includes habits of the mind "that allow the practitioner to be attentive, curious, self-aware, and willing to recognize and correct errors" (Epstein and Hundert, 2002, p. 228).

Competence as an ontic virtue, along with care completes—as mentioned above—the instantiation of caring. Like care, it is not one of the cardinal or theological virtues per se, nor is it like caring in making possible these traditional virtues. Rather, competence—along with care—is a mediating or intermediate virtue. In other words, it helps its possessor to obtain or express the cardinal and theological virtues. Competence then is an important character trait that helps a caring person obtain, develop, or express virtues like courage, justice, and especially prudence. In fact, it is crucial for developing the virtues that underlie the general competencies. According to Peter Angelos and colleagues, a virtue-based approach "can be put to new use to inform our understanding, interpretation, and implementation of the general competencies" (Larkin et al., 2005, p. 491). For example, competence is important in obtaining or expressing the virtues, such as carefulness or conscientiousness, fidelity, manual dexterity, and wisdom, associated with the general competency of patient care. It is also crucial for virtues, like compassion, empathy, honesty, and tolerance, associated with the general competency of interpersonal and communication skills.

4.1.6 Care-Competence Relationship

Finally, to understand care and competence fully requires knowledge of their structural relationship. That relationship, as schematized in Fig. 4.1, is cyclical and contains feedback information. The first segment of the cycle ($care_1 \rightarrow$ competence) refers to a person's motivation to care about ($care_1$) another person. That

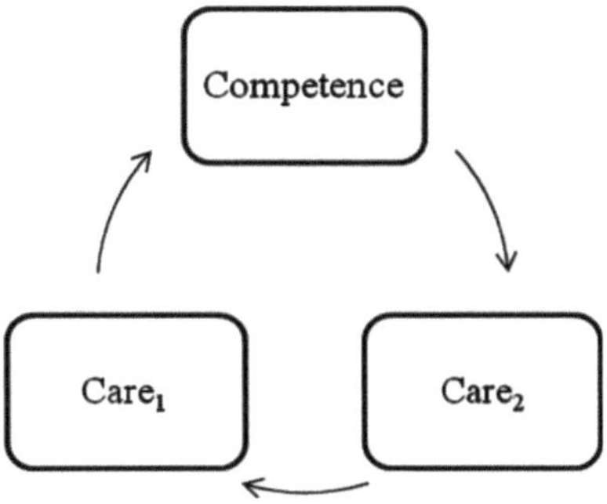

Fig. 4.1 Relationship between care and competence

motivation reflects an authentic impulse or desire to care about the other person and, in being so motivated, a caring agent desires not only to help a person needing care but ultimately to do so competently, i.e. to take care of that person ($care_2$). To care about another person to advance one's own agenda or biases generally does not lead to a genuine caring relationship. The reason is that the caring agent may not have the ethical competence to take care of ($care_2$) the other person, especially the other's existential or emotional needs. In addition, an agent who cares about ($care_1$) another person might be unable to form a caring relationship if the agent is technically incompetent to take care of the one in need. In other words, a person needing care may not trust a caring agent if that agent is incompetent to take care of ($care_2$) the person needing care (and the person needing care suspects or knows that the caring agent is incompetent).[11] Thus, to forge a genuine caring relationship requires at least two dimensions of the second type of care, a technical dimension ($care_{2a}$) and an ethical dimension ($care_{2b}$). In other words, to take care of another person comprehensively requires a caring agent to take care of that person competently in terms of correct technical actions ($care_{2a}$) and right ethical behavior ($care_{2b}$), which represents the second segment of the cyclic relationship between care and competence (competence → $care_2$). Specifically, a caring physician must be competent in the practice of medicine as both evidence–based science (technical competence) and patient–centered art (ethical competence), in order to take care of an individual patient's bodily and existential needs.

In the final segment of the care-competence cyclic relationship ($care_2$ → $care_1$), $care_2$ feeds back to reinforce the motivational dimension of $care_1$. As a caring agent takes care of a person in need competently, the caring agent experiences the satisfaction of meeting the needs of the cared-for person and in turn feels needed by that person, i.e. Mayeroff's reciprocal dimension of caring. This reciprocal connection between $care_2$ and $care_1$ not only establishes the caring relationship but also enhances it by expanding the potential of a caring agent to reach out to others in

[11] Trust here refers to more than simple reliance upon a physician's skill, but more importantly, it represents an attitude of deeply felt faith in the physician as a caring and competent person who can meet a patient's overall medical needs.

need of care—the caring agent also grows as a person and realizes his or her fulfillment as a caregiver. Not only does this feedback relationship between the two types of care reinforce a caring agent's motivation to care about ($care_1$) others, but it also strengthens a caring agent's desire to take care of ($care_2$) others competently. Thus, a caring agent makes every effort to maintain a level of competence that is current with community standards and expectations. Caring physicians, in van Hooft's sense of deep caring, ensure that the medical care they provide for patients is the most current and effective care possible and are at their professional acme—in that a genuinely professional physician is one who benefits patients, and does not harm them, through medical care. Lastly, in the absence of this feedback between $care_2$ and $care_1$, any natural desire to care about others in need atrophies—which thereby reduces the chance of an agent to take care of others in need and may lead to uncaring.

4.2 Uncaring

What is uncaring? Just as caring is difficult to define, so too is uncaring. In general, uncaring is a disposition or an attitude of a person who is unable or incapable of feeling any concern or empathy for another person. Furthermore, just as caring is the ontological virtue of the virtuous clinician that makes possible the traditional virtues, so too uncaring is the ontological vice of the unvirtuous or vicious physician that makes possible the traditional vices.[12] A distinction must be drawn between an unvirtuous person and a vicious person vis-à-vis uncaring. The former is a person who lacks the virtue of caring and exhibits uncaring by default, whereas the latter is a person who willingly rejects caring and chooses uncaring. The vicious person may reject caring and embrace uncaring because of an organic pathology or a mental condition, while the unvirtuous person is uncaring because something or someone distracts him or her from caring. For the present discussion, I will not entertain the notion of the vicious physician since it raises too many biological, ethical, and legal issues. Moreover, the number of vicious physicians is rather small, since the majority of people generally entering the healthcare professions do so with a genuine desire to help others. Finally, uncaring—like caring—exhibits both motivational and behavioral dimensions. Uncaring is motivational in that, at times, a physician may for some reason be unable to care about another person, while it is behavioral in that a physician is incapable to take care of another.

Sigrídur Halldórsdóttir (1996) advances a bridge/wall model of caring and uncaring for the health professions. According to Halldórsdóttir, caring represents a bridge for the healthcare professional to meet the patient's needs, either physical or psychological. As a bridge, caring allows the healthcare professional to be open

---

[12] In general, the unvirtuous person *vis-à-vis* the virtuous person simply lacks virtues, i.e. non-virtuous, and may not exhibit a vice, while the vicious person *vis-à-vis* the virtuous person not only lacks virtue but exhibits vice.

to and sincerely concerned about the patient's needs. It also permits the healthcare provider to be genuinely present and authentically committed to the patient and ultimately to be responsible morally in the face of the patient's needs. As a bridge, caring connects the healthcare professional and the patient at a fundamentally existential level. Uncaring, as a wall, represents indifference on the part of the healthcare provider to the patient's needs. This indifference exhibits several stages, developmentally. The first is disinterest in which the healthcare provider is inattentive to the patient's needs. The next stage is insensitivity to the patient's needs and especially feelings, which leads to the following stage of coldness in which the healthcare provider acts in a mechanical and business-like manner. The final stage involves inhumane acts, for example, when a healthcare provider completely ignores a patient to the patient's harm either physically or psychologically. The net result of this uncaring indifference is instrumental behavior, with almost, if not, total detachment on the healthcare provider's part for the patient's physical and existential needs.

Reported in the literature are several clinical studies of uncaring treatment of patients by healthcare professionals (Eliasson et al., 2008; Quirk et al., 2008; Widar et al., 2007; Wiman and Wikblad, 2004). For example, Marita Widar and colleagues report results from a study on the healthcare experiences of stroke patients (Widar et al., 2007). They find that often patients experienced uncaring attitudes from healthcare providers vis-à-vis pain associated with a stroke. These healthcare professionals ignored patient questions concerning pain or gave technical answers to questions asked by patients about their pain. "So then he reeled off a lot of statistics," recounted one patient, "about how many get well and how many need rehab and how many go away and die. I couldn't make head or tail of it. Goodness knows what it was for. Pointless" (Widar et al., 2007, p. 44). At times, patients felt that the healthcare professionals disbelieved them with respect to the veracity of their pain experience. Worse yet, patients experienced a "superiority" attitude from their healthcare providers concerning alternative treatments for pain.[13]

4.2.1 Carelessness

Just as the virtue of care is an ontic derivative of caring, so too the vice of carelessness is an ontic derivative of uncaring. Moreover, just as care partially instantiates caring, so too carelessness partially instantiates uncaring (incompetence is the other

[13] Interestingly, however, Mark Quirk and co-workers find that whether the healthcare provider was caring or uncaring depended largely upon the patient's perception (Quirk et al., 2008). For instance, a clinician's inquiry whether patients has someone they could discuss a cancer diagnosis was viewed as uncaring by some patients. Whether a healthcare worker is caring or uncaring is, they conclude, "in the eye of the beholder" (Quirk et al., 2008, p. 364). However, this conclusion does not exclude the reality that uncaring healthcare providers often do cause substantial physical and mental harm.

ontic vice that completes uncaring's instantiation). Carelessness is the lack of concern or consideration for another person and for his or her needs, whether physical or existential. As ontic, it pertains to a person's doing-in-the-world rather than fundamentally being-in-the-world. Carelessness often manifests itself as neglect or indifference to another's needs. Like care, it too exhibits motivational and behavioral dimensions. Motivationally, carelessness reflects a person's incapability to care about another person (carelessness$_1$), i.e. a person could care less about another. Behaviorally, it is a person's inability to take care of another person (carelessness$_2$), i.e. a person is careless in taking care of another. Whereas carelessness$_1$ is an incapability or insufficiency in caring about another person, i.e. not being ready or motivated to care about another, carelessness$_2$ is an inability or a failure to take care of another, i.e. to provide the necessary care for another. Finally, carelessness is not reducible to the traditional vices but serves as an intermediate vice that mediates between uncaring and those vices.

Carelessness is a common vice in healthcare, from mistakes or errors in diagnoses to technical oversights, such as leaving surgical instruments inadvertently inside patients. Such careless behavior often results in patient anxiety, emotional distress, physical harm, or even death. For example, a patient underwent surgery to remove a 6 kg tumor from the stomach (Anonymous, 2006). After discharge, the patient frequently complained of acute abdominal pain. An X-ray of the patient's abdomen revealed a pair of surgical scissors inside the patient, which the surgeon had accidently forgotten to remove. Another surgeon performed an emergency operation on the patient to remove the scissors. Another example of the vice of carelessness that resulted in serious harm involved a patient who underwent a radical double mastectomy for breast cancer (Celizic, 2007).[14] After the surgeon removed the breasts, routine pathological examination of the breast tissue reveled the patient did not have breast cancer. The error occurred earlier when a careless technician mishandled tissue samples by mixing up the patient's tissue sample with another patient's sample that was cancerous. The technician used a technique called "batching" in which the technician processed several tissue samples simultaneously instead of processing each sample separately. These examples reveal that carelessness as a vice may result for a variety of reasons, from simple distraction (in the case of the surgeon) to trying to economize (in the case of the technician).

4.2.2 Incompetence

Just as a gap exists between care$_1$ and care$_2$, so too one exists between carelessness$_1$ and carelessness$_2$. In other words, even though a person is not motivated to care about another person or is simply unable to care about another's welfare (carelessness$_1$) he or she cannot really be careless or incapable of taking care of another's welfare (carelessness$_2$) unless that person is incompetent to take care of

[14] I would like to thank Rachel Sherhart for drawing my attention to this clinical case.

another in terms of physical or existential needs. As competence is the ability to accomplish a task according to accepted professional standards or criteria, so incompetence is an inability to do so. Incompetence is simply more than making a mistake, it involves persistent errors in both medical judgments and practice (Good, 1998). Incompetence can be technical in that a person is incapable of performing a mechanical task effectively or efficiently, or ethical in that a person is unable to demarcate between right and wrong. Incompetence then is often the inability to comply with standards of practice or codes of behavior. Finally, incompetence is the foundation for the vices associated with the general incompetencies (Larkin et al., 2005). For example, incompetence is the basis for the expression of vices associated commonly with patient care, such as disrespect, inattention, negligence, recklessness, sloppiness, or with interpersonal skills, such as indifference, bigotry, callousness, insensitivity, and insincerity.

Incompetence is a frequent vice in healthcare delivery that often results in serious patient harm and distress, as well as death (Bark et al., 1994; Caulford et al., 1994; Rhodes, 1986). Technical incompetence, ranging from faulty medical knowledge to an inability to perform a mechanical or surgical procedure properly, is certainly one of the more prevalent vices is healthcare. For example, a 51 year-old female patient underwent surgery to repair a herniated disc. During the surgery, the surgeon accidently cut the major arteries and the patient died from the surgeon's incompetence to perform the operation safely (Daily Mail Reporter, 2009). Although technical incompetence is common, ethical incompetence on the part of physicians is often a chief complaint among patients. The following is an example of such incompetence,

> A dying patient was referred to hospital by his G[eneral] P[ractitioner] as his wife was no longer able to cope. The doctor was rude and demanded to know exactly what the patient thought he could do for him as he was not a casualty case. He shouted outside the cubicle, and a nurse intervened. Both the patient and his wife were very distressed. The patient died shortly afterwards (Bark et al., 1994, p. 125).

Unfortunately, many people are ignorant or unaware of their incompetence simply because they overestimate their own competence (Dunning et al., 2003). Ignorance of one's own incompetence is particularly common among medical students and novice physicians (Hodges et al., 2001). This ignorance is often an unintended outcome of medical school curricula. These curricula promote incompetence through an emphasis on general training, such as teaching pure or domain-specific medical knowledge particularly in terms of textbook and manuals, and not on context-specific training of skills necessary for clinic practice (Hodges, 2006). In other words, medical faculty do not teach students how to conduct a procedure in detail, especially in terms of individual patient's need, but teach only the broad outline of therapeutic protocols applicable to any patient.

4.2.3 Carelessness-Incompetence Relationship

Carelessness, along with incompetence, fully instantiates uncaring—in terms of the cyclical structure represented in Fig. 4.2—as care and competence instantiate

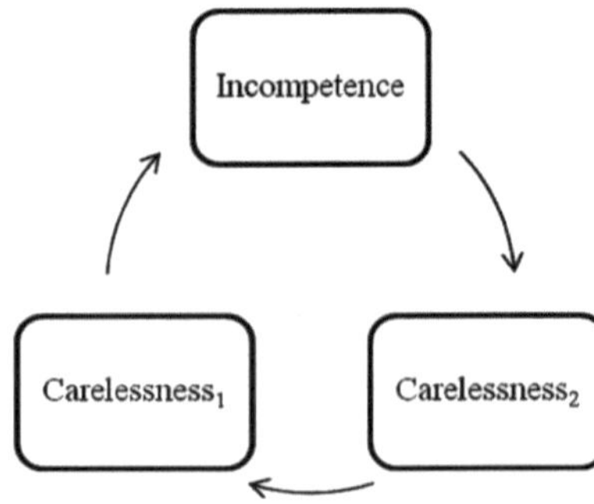

Fig. 4.2 Relationship between carelessness and incompetence

caring. The first segment of the cycle (carelessness$_1$ → incompetence) refers to an agent's lack of motivation to care about (carelessness$_1$) another person. That lack originates from an incapability to have an authentic impulse or desire to care about another, i.e. care less. Being unmotivated, an uncaring agent not only does not want to help a person needing care but ultimately does not do so competently, i.e. not to take care of that person (carelessness$_2$). Often, genuine caring relationships are not possible because of an agent's particular agenda or biases. Moreover, the uncaring agent may not have the ethical competence to take care of (carelessness$_2$) another person, especially another's existential or emotional needs. In addition, an agent who does not care about (carelessness$_1$) another person might be unable to form a caring relationship if that agent is technically incompetent to take care of the one in need. Inauthentic or uncaring relationships are the product of at least two dimensions of the second type of carelessness, a technical dimension (carelessness$_{2a}$) and an ethical dimension (carelessness$_{2b}$). The second segment of the cyclic relationship between carelessness and incompetence (incompetence → carelessness$_2$) represents not taking care of a person competently with respect to incorrect technical actions (carelessness$_{2a}$) and wrong ethical behavior (carelessness$_{2b}$). Specifically, an uncaring physician is incompetent in the practice of medicine as both evidence–based science (technical incompetence) and patient–centered art (ethical incompetence), and is unable to take care of an individual patient's bodily and existential needs.

In the final segment of the uncaring-incompetence cycle (carelessness$_2$ → carelessness$_1$), carelessness$_2$ feeds back to reinforce carelessness$_1$. In other words, as an uncaring agent does not take care of a person in need competently, carelessness$_2$ reinforces carelessness$_1$ or the inability of the careless agent to want to care about helping others in need. Operationally, a person who requires competent care may sever the relationship with a careless$_2$ agent, who then experiences the disappointment of not meeting the needs of the person and, in turn, feels affronted by him or her. This reciprocal connection between carelessness$_2$ and carelessness$_1$ then inhibits the formation of further caring relations with others because the careless or uncaring agent retreats from the pain associated with being unable to meet the needs of others. Not only does this feedback relationship between the two types of carelessness reinforce an uncaring agent's desire not to care about (carelessness$_1$) others but it also weakens an uncaring agent's capability to take care of (carelessness$_2$)

others competently. Eventually, a careless physician, for example, makes little effort to maintain a level of competence that is current with medical standards and expectations, and often harms patients, either physically or psychologically. Lastly, because of the feedback between carelessness_2 and carelessness_1, any natural desire to care about others in need atrophies and eventually disappears, thereby reducing the possibility of an agent to take care of others in need.

Finally, uncaring in terms of carelessness and incompetence, whether technically or ethically, can take a variety of forms. The first form of uncaring involves carelessness_1. In this form, a healthcare provider, for instance, is simply not motivated to care about the patient's welfare. The result can be devastating in that the provider is not motivated to care, or could care less, about the patient so that the provider is incompetent both technically and ethically to meet the patient's physical and existential needs. In some sense, this form of unvirtuous behavior is almost vicious—given the extent of possible harm to the patient. The other two forms involve either technical incompetence and ethical competence or technical competence and ethical incompetence. In the former, the healthcare provider takes care of the patient's physical needs but not the existential needs, while in the latter existential but not physical needs are taken care of.[15] Other forms of uncaring also exist. For example, a healthcare provider might care_1 about a patient but the provider is simply incompetent technically to take care_2 of the patient. In this example, however, the provider would seek someone who can supply such technical care_2. Or, the provider might care_1 about a patient but is simply incompetent ethically to take care_2 of the patient. This latter case is probably rather common in medicine and is often the result of the overemphasis of medical or healthcare education on the technical or scientific and bracketing the ethical or moral dimensions of healthcare.[16]

4.3 Summary

By way of summary for this chapter and as an introduction to the next, I would like to examine Howard Curzer's provocative assertion that care is not an appropriate virtue for healthcare providers but rather "it is a vice" (1993, p. 55).[17] For Curzer,

---

[15] A fourth possible form of carelessness_1 exists. The healthcare provider is competent both technically and ethically. In this case, the provider meets both the patient's physical and existential needs but at some point he or she is unable to sustain that competence in order to meet those needs.

[16] A rare form is the healthcare provider who cares about the patient but is incompetent both technically and ethically to care for the patient. Such a provider, if such a person could be called that, may be the result of a terribly faulty education. Of course, the goal is a healthcare provider who cares about the patient's physical and existential needs so that he or she is fully competent both technically and ethically to care for those needs.

[17] Robert Veatch (1985) also contends that virtues are not required for medical practice, under certain—if not most—circumstances. He claims that technical expertise is what is required, especially with respect to what he calls "stranger medicine," i.e. medicine practiced among strangers. Daniel Putman (1988), in critiquing Veatch, argues that even for stranger medicine virtues are

the requirement that these providers care for patients simply results in too many undesirable consequences, such as inadvertent patient harm, healthcare provider burnout, compromised objectivity, and paternalism, along with a variety of other "isms" like favoritism, sexism, and racism. "Expecting contemporary HCPs [heath care professionals] to care for their patients," charges Curzer, "is as unreasonable as expecting love from a prostitute" (1993, p. 66). Rather than care, he proposes benevolence as the apposite virtue for HCPs. In addition, maintains Curzer, "HCPs should act *as if* they cared for patients as individuals, but it is not necessary or even desirable for them really to care for patients" (1993, p. 62). He assures the reader that he is not advocating uncaring behavior but caring behavior in which HCPs treat their patients the same, as they would treat anyone off the street.[18] He champions substituting an ethics of care behavior for an ethics of care. In other words, as long as a HCP engages in benevolent behavior, such professionals should not be required to like—or to connect emotionally with—patients.

To defend the benevolence thesis, Curzer first divides virtues into general and role virtues. The former represent character traits that make a person good in general terms, while the latter are a specific set of traits that makes a person good with respect to a specific role and allows that person to accomplish the goals of that role. Importantly, general and role virtues do not necessarily overlap. In fact, a general virtue may not be a virtue for some roles. He gives the example of competiveness, which might function as a role virtue in the business arena but not in one's personal life. He acknowledges that care is certainly a general virtue and even a role virtue for some roles, like parenting. What Curzer disapproves of is the following thesis, "Care is a role virtue for HCPs" (1993, p. 53). To critique this thesis, he identifies three components of care as a role virtue for HCPs. The first is to minister to the patient, in which the HCP takes care of the patient; the next is to take interest in the patient, in which the HCP cares about the patient; and, the final is to like the patient, in which the HCP cares for the patient. According to Curzer, care as a role virtue must include all three components, especially the last one, on pain of being trivial. In other words, care of the patient involves not only ministering to and taking an interest in but also liking the patient. For Curzer, the objectionable part of care as a role virtue for HCPs is the requirement that HCPs like their patients, since this often leads to an inappropriate emotional attachment that yields the undesirable consequences listed above.

I would like to respond to Curzer's thesis in terms of virtuous and unvirtuous physician, by first returning to Peabody and a case study he presents in his classic paper to illustrate the importance of caring for the patient, especially in the clinical setting. In the case study, the patient, Mrs. Brown, is a young woman who—after meals—is suffering from nausea and upper digestive discomfort. After consulting

---

necessary for consistent medical practice—particularly when practice challenges a physician's self-interests.

[18] Curzer cites with approval Veatch's claim that contemporary medicine is "stranger medicine," since patients and physicians are generally strangers to one another.

a number of physicians, she is currently on a diet of milk and crackers; however, she is still suffering from her symptoms. She comes to the hospital with the hope of being properly diagnosed and treated for her symptoms. Unfortunately, various tests reveal no underlying anatomical pathology to account for the symptoms. Frustrated and eager to discharge the patient in order to fill the hospital bed with more clinically interesting patients, the attending clinician tells Mrs. Brown,

> you can send for your clothes and go home tomorrow. There really is nothing the matter with you, and fortunately you have not got any of the serious troubles we suspected. We have used all the most modern and scientific methods and we find that there is no reason why you should not eat anything you want to. I'll give you a tonic to take when you go home (Peabody, 1927, p. 878).

But upon arriving home, her symptoms return and she then tries alternative forms of medicine like chiropractic or Christian Science.

Peabody begins the analysis of this case study questioning whether Mrs. Brown's treatment by the attending clinician and hospital staff with all its technical probing was "too scientific." His answer is "not at all." After all, Mrs. Brown could be suffering from an unknown anatomical pathology, such as gastric ulcer, and the burden upon the staff is to determine what that pathology is. Peabody then asks a rather startling question of whether the treatment was not scientific enough. Crucial for understanding this question is what he means by a scientist and the scientific method. For Peabody, "a scientist is known, not by his technical processes, but by his intellectual processes; and the essence of the scientific method of thought is that it proceeds in an orderly manner toward the establishment of a truth" (1927, p. 879). Given this conception of a scientist and the scientific method, Peabody makes a bold assertion that the attending clinician and hospital staff were unscientific, since they did not utilize the scientific method properly and settled not for the truth but for a half-truth. A truly scientific examination of Mrs. Brown would not have stopped after the technical tests did not reveal any anatomical pathology and would have proceeded until the staff discovered the "real cause" of Mrs. Brown's symptoms.

To conclude his analysis, Peabody states, "Speaking candidly, the case was a medical failure in spite of the fact that the patient went home with the assurance that there was 'nothing the matter' with her" (1927, p. 879). Why was the case a failure? Was it simply because the attending clinician and hospital staff were unscientific? Or, did these HCPs not have sufficient interest in the patient as a person? The answer to these questions lies in Peabody's apt articulation for the role of care in treating patients, as quoted earlier, "One of the essential qualities of the clinician is interest in humanity, for the secret of the care of the patient is in caring for the patient" (1927, p. 882). The case was a failure because the attending clinician and hospital staff cared more for the patient's disease than for the patient. Once they could not identify an organic cause of the patient's complaint, they simply dismissed and then discharged her. They allowed the limits of their medical technology to constrain their ability to take care of the patient. And, so they accepted only a partial, not the full, fact of the patient's condition, i.e. the attending clinician's objectivity was compromised by not caring about the patient. Consequently, by not taking an interest in or caring about ($care_1$) the patient the attending clinician was unable to minister to or to take care of ($care_2$) the patient. These HCPs may be benevolent in their behavior, but, according to Peabody, they would not be caring.

How would Peabody respond to Curzer? Peabody would certainly agree with Curzer that inappropriate or excessive emotional attachment of a HCP to the patient might lead to undesirable consequences identified by Curzer. For example, an undue liking of Mrs. Brown might cause a clinician to withhold clinical test results that are important for her to maintain autonomy in deciding the best course of treatment, simply because the results presage a bad prognosis. Or, the clinician withholds the information from Mrs. Brown not wanting to hurt her because of an acute emotional attachment. However, Peabody would caution Curzer not to throw the proverbial baby out with the bath water. In other words, care is multidimensional and complex, as Curzer acknowledges, and so one must not eliminate care as a role virtue for clinicians simply because of inappropriate or excessive emotional attachment is possible. After all, for Peabody the secret of taking care of ($care_2$) or ministering care to the patient is in caring about ($care_1$) the patient. Importantly, he is not saying that the clinician must like the patient. Rather, for Peabody, the foundation of patient care is not care qua liking but rather care qua general interest in humanity, which translates into $care_1$ for the patient. This distinction is critical, Peabody would maintain, since genuine caring has an emotional component—interest in or concern for humanity—and without that component motivating clinicians to take care of ($care_2$) patients, especially patients they may not like, quality healthcare would be next to impossible. Peabody would challenge Curzer as to how the thesis of benevolence *sans* any emotional connection would motivate clinicians to be benevolent or to act for the good of their patients. For Peabody, Curzer's benevolent HCP is precisely the attending clinician who practiced a subpar technical medicine and failed to take care of ($care_2$) Mrs. Brown.

Following Peabody, I would stress that Curzer fails to appreciate the need for emotional attachment of the HCP to the patient in terms of motivating the HCP to minister care to or to take care of ($care_2$) the patient. In other words, Curzer identifies three different attributes of care when only two ($care_1$ and $care_2$) are sufficient, as articulated by Peabody and others.[19] A HCP need not like a patient but such a professional should be motivated to care about ($care_1$) or at least have a general interest in humanity and especially patients. I would argue that the reason for professional burnout is not that HCPs are emotionally attached to their patients but that they are attached inappropriately or have no emotional attachment, what is often called emotionally detached concern (Halpern, 2001). In fact, emotional detachment does not prevent burnout among HCPs (Huggard, 2003). Truly caring HCPs know the appropriate or proper boundaries for connecting to a patient emotionally, so to meet the patient's bodily and psychological needs without overextending themselves. Unfortunately, Curzer places too much emphasis on care qua liking to avoid

[19] For example, van Hooft (1996) identifies two attributes of care—although Noddings (1984) does list three. However, neither of them lists liking another person as a necessary attribute for caring.

what he considers a trivial notion of care, only to throw the baby (care) out with the bath water (emotional attachment). Care qua motivation and behavior suffices and does not depend on whether the caregiver likes the person needing care.

Although Curzer recognizes that care is motivational, he fails to appreciate its importance or to assign it a significant role vis-à-vis the other dimensions of care. The question he neglects to ask is why HCPs should be benevolent or behave in a caring fashion in the first place. His thesis of benevolence *sans* emotions is not robust enough to overcome the undesirable consequences he listed but rather would exacerbate them. Patients can surely tell when HCPs are simply going through the motions to behave *as if* they care. Such behavior is not only fake, as Curzer admits, but it is mechanical and it would thereby hinder forming a genuine patient-physician relationship, which would eventually have a debilitating consequence of the patient mistrusting the HCP. Bracketing or eliminating emotions from the health-care relationship runs the risk of emotions entering the back door and causing more damage than if emotions are initially incorporated appropriately into the relationship. I would argue that Curzer's thesis of benevolence *sans* emotional attachment is not a virtue but simply an ethical principle operating under either a deontological or consequential, but certainly not from a virtue ethics, perspective. In other words, HCPs are simply to follow the benevolence rule to do good and the nonmaleficence rule not to harm patients or to maximize the amount of patient good and to minimize patient harm.

The virtuous physician, on the other hand, is someone who genuinely achieves the patient's good from authentic behavior that is motivated by the compound or composite virtue of prudent love. Prudent love represents the transformation of care and competence specifically into the traditional virtues of love and prudence, respectively. In contrast, the unvirtuous physician does not achieve the patient's good because of the compound vice of loveless imprudence—which reflects the transformation of carelessness and incompetence into lovelessness and imprudence, respectively. I now turn in the next chapter to the notions of prudent love and loveless imprudence in the practice of contemporary healthcare.

References

ACGME, Accreditation Council for Graduate Medical Education. 2007. Outcome project: General competencies. Web site: http://www.acgme.org. Accessed October 15, 2009.

Allmark, P. 1998. Is caring a virtue? *Journal of Advanced Nursing* 28:466–472.

Anonymous. 2006. Careless doctors cause discomfort to patient. *Bio-Medicine* http://www.bio-medicine.org/medicine-news/Careless-Doctors-Cause-Discomfort-to-Patient-11148-1/.

Barbour, A. 1995. *Caring for patients: A critique of the medical model.* Stanford, CA: Stanford University Press.

Bark, P., C. Vincent, A. Jones, and J. Savory. 1994. Clinical complaints: A means of improving quality of care. *Quality in Health Care* 3:123–132.

Barker, P. 2000. Reflections on caring as a virtue ethic within an evidence-based culture. *International Journal of Nursing Studies* 37:329–336.

Benner, P. 1997. A dialogue between virtue ethics and care ethics. *Theoretical Medicine* 18:47–61.

Bevis, E.O. 1988. Caring: A life force. In *Care: An essential human need*, ed. M.M. Leininger, 49–59. Detroit, MI: Wayne State University Press.
Brody, J.K. 1988. Virtue ethics, caring, and nursing. *Scholarly Inquiry for Nursing Practice: An International Journal* 2:87–96.
Carraccio, C., R. Englander, S. Wolfsthal, C. Martin, and K. Ferentz. 2004. Educating the pediatrician of the 21st century: Defining and implementing a competency-based system. *Pediatrics* 113:252–258.
Cates, D.F., and P. Lauritzen (eds.) 2001. *Medicine and ethics of care*. Washington, DC: Georgetown University Press.
Caulford, P.G., S.B. Lamb, T.B. Kaigas, E. Hanna, G.R. Norman, and D.A. David. 1994. Physician incompetence: Specific problems and predictors. *Academic Medicine* 69:S16–S18.
Celizic, M. 2007. I don't want this to happen to anyone else. *Today* http://www.msnbc.msn.com/id/21127917/.
Clement, G. 1996. *Care, autonomy, and justice: Feminism and the ethic of care*. Boulder, CO: Westview Press.
Curzer, H.J. 1993. Is care a virtue for health care professionals? *Journal of Medicine and Philosophy* 18:51–69.
Daily Mail Reporter. 2009. Mother died after heart vessels severed during routine back operation. *Mail Online*: http://www.dailymail.co.uk/news/article-1221801/Mother-died-heart-vessels-severed-routine-operation.html
Duffy, J.R. 2009. *Quality caring in nursing: Applying theory to clinical practice, education, and leadership*. New York, NY: Springer.
Dunning, D., K. Johnson, J. Ehrlinger, and J. Kruger. 2003. Why people fail to recognize their own incompetence. *Current Directions in Psychological Science* 12:83–87.
Eliasson, M., G. Kainz, and I. von Post. 2008. Uncaring midwives. *Nursing Ethics* 15:500–511.
Epstein, R.M., and E.M. Hundert. 2002. Defining and assessing professional competence. *Journal of American Medical Association* 287:226–235.
Eriksson, S., G. Helgesson, and A.T. Höglund. 2007. Being, doing, and knowing: Developing ethical competence in health care. *Journal of Academic Ethics* 5:207–216.
Fry, S.T. 1990. The philosophical foundations of caring. In *Ethical and moral dimensions of caring*, ed. M.M. Leininger, 13–24. Detroit, MI: Wayne State University Press.
Fry, S.T. 1991. A theory of caring: Pitfalls and promises. In *Caring: The compassionate healer*, eds. D.A. Gaut and M.M. Leininger, 161–172. New York, NY: National League for Nursing Press.
Galvin, K.T. 2010. Revisiting caring science: Some integrative ideas for the 'head, hand and heart' of critical care nursing practice. *Nursing in Critical Care* 15:168–175.
Gilligan, C. 1982. *In a different voice: Psychological theory and women's development*. Cambridge, MA: Harvard University Press.
Good, M.-J.D. 1998. *American medicine: The quest for competence*. Berkeley, CA: University of California Press.
Halldórsdóttir, S. 1996. *Caring and uncaring encounters in nursing and health care–developing a theory*. Linköping University Medical Dissertations No. 493.
Halpern, J. 2001. *From detached concern to empathy: Humanizing medical practice*. New York, NY: Oxford University Press.
Halwani, R. 2003. Care ethics and virtue ethics. *Hypatia* 18:161–192.
Hamington, M. 2004. *Embodies care: Jane Addams, Maurice Merleau-Ponty, and feminist ethics*. Urbana, IL: University of Illinois Press.
Heidegger, M. 1996. *Being and time* (trans: Stambaugh, J.). Albany, NY: State University of New York Press.
Held, V. (ed.) 1995. *Justice and care: Essential readings in feminist ethics*. Boulder, CO: Westview Press.
Held, V. 2006. *The ethics of care: Personal, political, and global*. New York, NY: Oxford University Press.

Hodges, B. 2006. Medical education and the maintenance of incompetence. *Medical Teacher* 28:690–696.

Hodges, B., G. Regehr, and D. Martin. 2001. Recognizing one's own incompetence: Novice physicians who are unskilled and unaware of it. *Academic Medicine* 76:S87–S89.

Hollingsworth, W. 1995. *Taking care: The legacy of Soma Weiss, Eugene Stead and Paul Beeson*. Charlotte, NC: Professional Press.

Huggard, P. 2003. Compassion fatigue: How much can I give? *Medical Education* 37:163–164.

Jecker, N.S., and W.T. Reich. 2004. Care: Contemporary ethics of care. In *Encyclopedia of bioethics*, 3rd edition, volume 1, ed. S.G. Post, 367–374. New York, NY: Thomson Gale.

Jormsri, P., W. Kunaviktikul, S. Ketefian, and A. Chaowalit. 2005. Moral competence in nursing practice. *Nursing Ethics* 12:582–594.

Knowlden, V. 1990. The virtue of caring in nursing. In *Care: An essential human need*, ed. M.M. Leininger, 89–94. Detroit, MI: Wayne State University Press.

Kohlberg, L. 1981. *The philosophy of moral development: Moral stages and the idea of justice*. San Francisco, CA: Harper & Row.

Larkin, G.L., M.P. McKay, and P. Angelos. 2005. Six core competencies and seven deadly sins: A virtue-based approach to the new guidelines for graduate medical education. *Surgery* 138: 490–497.

Larrabee, M.J. (ed.) 1993. *An ethic of care: Feminist and interdisciplinary perspectives*. New York, NY: Routledge.

Ledesma, C.R. 2011. Relationship-based care: A new approach to caring. *Nursing Management* 42:40–43.

Leininger, M.M. 1988. Care: The essence of nursing and health. In *Care: The essence of nursing and health*, ed. M.M. Leininger, 3–15. Detroit, MI: Wayne State University Press.

Marcum, J.A. 2008. *An introductory philosophy of medicine: Humanizing modern medicine*. New York, NY: Springer.

Mayeroff, M. 1971. *On caring*. New York, NY: Harper & Row.

Noddings, N. 1984. *Caring: A feminine approach to ethics and moral education*. Berkeley, CA: University of California Press.

Parse, R.R. 1988. Caring from a human science perspective. In *Care: An essential human need*, ed. M.M. Leininger, 129–132. Detroit, MI: Wayne State University Press.

Paul, O. 1991. *The caring physician: The life of Dr. Francis W. Peabody*. Boston, MA: The Francis A. Countway Library of Medicine.

Peabody, F.W. 1927. The care of the patient. *Journal of American Medical Association* 88:877–882.

Peabody, F.G. 1933. *Francis Weld Peabody 1881–1927, A memoir*. Cambridge, MA: Riverside Press.

Pettersen, T. 2008. *Comprehending care: Problems and possibilities in the ethics of care*. Lanham, MD: Lexington Books.

Putman, D.A. 1988. Virtue and the practice of modern medicine. *Journal of Medicine and Philosophy* 13:433–443.

Quirk, M., K. Mazor, H.-L. Haley, M. Philbin, M. Fischer, K. Sullivan, and D. Hatem. 2008. How patients perceive a doctor's caring attitude. *Patient Education and Counseling* 72:359–366.

Reich, W.T. 2004. Care: History of the notion of care. In *Encyclopedia of bioethics*, 3rd edition, volume 1, ed. S.G. Post, 349–364. New York, NY: Thomson Gale.

Rhodes, P. 1986. Incompetence in medical practice. *British Medical Journal* 292:1293–1294.

Roach, M.S. 1992. *The human act of caring: A blueprint for the health professions*, revised edition. Ottawa, ON: Canadian Hospital Association Press.

Salsberry, P.J. 1992. Caring, virtue theory, and a foundation for nursing ethics. *Scholarly Inquiry for Nursing Practice: An International Journal* 6:155–167.

Slote, M. 2007. *The ethics of care and empathy*. New York, NY: Routledge.

Tishler, P.V. 1992. "The care of the patient": A living testimony to Francis Weld Peabody. *Pharos* 55:32–36.

Tishler, P.V. 2010. Soma Weiss, Alfred S. Reinhart, and the care of the patient. *Perspectives in Biology and Medicine* 53:75–86.
Tong, R. 1998. The ethics of care: A feminist virtue ethics of care and healthcare practitioners. *Journal of Medicine and Philosophy* 23:131–152.
van Hooft, S. 1995. *Caring: An essay in the philosophy of ethics*. Boulder, CO: University Press of Colorado.
van Hooft, S. 1996. Bioethics and caring. *Journal of Medical Ethics* 22:83–89.
van Hooft, S. 2006. *Caring about health*. Burlington, VT: Ashgate.
Veatch, R.M. 1985. Against virtue: A deontological critique of virtue theory in medical ethics. In *Virtue and medicine: Explorations in the character of medicine*, ed. E.A. Shelp, 329–345. Dordrecht: Reidel.
Veatch, R.M. 1998. The place of care in ethical theory. *Journal of Medicine and Philosophy* 23:210–224.
Warelow, P.J. 1996. Is caring the ethical ideal? *Journal of Advanced Nursing* 24:633–661.
Watson, J. 2008. *Nursing: The philosophy and science of caring*, revised edition. Boulder, CO: University Press of Colorado.
Widar, M., A.-K. Ek, and G. Ahlströöm. 2007. Caring and uncaring as narrated by persons with long-term pain after a stroke. *Scandinavian Journal of Caring Science* 21:41–47.
Wiman, E., and E. Wikblad. 2004. Caring and uncaring encounters in nursing in an emergency department. *Journal of Clinical Nursing* 13:422–429.

Chapter 5
On Prudent Love and Imprudent Lovelessness

I initially analyze the notion of virtuous physician with respect to the transformation of the ontic virtues of competence into prudent wisdom and of care into personal radical (including compassionate, empathic, and altruistic) love. These virtues serve to identify and define a virtuous physician. As Charles Bryan concludes a series of articles on the role of virtues in contemporary medicine, "wisdom and love are, respectively, the preeminent virtues for competence and caring, the twin pillars of medicine" (2007b, p. 135). However, the virtues do not function simply as single virtues isolated from one another to define the virtuous physician but as a compound or composite virtue of prudent love in which the two virtues are intimately integrated with one another. As a result of this intimate relationship, the compound virtue operates in a holistic, synergistic fashion, especially in terms of its relationship to the caring cycle and the care-competence cycle. I then examine the notion of the compound vice, imprudent lovelessness, which is derived from the transformation of incompetence and carelessness, as well as its relationship to the uncaring cycle and the carelessness-incompetence cycle. This compound vice identifies and defines an unvirtuous physician.

5.1 Prudent Love

A full explication of the notion of virtuous physician *qua* authentic professional requires transformation of both competence into prudence and care into love, along with their integration into the compound or composite virtue of prudent love. To that end, I first examine the transformation of technical and ethical competence into prudent wisdom. This wisdom entails knowing what diagnostic test to run or therapeutic modality to prescribe, not in terms of practicing defensive medicine but rather with respect to practicing offensive or effective medicine. Today, fear often motivates the unvirtuous physician to practice defensive medicine, especially given the multitudes of uncertainties facing the physician.[1] The virtuous physician, however,

[1] These uncertainties range from technical uncertainties like not knowing the etiology of a particular disease to personal uncertainties like not knowing a patient's preferences in terms of disease treatment (Ghosh, 2004).

J.A. Marcum, *The Virtuous Physician*, Philosophy and Medicine 114,
DOI 10.1007/978-94-007-2706-9_5,

practices not from such a dysfunctional attitude or stance but from one that has the patient's good in sight and not the bad that can happen to the physician. This prudent wisdom also involves knowing how to apply the theoretical in a practical manner to an individual patient who stands before the physician. Such wisdom requires not only an intimate knowledge of the patient, but also a physician's clinical experience and the ability to reflect on that experience.

The possibility for intimate knowledge of the patient is not simply the care a physician exhibits for the patient but also the love the physician has for that patient. This love is not just compassionate, empathic, or even altruistic in nature, but it also represents an attitude or a stance that is potentially radical or unrestricted vis-à-vis a patient's medical needs (Post, 2003; Toner, 2003). It is also not limited to a clinician's emotional response but includes the prudent wisdom of knowing how to engage the affective aspect of medical care. Together, the two traditional virtues of prudence and love compose a compound or composite virtue, prudent love, which empowers the virtuous physician *qua* authentic professional to make wise and caring clinical decisions in order to provide the quality-of-care a patient expects and deserves. In this section, I first discuss prudent wisdom and then radical love, concluding with an exposition of prudent love's nature and its relationship to caring.

5.1.1 Prudent Wisdom

Traditionally, as noted in Chapter 3, prudence is the ability to make decisions often in the face of uncertainties. As noted then in Chapter 4, such decisions depend upon skills learned while being trained to achieve a specific level of competence, particularly in a profession like medicine. However, as hinted in that chapter, competence, although necessary, is insufficient. What is required further is a capacity to make wise and prudent decisions. Such wisdom weds both the theoretical and practical dimensions of traditional wisdom and prudence, i.e. modern prudence is not simply limited to a traditional understanding of Aristotelian *phronesis* or practical wisdom.[2] For example, Paul Baltes and colleagues of the Berlin wisdom project, founded by Baltes in the 1980s, propose a comprehensive notion of wisdom that not only combines the theoretical and practical but also the specific dimensions of wisdom pertinent to the particularities of life. "The Berlin Paradigm," explains Baltes and his colleague Jacqui Smith, "combines a broad definition of wisdom as excellence in mind and virtue with a specific characterization of wisdom as an expert knowledge system dealing with the conduct and understanding of life" (2008, p. 58). To

---

[2] Aristotle himself champions a larger notion of phronesis or prudence *vis-à-vis* the other intellectual virtues than ascribed to him traditionally, when he writes, "for with the presence of the one quality [virtue], practical wisdom [phronesis], will be given [or possess by the knower] all the virtues" (1998, p. 158). See Flyvbjerg (2001), for further discussion of the relationship of phronesis to episteme and techne.

that end, Baltes and Smith identify five components of wisdom as expert knowledge system, which they divide into two separate strata.

The first stratum combines the two components of theoretical and practical wisdom, or what Baltes and Smith (2008) call factual and strategic knowledge. Factual knowledge pertains to understanding the underlying reality behind the phenomenal and is similar to Aristotle's notion of theoretical wisdom. According to Baltes and Smith, it is also concerned with understanding the "fundamental pragmatics of life." Importantly, individuals utilize this knowledge to construct their own unique lives. "It is applied," insists Baltes and Smith, "to life planning (e.g., which future life goals to pursue and how?), life management (e.g., how to deal best with critical problems such as suicide or family conflict?), and life review (e.g., how best to make sense of our life history and past experiences?)" (2008, p. 58). The understanding associated with the pragmatics of life includes the nature of human existence and development, social mores and contexts, and relationships among social members. Strategic knowledge is procedural and practical in nature and involves how and when to apply one's knowledge of life's pragmatics.

In the second stratum, Baltes and Smith combine the final three components of their notion of wisdom. The first is "lifespan contextualism" or "knowledge about the normative and nonnormative influences on an individual's life" (2008, p. 58). These influences can be biological, social, or cultural in their nature. The next component is value relativism, which pertains to knowledge about values and goals, along with their prioritization, held by an individual. This component is crucial for developing respect for values and goals contrary to one's own, although one need not adopt personally those contrary values and goals. The final component involves life's uncertainties and contingencies and how best to manage those uncertainties and contingencies. The Berlin notion of wisdom, with its emphasis on the life's pragmatics, pertains to how best to plan one's life in terms of future goals and aspirations, manage one's life with respect to its manifold uncertainties and contingencies, and review one's life in order to find meaning in it. Such wisdom is not only advantageous for the individual but also for others in that "it contributes to the construction of the lives of others in the form of good advice, exceptional judgment, excellent mentoring, or insightful organization of education and society" (Baltes and Smith, 2008, p. 58).

Although Monika Ardelt (2004a) does not challenge the theoretical dimensions of the Berlin notion of wisdom, Ardelt does challenge its seemingly exclusive focus on wisdom per se as expert knowledge. Rather than focus on wisdom per se and limit it to expert knowledge, she focuses on the wise person and identifies three fundamental characteristics of that type of person, based on earlier work by Vivian Clayton and James Birren (1980). The first characteristic is cognitive, which "refers to the desire to know the truth and attain a deeper understanding of life, particularly with regard to intrapersonal and interpersonal matters" (Ardelt, 2004a, p. 275). She acknowledges that this characteristic is similar to the Berlin definition of wisdom but maintains that it reflects the wise person and not just expert knowledge. The next characteristic is reflective and, according to Ardelt, "represents self-examination, self-awareness, self-insight and the ability to look at phenomena and events from different perspectives" (2004a, p. 275). This characteristic helps the wise person to understand a personal and subjective viewpoint in order to transcend it and to understand and appreciate another's viewpoint. The final characteristic is affective and "consists of a person's sympathetic and compassionate love for others" (Ardelt, 2004a, p. 276). This characteristic accounts for the ability of wise people to engage others in a genuinely heartfelt manner in order to respond wisely to them. Ardelt admits that her notion of the wise person is a "Weberian ideal" but she argues that wisdom represents a range from high to low on which a wise person approximates the ideal.[3]

Andrew Achenbaum (2004) utilizes but extends both the Berlin notion of wisdom and Ardelt's notion of wise person by focusing on wise acts and three key features that characterize them. The first is objectivity in that wise acts are impartial. "Wise people choose," according to Achenbaum, "their best course of action in an objective manner, unconstrained by conventional ways of expected behavior" (2004, p. 302). A wise person acts with a vision that sees not only the potential within given circumstances or a situation but also the limitations. To achieve such objectivity, a wise person must act rationally—the next feature of the wise act. "Rational thinking," claims Achenbaum, "informs how wise people treat others so that they act in a just, compassionate way" (2004, p. 302). In other words, a wise person thinks deeply through introspection about the potential and limitations of the action before acting. The final key feature of a wise act is transcendence. "Wisdom's vision is a worldly wise perspective," stresses Achenbaum, "which reaches for higher stakes" (2004, p. 302). The wise act requires this transcendence because it strives for universality and not simply for the parochial. As wise people act objectively and rationally, according to Achenbaum, they transcend the *Angst* associated with the here and now to enlarge their vision of reality so to engage reality as people of integrity.

The role of wisdom in clinical medicine is the subject of several scholarly examinations and analyses, besides those cited previously in Chapter 3 (Egonsson, 2007; Henry, 1993; Parker, 2002; Szawarski, 2004). Ricca Edmondson and Jane Pearce (2007), for example, develop a triadic model for the role of wisdom in clinical decision-making. Briefly, the three components of their model consist of the capacities of the self as deliberator, the other as beneficiary of the deliberations, and finally the deliberated problem. Although this model goes a long way to incorporating wisdom into clinical practice, a more robust model includes additional components of wisdom as delineated by the Baltes-Ardelt-Achenbaum exchange. I propose to converge the Berlin notion of wisdom as expert knowledge, Ardelt's notion of the wise

[3] Berlin's reaction to Ardelt's alternative notion of wisdom as the wise person in terms of the cognitive, reflective, and affective is to challenge as inaccurate her characterization of its notion of wisdom. Baltes and his colleague Ute Kunzmann (2004) argue that the Berlin notion includes the reflective and affective dimensions of wisdom. In critique of Ardelt's notion, they contend that wise people are not wisdom itself but only approximations. In response, Ardelt emphasizes that wisdom is not wise knowledge found in texts but wise people and that "the wisdom-related knowledge that is written down in texts remains theoretical or intellectual knowledge until a person re-transforms it into wisdom" (2004b, p. 305).

person as cognitive, reflective, and affective, and Achenbaum's notion of wise action as objective, rational, and transcendent, in order to explicate the traditional intellectual virtue of prudence vis-à-vis the notion of virtuous physician—even though these notions are apparently exclusive of each other for accounting for the notion of wisdom and often presented as such by their proponents. To that end, I first appropriate the Berlin notion of wisdom for clinical medicine and then utilize it as the foundation for Ardelt's notion of the wise person and for Achenbaum's notion of wise action to discuss what the prudent physician is and does in terms of clinical practice.

For the prudently wise physician in terms of the Berlin notion of wisdom, the first stratum is factual and strategic clinical knowledge. In terms of factual knowledge, the prudent physician must be knowledgeable about the contemporary state of medical facts and information. That physician must not only know what is current but must also be well-trained, in terms of skillfully applying that knowledge to help the patient recover or heal. In other words, the prudent physician must be technically competent. Besides being technically competent, the prudent physician must also be morally or ethically competent in distinguishing between good and bad or right and wrong, respectively. Simply being technically competent may be insufficient in the clinic or hospital, especially when the patient's values or preferences conflict with those of the physician or the healthcare system. In terms of strategic knowledge, the prudent physician must know best when and how to bring factual knowledge to bear on a patient's case. The procedural application of factual knowledge can have a significant impact clinically on a patient's outcome, given the veracity of that factual knowledge. Misapplication of such knowledge can be devastating for the patient's health. Importantly, practical and theoretical medical knowledge go hand-in-hand and the prudent physician understands the significance of this relationship. For, to act without truly knowing the underlying causes could ultimately harm the patient.

As for the second stratum, the prudently wise physician through a careful medical history accesses the vital details of the patient's illness story, i.e. the lifespan context. Such details are important for understanding the context in which the patient suffers vis-à-vis the disease. Without this context, the physician is unable to identify and factor into both diagnosis and therapy crucial elements that may have a significant impact on clinical outcome. Obtaining this context also provides the prudent physician knowledge of the patient's value system or preferences, which may be germane to treating the patient successfully. The patient's values are essential not only for understanding the patient's overall lifespan context but also for navigating problem areas that arise if values conflict between physician and patient. The prudent physician must be able to appreciate and respect the patient's value system and the role it plays in the patient's life and health. Finally, the prudent physician can negotiate the uncertainties and contingencies that often accompany patient treatment. Through life experience, the physician can assist the patient in adjusting to these uncertainties and in using them to achieve a positive medical outcome, whether curing the disease or simply managing it effectively. In sum, the Berlin notion of wisdom as expert knowledge provides an epistemic basis for the prudent physician's clinical practice. Through such wisdom, the prudent clinician has the knowledge and intellectual resources to help and not to harm the patient.

Although expert knowledge is vital for wise deliberation on the part of the prudently wise physician, such a physician must be wise per se, i.e. a wise person. Ardelt's characterization of the wise person as cognitive, reflective, and affective is an apt characterization of the prudent physician and rounds out the personal dimension of the virtuously wise physician vis-à-vis the Berlin notion of wisdom. First, as cognitive, the prudent physician knows not only the most current medical facts but also what those facts mean for the individual patient and his or her illness story. As reflective, the prudently wise physician embeds knowledge of both medicine and the individual patient within a larger context that maintains the integrity of both patient and physician. Lastly, as affective, the prudent physician displays the appropriate concern and compassion not only for the patient's pain and suffering but also for the existential *Angst* experienced during illness. Ardelt's wise person may represent a Weberian ideal but that person *qua* physician represents an Oslerian ideal as well, for "wise elders tend to be satisfied with their life because they are able to accept the reality of the present moment with equanimity, which helps them to deal with life's uncertainty and the physical, social, and emotional losses that often accompany old age" (2004a, p. 281). Such characterization represents the ideal Oslerian prudent physician, who in the face of clinical uncertainties and challenges faces them with a similar equanimity to treat the patient wisely.

The prudently wise physician must not only have wisdom as expert knowledge and be a wise person, but that physician must also act wisely or be the agent of wise actions. To that end, Achenbaum's characterization of wise acts as objective, rational, and transcendent, is certainly applicable to a prudent physician's wise acts. With respect to objectivity, the prudent physician's wise acts are impartial and neutral with respect to either the physician's or the medical community's agenda as to what is best for the patient. The prudent physician includes the patient's illness story in both diagnosis and therapy, in order to be as objective as possible and to avoid imposing the physician's subjective concerns and biases onto the patient. The prudent physician's wise acts are also rational and conform to standards for attaining valid and sound clinical judgments upon which the clinician in consultation with the patient makes decisions that are best for the patient's wellbeing. Lastly, the prudent physician's wise acts are transcendent in that the acts are the result of a wider perspective than simply a local perspective in which the physician operates. As such, these acts are applicable to patients other than those the physician is treating and may serve to help stimulate the physician to further wise acts.

Importantly, Malcolm Parker (2002) argues that EBM is not contrary fundamentally to traditional clinical wisdom. Parker addresses the fears many clinicians have concerning EBM, especially fears about reducing the medical world to just the measurable or quantifiable and restricting clinical judgments to simply empirical evidence and eliminating the clinician's experience entirely. "EBM does not privilege evidence over the overall judgment of the clinician," Parker assures the fearful clinician, "but it does privilege systematic knowledge claims over the more fallible claims of experience. Nor does it suggest," he adds, "that clinical wisdom does not involve much more than 'hard' empirical data" (2002, p. 275). However, that "more" required for clinical wisdom is not a mystical element unassailable to valid reasoning, but rather it is the cognitive ability of the clinician to apply empirical evidence obtained from clinical studies to the individual patient's medical needs. Certainly, the art of medicine is important in clinical practice but an art informed by the technical and scientific dimensions of medicine. Lastly, clinical decision-making is not simply the sum of EBM and clinical wisdom. "EBM ought not be conceived as *limited* or *balanced* by clinical wisdom," concludes Parker, "but as increasingly coming to constitute its scientific component" (2002, p. 279). Clinical wisdom, in other words, is an inherent component of the EBM project.[4]

Finally, as Zibigiew Szawarski (2004) points out, wisdom itself can also have a healing power in treating patients, which is separate either from the healing power of nature itself or from the advances in medical science to cure disease. Szawarski identifies three key features of the healing power of wisdom: knowledge, clinical judgment, and self-trust. Knowledge can be (1) universal knowledge about the world in general, especially about the human predicament, (2) broad medical knowledge about diseases and how to treat them, and (3) specific clinical knowledge about the individual patient consulting the physician, especially the values or preferences a patient may hold about life and health. Next, clinical judgment pertains to the art of medicine in which a clinician can call upon a tacit dimension of the decision-making process to make sound diagnostic, prognostic, and therapeutic decisions. Finally, self-trust refers to the capacity of clinicians not to second-guess themselves and to trust in their clinical abilities. "If you do not trust yourself," concludes Szawarski, "you cannot expect that your patient will trust you" (2004, p. 192). The prudently wise physician, then, is the person who can bring these features to bear in treating the individual patient efficiently, effectively, and safely.

5.1.2 Radical Love

Besides prudence, the other chief traditional virtue for explicating the notion of virtuous physician is love, which healthcare practitioners often identify as compassion, empathy, or altruism. This love represents a transformation of the ontic

[4] For Parker, as for Hofmann (2002), Aristotle's notion of *techne* captures more adequately the role of wisdom in clinical decision-making, than Aristotle's notion of *phronesis* or practical wisdom. However, *techne* has its limitations, especially with respect to clinical decision-making and the practice of medicine. Although technical application of clinical information and knowledge is important in providing quality healthcare, it may lead to reduction of patients to simply their disease conditions—a reduction often exemplified by statements like, "The congestive heart failure in room 221." Besides the limitations of *techne*, practical wisdom and theoretical wisdom also exhibit grave limitations. Practical wisdom in the clinic without some understanding of the causes underlying illness can be blind, while theoretical wisdom without some pragmatism can be sterile. Hence, the notion of prudent wisdom advocated herein provides not only the necessary but also the sufficient cognitive faculties to account for sage clinical judgment and decision-making.

virtue of care. Care, as both motivational ($care_1$) and behavioral ($care_2$), is critical for delivering quality healthcare in a professional manner. However, to deliver such healthcare requires a dimension of intensely felt commitment and solicitation on the healthcare provider's part. To bracket that dimension or to deny its importance is to run the risk of eliminating a vital human component in healthcare. Deep down love predicates the very nature of human relationships, and the nature of patient-physician relationships is no different. That love, as noted in Chapter 3, is cognate with the virtues of compassion and empathy. It also includes altruism, as discussed in this chapter. In terms of explicating the virtue of love for prudent love vis-à-vis the notion of virtuous physician, love is first explored in terms of compassion as compassionate love and then in terms of empathy and altruism as empathic and altruistic love, respectively. For the virtuous physician, compassionate, empathic, and altruistic love provides the emotional foundation and complements the cognitive foundation that prudent wisdom provides for practicing quality healthcare. Although these forms of love are necessary for understanding the nature of love required for the prudent love of the virtuous physician, they are insufficient to capture its essence. To that end, I utilize Jules Toner's notion of personal radical love and communion to complete the transformation of care into love.

5.1.2.1 Compassionate Love

The professional literature is replete with the recognition of the need for compassionate or deeply felt love in delivering quality healthcare professionally (Fehr et al., 2008; Graber and Mitcham, 2004; Post, 2003, 2010; Sprecher and Fehr, 2006; Taylor, 1997). For example, Lynn Underwood (2004, 2008) champions the benefits and advantages of compassionate love for healthcare practice. According to Underwood, compassionate love is distinct from both empathy and altruism, as well as from romanticism. She defines compassionate love as "a self-giving, caring love that values the other highly and has the intention of giving full life to the other" (Underwood, 2008, p. 4). A vital component in the definition is valuing another person because the other person deserves respect as a human being. Compassionate love then requires comprehending and assenting intellectually to that valuing or to the inherent value of another person. In addition, it is heart-felt at a fundamental level. "Not that everyone will feel gushing emotion when giving compassionate love to another," admits Underwood, "but some sort of emotional engagement and understanding seem to be needed to love fully in an integrated way" (2008, p. 8). Lastly, she also stresses the volitional nature of compassionate love. In other words, the compassionate person freely chooses to care for another person in order to promote or realize another's good or wellbeing.

Underwood (2008) proposes a research model to investigate empirically the notion of compassionate love, to explicate precisely it meaning, and to enlarge broadly its scope. The model is composed of three modules. The first is the context or environment, what Underwood calls the "substrate," in which a compassionate person loves. The contextual or environmental elements influencing a compassionate person include cultural, social, and physical factors, as well as emotional, cognitive, and personality traits. She gives the example of an extrovert as more compassionate than an introvert—or at least exhibiting compassion more freely. In addition, the substrate can develop or change over time, such that what was once an important factor or trait for expressing compassionate love is no longer important or even necessary. Also, different professional situations often require different contexts or environments for expressing compassionate love. The next module represents the motivation for initially expressing compassionate love and the discernment on when and how to express it. Because motives are often mixed, Underwood encourages self-reflection on motives to overcome or eliminate self-centered ones. Discernment is critical in self-reflection and relies on the intellectual ability to evaluate critically what is required to love another person compassionately. The last module pertains to a compassionate person's attitudes and actions, in which the person expresses compassionate love fully and completely. Finally, the full expression of this love has a positive impact on the first module, supporting and promoting the context or environment for further expression of compassionate love.[5]

Stephen Post also advocates a notion of compassionate love, especially for the healthcare professions. "Love can take the form of active *compassion*," insists Post, "when someone is suffering and needs support. Compassion includes responsive helping behavior. It is an emotional state with practical consequences. In times of compassion," he adds, "we give ourselves and we discover ourselves" (2003, p. 5). His notion has several important implications for healthcare providers. First, compassionate love is a response that requires cognitive assessment of the other's need in order to respond appropriately. Next, it also represents a native or innate affective desire to help others who are suffering or in pain. Third, it is not only cognitive and affective in nature but also functional. In other words, the compassionate healthcare provider responds from the heart to the call of the person in need through the provider's hands-on or practical help. Moreover, the compassionate healthcare provider, in loving patients, not only discovers who patients are as persons but also who the provider is as caregiver. Finally, Post (2010) unites compassionate love with human dignity. When we treat each other with compassion, he argues, we acknowledge each other as people of worth and value.

Charles Dougherty and Ruth Purtilo (1995) claim that compassion is not only critical for quality healthcare delivery but that it is also the physician's duty. Dougherty and Purtilo define compassion as "an ability to identify with another's experience of suffering" and "the disposition to want to alleviate the other's suffering" (1995, p. 427). Hence, they identify two key features of compassion. The first is a capacity to pull alongside of the person suffering or in pain and to connect with that person in order to understand and possibly experience to some extent that suffering or pain. They utilize Warren Reich's three phases of patient suffering to provide opportunities for physicians to connect with the patient's

[5] Underwood (2008) also discusses the contrasting negative behavior in which compassionate love is not given or expressed and its impact on human relationships that require care.

suffering.[6] The second key feature is its virtuous nature, i.e. the desire and ability of the physician to affect change for the patient's good, whether in terms of curing the patient's illness or helping the patient manage it. Dougherty and Purtilo defend the claim that compassion is a physician's duty based on three professional responsibilities of the medical profession. The first is the physician's fiduciary responsibility to place the patient's interests first. The next professional responsibility is due care, which, according to Dougherty and Purtilo, "requires physicians to maintain a reasonable range of professional skills and to use them with appropriate diligence" (1995, p. 429). The final responsibility is confidentiality in which the physician must protect the vulnerable patient from undue harm associated with the privileged knowledge shared in the confines of the patient-physician relationship.[7] Although Dougherty and Purtilo realize that some physicians exceed the compassion duty, they conclude, however, "a basic level of compassion is a duty for all physicians" (1995, p. 432).

5.1.2.2 Empathic Love

Empathy, as seen in Chapter 3, is distinct from compassion as a cognate virtue of love; however, with respect to the notion of prudent love it complements compassion.[8] Whereas compassion stresses the ability of a loving person to pull alongside the suffering person, empathy affords the opportunity to enter the suffering person's world and eventually to experience it. For healthcare professionals, empathy is an important element in providing quality healthcare (Larson and Yao, 2005; Moore, 2010; More and Milligan, 1994; Neuman et al., 2009; Reynolds and Scott, 1999; Spiro et al., 1993). For example, in a review of the literature on the role of empathy particularly in primary care delivery, Stewart Mercer and William Reynolds (2002) report that empathic healthcare practice overwhelmingly improves clinical outcomes. Although measurement of empathy varies from study to study, empirical research does confirm the positive benefits of empathic consultation in a primary healthcare setting. Moreover, teaching empathy to medical students in the clinic enhances their empathic stance towards patients. Based on their literature review, Mercer and Reynolds conclude, "Empathic consulting in primary care should be

---

[6] Reich (1989) identifies three phases during the development of patient suffering. The first is mute suffering in which the patient cannot articulate the suffering completely. A physician must attend to clinical clues to aid the patient to give voice to that suffering. The next phase is expressive suffering in which the patient does begin to voice the suffering, and the physician's role is to guide the patient towards healthful articulations. The last phase is new identity in which the suffering transforms the patient's life in terms of meaning. The physician must continue to aid and support this transformation process.

[7] Dougherty and Purtilo also address several objections to the duty of physicians as compassionate caregivers, including compassion as simply or exclusively a personality trait or it leads to burnout and not all patients want or deserve compassion.

[8] To remind the reader that in Chapter 3, I identify compassion with sympathy. So, what is true for the relationship between compassion and empathy is also true for the relationship between sympathy and empathy.

encouraged and the tradition of holism in general practice is a strong foundation. Methods of assessment of quality of care in general practice," they add, "should include the human dimension of the clinical encounter, of which empathy is a key part" (2002, p. S11).

To date, healthcare pundits have yet to reach consensus in defining empathy. Definitions often range from simply borrowing the patient's feelings about the illness experience to actually possessing those feelings, even if only vicariously (Halpern, 2003; Hojat, 2007; Kalisch, 1973; Mercer and Reynolds, 2002; Spiro et al., 1993). Jodi Halpern, for example, defines empathy in terms of the clinical encounter as an emotional attunement with the patient's world of suffering. "Emotional attunement," according to Halpern, "operates by shaping what one imagines about another person's experience. In trying to imagine what the patient is going through," she adds, "physicians will sometimes find themselves resonating" (2003, p. 671). Through resonating with the patient's suffering, the healthcare worker understands dimensions of that suffering that allow for effective treatment. Halpern explains how empathic physicians achieve effective treatment through "associative reasoning," which involves physicians connecting not only emotionally with a patient but also cognitively. Through this connection or association, physicians can focus on the patient's illness story and come to understand what that story signifies or means to the patient.

Janice Morse and colleagues (1992) identify four components to the notion of empathy.[9] The first is moral, which reflects a personal internal motivation to enter into another person's world and to experience it and which represents a receptive and an at-hand attitude towards another. The next component is emotive, in which the empathic person shares in the other person's subjective world of psychological feelings, emotions, and concerns. "The emotive component of empathy," to quote Morse and coworkers, "refers to one's ability to subjectively perceive and share another person's psychological state or intrinsic feelings—that is to feel what others feel" (1992, p. 275). The third component is cognitive, which pertains to understanding another person's feelings from an objective and rational perspective or stance. "The cognitive component of empathy," according to Morse and colleagues, "includes perspective taking, that is, the intellectual ability to understand another's perspective and predict their thoughts" (1992, p. 275). The final component is behavioral, in which the empathic person communicates understanding of the other person's inner world of feelings and concerns in order to assure the other that he or she genuinely knows about and understands the person's inner world.

Empathy then is a complex, multidimensional notion that defies simplistic definitions such as Howard Spiro's oft-quoted definition, "Empathy is the feeling that persons or objects arouse in us as projections of our feelings and thoughts. It is evident when 'I *and* you' becomes 'I *am* you,' or at least 'I might be you'" (1992,

[9] Interestingly, Morse and colleagues—based on their notion of empathy—do not find that empathy is appropriate for nurses in a clinical setting (Morse et al., 1992, 2006).

p. 843). However, utilizing Morse and colleagues' components of empathy, Mercer and Reynolds define empathy comprehensively as an ability to

(a) understand the patient's situation, perspective and feelings (and their attached meanings);
(b) to communicate that understanding and check its accuracy; and,
(c) to act on that understanding with the patient in a helpful (therapeutic) way (2002, p. S11).

Their definition presumes the moral and emotive components, while explicitly stressing the (a) cognitive and (b) behavioral components. Moreover, they add another component (c)—acting on behalf of the other. This is an important component, especially for clinical practice, in that an empathic response must lead to some benefit for the patient—particularly in terms of either curing or managing a patient's illness effectively. Without this final step, empathic behavior is almost worthless to the patient's situation.[10]

Besides Morse and colleagues' four components of empathy, Mercer and Reynolds also utilize Godfrey Barrett-Lennard's empathy cycle to formulate their definition. The Barrett-Lennard cycle consists of three phases, with an initial phase in which a person resonates with another person's feelings through listening to that person's concerns and understanding them ("resonated empathy"). The next phase involves communication of those feelings as understood by the empathic person back to the person in need ("communicated empathy"), while the last phase pertains to the reception of the understood empathic feelings by the person in need ("received empathy"). The last phase loops back onto the first to reinforce or validate (or invalidate) the resonating or empathic feelings. "The total interactive sequence within which these phases occur," concludes Barrett-Lennard, "begins with one person being self-expressive in the presence of an empathically attending other, characteristically leading to further personal expression and feedback to the empathizing partner" (1981, p. 95). Again, parts (a) and (b) of Mercer and Reynolds' definition represent the three phases of the empathy cycle, while part (c) is a novel addition.

William Zinn (1993) advocates a role of empathy in healthcare through the notion of the "empathic physician."[11] According to Zinn, empathy represents "a process for understanding an individual's subjective experience by vicariously sharing that experience while maintaining an observant stance" (1993, p. 306). The empathic process includes an initiation component in which "a behavior caused by an internal emotional state can arouse in another an internal emotional state that is outwardly manifested by the same behavior" (Zinn, 1993, p. 308). Through such awakening, the physician can connect with the patient at a fundamental experiential level and

[10] Recently, Melanie Neumann and coworkers use Mercer and Reynolds' definition to develop a model for clinical empathy (Neuman et al., 2009).

[11] Shimon Glick (1992) also develops a notion of "empathic physician" similar to Zinn and offers pedagogical means to achieve such a physician.

share in the patient's world. The process also involves, as it does for Halpern, a resonating component on the physician's part that is not only emotional but also cognitive. This component originates in the physician's relevant previous experience, as well as in the capability to imagine and fantasize what the patient is experiencing and the world in which the patient exists. Through the empathic process, not only does the physician enter the patient's world, but the patient can share that world and its burdens with another—the empathic physician. This sharing of the world by both the patient and physician, claims Zinn, is the root of empathy's therapeutic potential. The physician may realize that potential through use of appropriate language to engender in the patient deeper awareness of the illness and its meaning and impact on the patient's life.

Besides the philosophical examination of empathy's role in healthcare, clinical researchers are investigating its role through measuring and quantifying empathy of healthcare students and providers (Hemmerdinger et al., 2007; Hojat, 2007; Pedersen, 2009). For example, Mohammadreza Hojat and colleagues at the Jefferson Medical College in Philadelphia employ the Jefferson Scale of Physician Empathy (JSPE), which they developed, as an instrument to quantify and study empathy in clinical settings. JSPE is a self-report instrument in which healthcare providers answer Likert-type statements on a seven point scale.[12] In a study of medical professionals, Hojat and colleagues demonstrate that women and psychiatrists score higher than men and other medical specialists, respectively (Hojat et al., 2002). Recently, from a critical review of over 200 publications on empirical empathy studies, Reidar Pedersen (2009) reports that empathy studies vary greatly as to methods employed and stress preferentially either the cognitive or emotive component of empathy. "In sum," concludes Pedersen, "the empirical studies of empathy tend to separate empathy from main parts of clinical perception, judgment, and communication" (2009, p. 318). He recommends that future studies include these elements of clinical practice into models for empathy.[13]

Clinical investigators propose various models to account for and explicate the notion of empathy (Barrett-Lennard, 1981; Squier, 1990). Recently, several investigators examining the role of empathy in healthcare put forward models to guide and conduct future research to overcome problems associated with past empirical studies (Neuman et al., 2009; Norfolk et al., 2007). For example, Tim Norfolk and colleagues advance a model for clinical empathy to establish rapport between patient and physician in terms of a therapeutic relationship. Their model comprises several components (Norfolk et al., 2007). The first involves the empathic motivation of the physician to be curious about and caring for the patient's illness, as well as over the illness experience. Along with this component are the specific skills necessary to obtain the patient's illness story in sufficient detail to empathize with the pain and

[12] An example of a statement is, "I believe that empathy is an important therapeutic factor in medical treatment" to which the subject either agrees (7 = strongly agree) or disagrees (1 = strongly disagrees).

[13] Pedersen also finds poor predictive value of empathy empirical studies for selecting medical students—a finding corroborating another literature review (Hemmerdinger et al., 2007).

suffering associated with the patient's illness. These skills include identifying cues from the patient to construct an accurate perception of the illness and its experience. The next component of their model is for the physician to share construction of the illness perception with the patient, in order to validate its accuracy. The outcome of the model is empathic understanding, which leads to greater rapport between the patient and physician and to a robust therapeutic relationship. Importantly, according to this model "the rapport established between doctor and patient is not a static moment or outcome, but rather a dynamic, iterative process in which the doctor attempts to reach an increasingly accurate understanding of the patient's thoughts, feelings and expectations" (2007, p. 693). Based on their dynamic model, a pilot study confirms that training in specific skills for empathic understanding significantly improves rapport between patient and physician (Norfolk et al., 2009).

Criticism and skeptics on the role and feasibility of empathy in healthcare abound in the literature (Edwards, 2001; Hirsch, 2007; Landau, 1993; Pedersen, 2008). Kelly Edwards divides the criticisms into five categories. The first is trivialization of the clinical exchange between patient and physician. The problem is that the dialogue can be strained and artificial, unless the healthcare provider genuinely attempts to enter into the patient's world in a fundamental way. The next category reflects the distraction of physicians from providing the objective care a patient needs and deserves. Edwards cites Richard Landau's well known criticism that finding a place for empathy in a physician's clinical practice is difficult, if not impossible, given its many technical requirements. "Physicians who deliberately cultivate empathy, who place themselves in the patient's position," frets Landau, "will not be able to reliably fulfill all of these requirements" (1993, p. 108). The third involves the problems associated with empathizing with another person, especially when the experience is foreign to the physician—unless one has had the same experience. Empathy, then, is not possible and unfair to demand of physicians. The next category refers to an inability to attune emotionally to a patient, particularly a difficult patient. Central to this criticism is that fulfilling the clinical role is demanding enough, let alone attempting to engage emotionally the patient's role. The last category pertains to the problem of a physician dominating patients by telling them what their feelings or emotions should be or are, thereby co-opting patients' illness experience.

In response to criticisms of clinical empathy, Pedersen (2008) acknowledges the various problems associated with it in terms of clinical practice. However, he offers an alternative approach to implementing empathy within the clinic—a hermeneutic approach based on Heidegger's and Hans-Georg Gadamer's phenomenological philosophy. According to Pedersen, attempts to define empathy often ignore the historical "situatedness" of the empathic relationship. Moreover, he asserts that proponents of empathy generally fail to notice the moral commitment associated with empathy. "The concept of empathy," argues Pedersen, "does not only refer to understanding another person, but also to understand that other person in a right or appropriate way, e.g. to be able to interact with that person in a good way. Thus" he concludes, "my tentative suggestion is to describe empathy as appropriate understanding of another human being" (2008, p. 332). Empathic understanding is not simply epistemic but incorporates the hermeneutical or interpretive with respect to historicity, horizon, dialogue, and the hermeneutical circle. Importantly, it must be appropriate, not in terms of being absolute but rather with respect to being sensitive to a particular pertinent context meaningful to the patient. Empathy is a process, for Pedersen, since "to achieve appropriate empathic understanding the subject and object have to participate in a dialogue and reflect on their understanding and experiences; and the intersubjective truths gained are never complete," he cautions, "but rather revisable results from an ongoing process" (2008, p. 333).

5.1.2.3 Altruistic Love

Altruistic love is the final dimension of love that requires discussion before explicating the notion of radical love and the compound virtue of prudent love in the following sections.[14] Although the notion of altruism is present in earlier Jewish and Christian thought, especially in terms of agape love, not until the nineteenth century does Auguste Comte introduce the term "altruisme" (Scott and Seglow, 2007). Altruism for Comte is part of his philosophy of positivism and represents the moral demands of that philosophy to live for others. According to Comte, "the expression, *Live for Others*, is the simplest summary of the whole moral code of Positivism" (1968, p. 556). He specifically contrasts altruism to egoism and the self-centeredness it entails. Since Comte's introduction of the term altruism, altruism has had a contentious history with two main trajectories (Scott and Seglow, 2007). The first pertains to biological evolution, and the question of whether altruism represents a selection force in Darwinian evolution. The second concerns the mechanism of altruism in human behavior, and the issue of whether reason or emotion is responsible for motivating altruistic acts. "It is important," note Niall Scott and Jonathan Seglow, "for both moral philosophers and the rest of us, to consider whether we act altruistically because of desires and sentiments or on the basis of reason (or perhaps both)" (2007, p. 21). Unfortunately, the solution to this quandary remains elusive and hence defining altruism is problematic.

Although no consensus exists for defining altruism, contemporary definitions do reflect—in spirit—Comte's notion of altruism (Krebs, 1970; Piliavin and Charng, 1990; Post et al., 2002; Scott and Seglow, 2007). For example, Kristen Monroe defines altruism as "behavior intended to benefit another, even when this risks possible sacrifice to the welfare of the actor" (1996, p. 6). Monroe goes on to identify several crucial points pertinent to her definition. The first is behavioral in that altruism involves acting or doing something for someone in need. In other words, good intentions are simply insufficient—one must do the good. The next point is the action's goal-directedness, which must be towards advancing the other's welfare. "If another's welfare is treated as an unintended or secondary consequence of behavior designed primarily to further my own welfare," argues Monroe, "the act is not altruistic" (1996, p. 6). Another point is that intention trumps outcome. If the outcome

[14] Whether altruism is a virtue is debatable, and so I elected not to include it in the discussion on cognate virtues of love in Chapter 3 but to reserve its inclusion until this chapter.

results in bad rather than good for the other, still the act is altruistic if the person acted with good intention. A further point is that the altruistic act must (or possibly) diminish the actor's welfare; otherwise, if the act increases the actor's welfare then the act is part of a "collective" welfare. Finally, the action must be unconditional, with no thought of reward.

Part of the problem in arriving at a consensus definition for altruism is the various types of altruism traditionally proposed in the literature, many of which are discipline-dependent. For example, economists, psychologists, biologists, sociologists, anthropologists, political scientists, etc., define altruism differently. However, since the 1980s—in which a "paradigm shift" occurs in studying altruism—two types of altruism predominate (Humphrey, 1997; Kitcher, 2010; Piliavin and Charng, 1990; Sober, 2002). The first is evolutionary or biological altruism. Philip Kitcher summarizes this type of altruism accordingly, "an agent *A* is said to act altruistically towards a beneficiary *B* when *A*'s action promotes the expected reproductive success of *B* at expected reproductive cost to *A*" (2010, p. 121). The agent of evolutionary or biological altruism need not be conscious or mindful of the altruistic act. Of course, the problem with this type of altruism is that natural selection should eliminate altruistic organisms since these organisms are less fit or able to compete with their rivals in terms of reproductive success.[15]

The second predominate type of altruism is psychological or behavioral. According to this type, the actor is conscious or mindful of the altruistic act and possesses a desire or intention to perform it. "An altruistic desire," explains Elliot Sober, "is an other-directed desire in which what one wants is that another person does well" (2002, p. 19). The problem associated with psychological or behavioral altruism is whether altruistic desires or intensions are ultimate, in the sense that the altruistic agent desires purely from nonself-directed goals, or whether the desires are instrumental, in that the agent's desires also include self-directed goals. Another important problem associated with this type of altruism is altruistic love. According to Sober, an altruistic desire can or cannot be attended with an emotional or a loving desire. The problem specifically is how that transformation from a non-loving to a loving altruistic desire occurs. To address these problems, scientists are undertaking empirical or experimental research to resolve them—which deserve brief consideration.

Empirical and experimental research on altruism and altruistic love begins in earnest with the 1980s paradigm shift (Batson, 2002; Krebs and Van Hesteren, 1992; Piliavin and Charng, 1990; Post et al., 2003). That research focuses not only on the problems identified in the previous paragraph but also on other problems associated with them, including the motivation for altruistic actions, the existence of an altruistic personality, the role of nature and nurture in altruism, the genetic basis of altruism, and the ontogenic development of altruism and the conditions and factors influencing that development. One of the major criticisms of altruism is that

[15] Darwin proposed solution of group selection in which an altruistic group is fitter than a non-altruistic group is an attempt to resolve this problem (Sober, 2002).

egoistic concerns are always present, no matter how altruistic an act may appear. Proponents of altruism have conducted research to investigate this criticism. For example, Batson (1991; 2002) examines whether egoistic concerns are responsible for motivating altruistic actions. From his experimental results, he concludes that the basis for altruistic actions is not egoistic. To quote Batson,

> Experimental research has tested the claim that empathic emotion evokes altruistic motivation—motivation with the ultimate goal of increasing another's welfare. Results of more than 25 experiments designed to test this empathy-altruism hypothesis against various egoistic alternatives have proved remarkably supportive of it, leading to the tentative conclusion that feeling empathy for a person in need does indeed evoke altruistic motivation to help that person (2002, p. 104).

However, Batson acknowledges that other motivations besides empathy, such as an altruistic personality, may be responsible for altruistic acts.[16] To resolve this problem and others confronting altruism requires, he contends, more research in almost every discipline (Post and Underwood, 2002).

Altruism has an interesting role in medicine (Anonymous, 2000; McGaghie et al., 2002; Pilowsky, 1977). Some medical professionals consider altruism an attribute or even a virtue of physicians. For example, the American Board of Internal Medicine identifies altruism as "the essence of [medical] professionalism" (1998, p. 5). Other medical professionals expand the role of altruism to include the larger medical community and its advancement. For example, in a 1927 editorial in the *Journal of the National Medical Association* an editor extols the altruism of past medical researchers and their sacrifices to make discoveries that benefit patients. The editor goes on to contrast these heroes to community members "who contribute nothing to its welfare...who take everything from the profession and add nothing to it. They might be defined as professional parasites—a burden and a menace to medical ideals" (Anonymous, 1927, p. 23). Besides physicians, patients can be altruistic. Notably, patients, who are generally terminally ill, often contract to donate organs upon their death. Medical professionals and ethicists recognize the role altruism plays in motivating patients and others to donate for the benefit of others—even those unrelated to the donor. "Altruistic beliefs and values," claims James Childress and Catharyn Liverman, "are one of the mainstays of the voluntary organ donation system" (2006, p. 70).[17]

Whereas compassion and empathy allow healthcare providers to enter the world of a suffering patient and to experience it, if only vicariously, altruism creates the opportunity not just to experience that world but also to live it in such a manner to expose physicians to the risk and dangers associated with a patient's suffering. In other words, physicians might have to give of themselves in such a manner that imposes some cost or loss for them. Situations or conditions may even sometimes

---

[16] In contrast to Batson, Elliot Sober and David Wilson (1998) propose a biological justification for psychological altruism. Their proposal has generated considerable controversy (Schulz, 2009; Sesardic, 1999; Stich, 2007).

[17] Interestingly, altruism has general positive health benefits for the patient and appears to increase overall wellbeing and even longevity (Post, 2005; 2007).

call physicians to risk their lives, given the dangers of their occupation. The question that beleaguers healthcare providers is whether physicians should be altruistic in the first place; and, if they should be altruistic what behaviors comprise altruistic acts. In commentary on the perceived decline in altruism in medicine, Roger Jones (2002) claims that such altruistic acts as providing free medical care for those who cannot pay or providing services outside contractual obligations—"a general willingness to go the extra mile in professional activities"—are no longer as prevalent among medical professionals. However, not all physicians agree that medical professionals should be altruistic. For example, Walter Glannon and Lainie Ross (2002) contend that altruism actually interferes with the fiduciary relationship with the patient and propose that beneficence better serves that relationship.[18] Jeffrey Bishop and Charlotte Rees (2007) also assert that altruism is inappropriate for medical professionals and argue that physicians should exhibit pro-social behavior predicated on practical wisdom or Aristotelian *phronesis*.[19]

5.1.2.4 Radical Love

Although the forms of compassionate, empathic, and altruistic love provide a fuller understanding of the nature of love required for explicating the prudent love of a virtuous physician, they are insufficient—as noted at the chapter's beginning—to capture its essence adequately and completely. In *Love and friendship*, Jules Toner (2003) develops a notion of personal radical love and communion sufficient to transform care into love with respect to the compound virtue of prudent love. Toner begins with a descriptive analysis of radical love—what he calls its "total concrete experience." "In the full concrete experience of love," as Toner articulates his experience of it, "our whole being, spirit and flesh, is involved: cognitive acts, feelings and affections, freedom, bodily reactions—all these are influencing each other and all are continually fluctuating in such a way as to change the structure and intensity of the experience" (2003, p. 65). From this experience, he derives love's crucial or radical element common to its various forms.

From the descriptive analysis of the total concrete experience of love, Toner distinguishes five essential components composing it. The first is cognitive, which includes a person's sensations and perceptions, memories, imagination, insights and conceptualizations, and judgments. The next component is affective and includes a person's emotions, feelings, and passions. The third component is what he terms the "act of affectivity" to distinguish the active role of affections from their passive role. This act represents the spontaneous response of a loving agent to love's object. Freedom is the next component and pertains to the volitional nature of love. "By *freedom*," to quote Toner, "I mean power of self-determination by choice which is

[18] Glannon and Ross' article is part of a symposium on the role of altruism and supererogation in medicine (Downie, 2002; McKay, 2002; McLean, 2002).

[19] Interestingly, a survey of students within professional schools reveals that medical students are more altruistic than either law or business students (Coulter et al., 2007).

not determined by any condition or cause whether extrinsic to the agent or intrinsic to the agent but extrinsic to the act of choosing" (2003, p. 66). The final component involves bodily reactions to love, which include the physiological, such as increased heart rate and blood pressure, and/or the behavioral, such as smiling and laughing.

In addition, Toner identifies three key features of the total concrete experience of love. First is the response of radical lover to the beloved. For Toner, radical love includes "my response to your total reality. It is directly and explicitly a response to your actuality, primarily and in every instance to your fundamental actuality as a personal act of being; secondarily, to your qualitative actuality revealed in your acts and partially revealing your act of being. It is indirectly and implicitly," he adds, "a response to your potentiality, dynamism, and need. This response is experienced as liberation of the subject's energy for love and liberation from the confinement of individual being. It is at the same time experienced as a willing captivity to the beloved" (2003, p. 101). The next feature of radical love, which emanates from the lover's response to the beloved, is union of lover and beloved. This union represents what Toner calls a "consonant presence" in which "the lover is in consonance with the loved one's reality as a person, not merely with the loved one's thoughts or affections" (2003, p. 112). The final feature is "the most fundamental aspect of radical love," which is "affective affirmation of the beloved for and in the beloved's own self, in the beloved's very act of personal being" (2003, p. 148). This "irreducible root" of radical love, then, is the comprehensive affirmative response of a lover for the beloved per se and, in turn, the beloved's comprehensive affirmative response to his or her lover per se, which ultimately results in communion of lover and beloved.

From these crucial features for the total concrete experience of radical love, Toner provides the following definition of it. For Toner, "radical love is a response in which the lover (I) affectively affirm the beloved (you) for yourself (as a radical end), in yourself (on account of your intrinsically lovable actuality), directly and explicitly in your personal act of being, implicitly in your total reality, by which affirmation my personal being is consonantly present to and in you and yours is present to and in me, by which I affectively identify with your personal being, by which in some sense I am you affectively" (2003, p. 163).[20] Radical love then is an affective act in the sense that feelings, although crucial, are not necessary. A lover may not at every moment feel loving. However, Toner insists that "integral" love always includes these feelings toward the beloved. Importantly, the affective act is affirmative. "One who radically loves," claims Toner, "approves, endorses, intentionally confirms, the loved one's whole actuality insofar as it accords with his true self, along with his potentiality and dynamism toward fuller life" (2003, p. 195). Next, the lover's affective affirmation of the beloved is per se or for the beloved in himself/herself for himself/herself. In other words, the lover affirms the beloved as

[20] Toner also provides a concise definition, "radical love is: 1. an affective act of affirming the loved one, 2. in herself for herself, 3. which act itself constitutes an actual union of the lover and beloved (rather than only tends toward such union)" (2003, p. 195).

a means towards his/her own ultimate or radical ends and not towards those of the lover or another person.

Importantly, radical love involves union of the lover and beloved. This union is more than a simple mutual presence "*to* each other" but constitutes "a mutual [consonant] presence *in* each other, a mutual living in" (Toner, 2003, p. 198). Such union involves two critical elements. The first is the "giving self." "In the act of radical love," notes Toner, "the lover renders himself a gift to the loved one and in doing so, *in the very same act*, opens himself to and accepts the loved one into himself" (2003, p. 198). Without the giving of self or rendering one's affirmation of the beloved, according to Toner, radical love is not possible. The second is "affective identification." Such identification pertains to the experience in which the lover identifies and shares in the beloved's life, without compromising the integrity of either's life. Rather than compromising the integrity of one's life, radical love enriches one's life and actualizes the potential inherent in one's being. Toner summarizes radical love accordingly, "a medium through which the lover intentionally joins his own ontic self-affirmation, his very act of being, to his loved one's act of self-affirmation or act of being" (2003, p. 199). As such, this love is the foundation for the different forms of love.

Toner identifies several subtypes of radical love, with personal love constituting its fullest and complete form of union between lovers. Personal love is radical love for another person and not for non-human objects. He divides personal love into three modes. The first pertains to loving and knowing "*that* the other is a person and perhaps know[ing] *that* she has some urgent need" (Toner, 2003, p. 207). The next mode involves loving and knowing such a person, especially the lovable qualities of that person, through one's own experience or through another's testimony. The final mode, Toner admits, is more difficult to define, given one's inability to apprehend or comprehend fully another person. But, Toner offers the following definition for what he calls personal radical love, "an affective affirmation informed through knowledge of the loved person in his or her total reality, primarily the unique personal self, but also and necessarily, in some way, by all else in her reality" (2003, p. 208). He identifies two key elements of this love: the integrally and the intentionally truthful. The first pertains to the intense feelings associated with personal love and the second to a sincere and honest response of the lover for the beloved. Moreover, he admits that the truthful nature of personal radical love represents perfection. "Actual perfectly truthful love is an ideal," Toner concedes, "that is not necessary to reach in order to love personally. If it were," he realizes, "who of us could lay claim to personal love for anyone. It is enough for human love to be genuinely personal," he admits, "if it is intentionally truthful, enough and necessary" (2003, p. 212).

The culmination of personal radical love is personal communion of beloved and lover. "Personal communion," according to Toner, "begins only when there is mutual personal love, mutual belief in each other's love, and awareness in each of the other's belief" (2003, p. 213). From this communion of two acts of radical love by the two agents experiencing personal love is the "one composite act of one composite agent." And the growth and actualization of each of the agents is directly proportional to the strength of their combined and integrated personal radical love.

Toner summarizes the notion of personal radical love with the following "formula:" "Communion of personal love is the relationship constituted by mutual personal love, integral and intentionally truthful, when the persons in the relationship both believe in each other's love and are aware of each other's belief and are responding each to the other as the other is actualized and revealed in his or her love, the mutual responses so interdependent as to form one composite agent" (2003, p. 215).

Toner's notion of personal radical love represents the needed resource to transform a physician's care of a patient into love for that patient, which fulfills and yet exceeds the compassionate, empathic, and even altruistic forms of love. Toner himself examines the nature of care vis-à-vis this radical love. He begins with the origins of care, with its naissance in the beloved's need(s). "When the object of radical love is in need," to quote Toner, "care arises" (2003, p. 82). Care is a response of the lover to the beloved and, hence, it is a type of affirmative affective act. According to Toner, care represents a desire to meet or fulfill a relative end, an end pertaining to the lover's need to realize or actualize the beloved's radical end. For example, a physician taking care of an ill patient represents a relative end for the physician while for the patient it, especially in terms of a cure, represents a radical end. The transformation of care to love requires a conversion of care from a derivative of desire, i.e. to care for only one's ends, to one of radical love, i.e. to care for another's ends as if those ends are one's own. Toner calls this type of care radical care, to distinguish it from relative care, and he defines it in terms "of a fulfilling end for the one radically cared for" (2003, p. 204). As such, radical care is a subtype of radical love, what he calls radical care love. According to Toner, "radical love and radical care are not fully distinct acts. They have," he explains, "the same object, the radical end, the person; and radical care is really only a qualified form of radical love" (2003, pp. 77–78).[21] What ultimately distinguishes care based on radical ends from care based on relative ends, according to Toner, is that once a caregiver satisfies the caregiver's relative end by realizing the cared-for person's radical end the caregiver no longer needs to care since no need is palpable. For the radical caregiver, however, once he or she meets the need of the cared-for person the caregiver still cares about the cared-for person and is ready to take care of additional needs that might arise.

Toner's notion of radical love, then, is apt for appropriating the virtue of love to clinical practice. The loving clinician affectively affirms a patient's healthcare need as a radical end in the patient's personal act of being and strives to meet that need caringly and competently. The basis of this loving response is the clinician's ability and willingness to enter into a patient's illness world and to be authentically present to the patient in order to alleviate the patient's suffering and to restore health. In turn, the patient affectively affirms the clinician's efforts thereby confirming the clinician's need to be perceived as a genuine healthcare provider. Radical love represents the loving clinician's whole-hearted affective interaction with patients, in

[21] Elsewhere, Toner writes, "Since the object of radical care is the one affectively affirmed in himself for himself, even though in need, care is only a modified form of radical love, not fully distinct from it" (2003, p. 200).

order to pull alongside them and to meet fully their ultimate or essential healthcare needs. Certainly, a loving clinician responds to patients not simply as pathological specimens but as whole persons in their lifespan contexts. For, a loving clinician is in consonance or intimate contact with patients in order to forge therapeutic unions or relationships with them. The "irreducible root" of radical love in medicine is the comprehensive affirmative response of a loving clinician for the patient and, in turn, the patient's comprehensive affirmative response to the clinician. Finally, the notion of union involved in radical love is germane to clinical practice in terms of forging an effective therapeutic relationship between clinician and patient. The loving clinician gives of self as gift *qua* healer to the patient who initially came to the clinician in an opened, vulnerable state, thereby giving to the clinician the patient's self in need of restoration. Through this mutual affective identification involving the clinician's realization of the patient as one in need of healthcare and the patient's acknowledgement of the clinician as genuine healthcare provider, a therapeutic relationship is not only possible but also actualized.

5.1.3 Compound Virtue of Prudent Love

The compound or composite virtue of prudent love defining the virtuous physician represents a combination of prudent wisdom and personal radical love. In this relationship, prudence and love complement one another as the prudently loving physician attends to the patient's medical needs.[22] As a prudent healthcare provider, on the one hand, the physician provides the best competent medical care currently possible for the patient, not only in terms of knowing why (theoretically) this is the best care but also how (practically) to utilize that care for the individual patient. By being a person who is prudent, the virtuous physician engages in prudently wise acts. As a loving healthcare provider, on the other hand, the virtuous physician provides competent and prudent healthcare in a caring and gracious manner, with respect to appreciating and understanding the patient's illness story. Through that story, the loving physician draws alongside the patient compassionately and enters empathetically into the patient's world of pain and suffering caused by the illness; and, when conditions demand it, the physician even sacrifices altruistically his or her comfort for the patient's welfare. By loving the patient because the patient is a person who is worthy per se of such love, the loving physician meets both comprehensively and holistically the patient's ultimate or radical needs and ends of healing—especially emotionally. Hence, the prudently loving physician employs the virtues of prudent wisdom and personal radical love simultaneously and harmoniously to complement or balance one another.

---

[22] Both prudent wisdom and radical love also supplement one another by supplying or adding important features of clinical practice that the other lacks, thereby rounding out that practice comprehensively. This supplementation differs from complementation of prudence and love as prudent love in that the former augments the practice with what the other lacks while the latter completes and thereby improves virtuous practice.

However, the compound virtue of prudent love is more than simply the combination or complementation of prudent wisdom and radical love. Rather, it also represents a species of these virtues that when in combination surpasses or transcends the properties of each virtue individually. The relationship between the two virtues is synergistic in that prudent wisdom sharpens and strengthens personal radical love and radical love in turn enlarges and deepens prudent wisdom. In other words, prudent wisdom vis-à-vis personal radical love empowers the virtuous physician to make the best and wisest decisions possible because it is the loving thing to do for the patient, and personal radical love vis-à-vis prudent wisdom enables that physician to care for the patient's needs affectionately because it is the wisest thing to do. The synergy between prudent wisdom and personal radical love then drives the virtuous physician to maintain and enhance his or her competence through continuing education even when no patient need is evident or immediate. Rather, the anticipation of such future need motivates the physician. In turn, synergy between the two virtues motivates the virtuous physician to love the patient even when the patient is in no need of competent medical care. Consequently, this synergy then makes prudence loving and love prudent. The clinical outcome is not just comprehensive healthcare but a full and rich holistic healthcare, which restores the patient's dignity and integrity regardless of whether the patient is cured or not.

Prudent love synergy is particularly critical for forming robust patient-physician relationships, which compose the virtuous practice of medicine. An important component of those relationships is the generation of appropriate professional boundaries for defining patient-physician dyads so that the relationships are therapeutic for and not harmful to patients (Epstein, 1994; Farber et al., 1997; Gabbard and Nadelson, 1995; Gutheil and Gabbard, 1993; Nadelson and Notman, 2002; Spruiell, 1983). Boundaries are important for instituting relationships and for determining when one party inappropriately crosses a boundary resulting in misconduct and particularly in patient harm or injury. The problem with establishing boundaries is that a tension or ambiguity exists between getting too close to the patient and remaining too distant. If physicians get too close, particularly emotionally, then inappropriate behavior, such as sexual relations, may occur. This is especially true in specialties such as psychiatry, where therapists unfortunately at times engage in sexual misconduct with patients.[23] However, if physicians are too distant or detached, then they may lose their moral compass and engage in such harmful activity as nonconsensual human experimentation. Emily Friedman (1990) cites the atrocities Nazi doctors committed during World War II, as an illustration of physicians who transgressed the boundary of "do no harm." These doctors were too detached from their Jewish test subjects because of the goals of the Nazi war effort, which blinded the doctors to their unethical and immoral behavior and to the harm they inflicted on their test subjects.

---

[23] In fact, most of the literature on boundaries in medicine addresses sexual transgressions of psychotherapists.

As many commentators on the notion of therapeutic boundaries recognize, no set protocols or algorithms are available for establishing these boundaries, making the task of defining them difficult. However, several definitions are available within the literature. For example, Glen Gabbard and Carol Nadelson define therapeutic boundaries as "the parameters that describe the limits of a fiduciary relationship in which one person (a patient) entrusts his or her welfare to another (a physician)" (1995, p. 1445). These parameters include, for instance, time, location, financial remuneration, and the physician's and patient's role in the therapeutic relationship (Gutheil and Gabbard, 1993). The last parameter, the role of the physician (or patient), is particularly crucial for forming beneficial boundaries to safeguard the therapeutic relationship. If the physician's role is ill demarcated or not demarcated at all, for instance, patients may place unrealistic expectations on the physician leading to boundary transgressions and misconduct—at least from the patient's perspective. Neil Farber and colleagues expand upon Gabbard and Nadelson's definition to explicate the limits that constitute boundaries in medical practice. "Boundaries in patient care," they write, "are mutually understood, unspoken, physical, and emotional limits of the relationship between the trusting patient and the caring physician" (Farber et al., 1997, p. 2291). Significantly, both the emotive and physical limits require demarcation and acknowledgment in order to prevent boundary transgressions. Finally, Richard Epstein (1994) points out the spatial metaphoric nature of boundaries, especially in terms demarcating self from non-self. "The concept of personal boundaries," notes Epstein, "employs a spatial metaphor that helps us to describe and define our relationships with other beings and objects in the external world" (1994, p. 15). Unfortunately, the metaphorical nature of boundaries makes them hard to define and explicate precisely, thereby compounding their practical application in clinical practice.

The difficulties associated with defining therapeutic boundaries aside, a critical issue facing the clinical relevance of these boundaries is how to generate or establish them—especially for pedagogical purposes. I argue that prudent love synergy serves to address this issue with respect to developing and determining appropriate boundaries for robust patient-physician relationships that are therapeutic for patients and gratifying for physicians. Neil Farber and colleagues propose that love is useful in forming therapeutic boundaries between patients and physicians. "This love," the authors note, "is not characterized by romantic feelings, but rather denotes a platonic relationship and one in which intimacy is used to benefit the patient" (Farber et al., 1997, p. 2291). Specifically, the authors identify agape or servant love as the type of love required for establishing patient-physician boundaries. Unfortunately, they do not demonstrate specifically how such love operates in establishing these boundaries. The compound virtue of prudent love addresses this problem. First, such love is wisely prudent in determining what course of action is best and appropriate for treating a patient. Such action is not just practical in nature but it also depends on the etiological factors causing the patient's illness. Secondly, such prudence is loving in that the physician embraces the patient's illness experience in order to enter the patient's world of pain and suffering to understand and meet the patient's physical and emotional needs. A physician who is both prudent and loving knows best how to establish the professional boundaries for therapeutic success.

Besides generating and establishing robust therapeutic patient-physician relationships, prudent love synergy is also critical for addressing the possibility of abuse because of the power differential or inequity inherent in the patient-physician relationship (Marcum, 2008).[24] Traditionally, the locus of power in the patient-physician relationship is the physician. The reason is chiefly that the physician has the attributes of knowledge, training, and skill to help the patient, while the patient often lacks these attributes. Moreover, even physicians who are sick are ill advised to treat themselves, since disease often weakens a person not only physically but also mentally. With the rise of patient autonomy or respect for patient autonomy on the part of healthcare providers, the power within the therapeutic relationship often shifts to the patient; however, the physician is mainly responsible for delivering the clinical goods. Given the technical nature of modern medicine, patients find it difficult, if not impossible, to navigate the different therapeutic options available, let alone diagnose their illness with any accuracy or certainty. Consequently, opportunities for abuse given the power inequity in the therapeutic relationship still exist, even with contemporary emphasis on patient autonomy.

The virtuous physician does not abuse the power differential or inequity of the therapeutic relationship but rather uses the power entrusted to him or her by both the patient and society to the patient's and society's benefit and welfare. For, virtues ensure that a physician does the good because the physician is good. Part of the reason for the crises of quality-of-care and professionalism is that the patient-physician relationship may be dysfunctional, in that the physician, who holds most of the power in the relationship, may have little concern for the patient's existential needs and fears. In a dysfunctional relationship, the physician is the active agent dictating therapy with little regard for the emotional trauma of the patient's illness story while the patient is a passive recipient often suffering in silence and too often alone. The virtuous physician, as a genuinely caring and competent professional, strives to provide healthcare that meets the patient's needs not only in terms of a cure but also in terms of healing the patient or of assisting the patient obtain some sense of meaning and wholeness even when no cure is available. Although the virtuous physician represents a professional ideal for providing the quality-of-care patients expect and deserve, still it is an ideal that is worth striving to achieve and, to some extent, is achievable.

Just as the cyclic interaction defines the relationship between care and competence, so too, it defines the synergistic relationship between prudent wisdom and personal radical love. The love-prudence cycle (Fig. 5.1) is similar in structure to the care-competence cycle (Fig. 4.1). Beginning with $love_1$, particularly a personal radical $love_1$, the virtuous physician is motivated to connect with the patient in order to achieve the individual patient's radical ends, especially cure of disease and restoration of health. That connection involves not just a desire to care about the patient

[24] Importantly, philosophers of medicine recognize several different types of patient-physician relationships, based on their power differential. Moreover, the type of relationship may even change as the relationship develops, during course of treatment—hopefully, form thin to thick care.

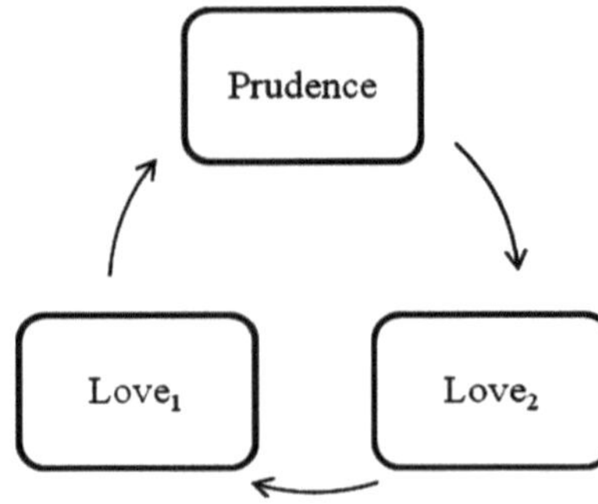

Fig. 5.1 Relationship between love and prudence

but a deep-seated love that provokes the physician to care about the patient in a compassionate, empathic, and possibly altruistic loving manner. But, such love is not undirected vis-à-vis prudent wisdom. Rather, such wisdom informs and shapes love$_1$ so that the physician can meet the patient's medical needs astutely and competently. Prudent wisdom operates not only in terms of the technical dimension of virtuous medical practice in judiciously utilizing medical technology correctly but also with respect to its ethical dimension in discerning the right or good way of engaging that technology. By informing the physician's love$_1$ for the patient, the physician can love$_2$ the patient concretely in meeting the patient's medical needs in a technically correct and morally good manner. The outcome is not only a patient who feels truly taken care of but also a physician who feels that he or she is fulfilling his or her duty as a physician. Finally, love$_2$ feeds back onto love$_1$ stimulating the physician to act more prudently, making ever wiser clinical decisions in the care of the patient and leading to a deeper love$_2$ for the patient. Moreover, this feedback results in expanding the physician's love$_1$ for more or demanding patients. Thus, prudent love synergy is an outcome of this cyclical, feedback relationship in which prudence and love enhance each another's properties to achieve what each cannot do separately.

Finally, the care-competence and the love-prudence cycles are part of a larger structure that includes caring (Fig. 5.2). In this larger structure, caring—as the

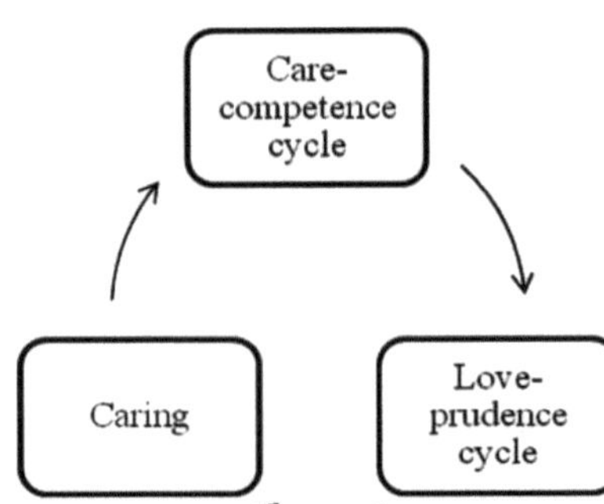

Fig. 5.2 Relationship between caring and the care-competence and love-prudence cycles

onotological virtue—makes possible the care-competence cycle, which in turn drives the love-prudence cycle. In other words, care motivates the physician to meet the patient's medical needs in a loving manner while competence in a prudently wise way. The virtuous physician's caring for the patient influences the patient positively, which in turn encourages the virtuous physician's caring through the feedback loop, love-prudence cycle → caring. This encouragement of caring leads to further motivation to care and competence, which in turn leads to deeper and richer lovingly wise decisions and acts in treating the patient. The overall outcome of this synergistic feedback is the virtuous physician's and other healthcare providers' ability to provide quality healthcare for the patient, both physically and existentially, which continues to escalate the quality healthcare provided by the virtuous physician and other healthcare providers in a professional manner. The patient experiences quality healthcare provided by a competent and prudent physician who deeply cares for and loves the patient. Peabody certainly exemplifies the notion of virtuous physician, which his peers recognized. In Peter Tishler's assessment of Peabody as a clinician, for example, he writes, "Peabody was intensely dedicated to the care of patients, to the training and nurturing of other professionals, and to the development of clinical investigation. . .he was the charismatic embodiment of the virtues [especially caring] about which he wrote" (1992, p. 35).

5.2 Imprudent Lovelessness

Just as the compound virtue of prudent love undergirds the virtuous physician's wise and caring decisions and actions that yield quality healthcare in a professional manner, so too, the compound vice of imprudent lovelessness undergirds the unvirtuous physician's unwise and uncaring decisions and actions that result in the delivery of poor quality and unprofessional healthcare. The notion of unvirtuous physician *qua* inauthentic and unprofessional healthcare provider entails a transformation of incompetence into imprudence and carelessness into lovelessness, yielding an imprudent lovelessness that represents the compound vice identifying and defining an unvirtuous physician or healthcare provider. As noted at the beginning of the chapter, these physicians and providers operate from a dysfunctional attitude or stance—often practicing defensive medicine to protect themselves from legal action by patients for possible bodily harm and personal pain and suffering. Generally, the patient's good or agenda does not drive or motivate the unvirtuous physician but rather the physician's own good or agenda. Such a physician is often not knowledgeable of current medical practices or more importantly not interested in the patient's personal illness story. Consequently, the unvirtuous physician is usually unable to form a patient-physician relationship that is therapeutic for the patient or even satisfying personally and professionally for the physician. In this section, I first examine the transformation of incompetence into imprudence and then carelessness into lovelessness, followed by explication of the compound vice, imprudent lovelessness, and finally by discussion of its relationship to uncaring.

5.2.1 Imprudence

As discussed in Chapter 3, imprudence consists of rash and incautious behavior particularly in forming judgments and in making decisions. Imprudent behavior can also result from fallacious and illogical reasoning and arguments. The outcome of this behavior is generally poor judgments and unsound decisions, leading to injudicious and, at times, stupid actions. In addition, imprudent people are often unable to anticipate or presage the fallout of their behavior or, if they can, are incapable of doing anything about it—which is symptomatic of the imprudent person's moral weakness. Moreover, while prudent people can make decisions in the face of uncertainties and act accordingly, they often fail to make decisions—or at least good decisions—under these conditions and either do not act or—if they do act—act poorly or improperly. Imprudent people, consequently, come across as irresponsible and unwise or foolish. Finally, just as competence is insufficient to explicate adequately the notion of virtuous physician and requires transformation into prudence, so to, incompetence is insufficient to explain fully the notion of unvirtuous physician and necessitates transformation into imprudence. In this section, I explore the imprudent physician as the antithesis of the prudent physician vis-à-vis the analysis provided earlier in terms of the Berlin wisdom project and Ardelt's and Achenbaum's responses to that project.

The two strata approach of the Berlin wisdom group for explicating wisdom is also adaptable or applicable for approaching and explaining imprudence (Baltes and Smith, 2008). For the first stratum, which consists of theoretical and practical wisdom, imprudence involves faulty factual or theoretical knowledge about the underlying reality of phenomena, which may lead to a distorted view of the world. Such a distorted worldview has serious consequences for the ability—or inability—of the imprudent person to investigate and discover—and ultimately to understand—unknown phenomena. With respect to practical wisdom, imprudence also involves a distortion but a distortion of life's basic pragmatics or principles. The distorted practical worldview often leads to an inability to utilize common sense to construct a viable and rewarding life. Imprudent people specifically are unable to plan, manage, and review their lives adequately and honestly, resulting in failure of their cognitive faculties to achieve life's goals and aspirations. For the second stratum, imprudence consists of a powerlessness to grasp the context or situation of a person's life with respect to the physically and socially normative factors that shape the life. The next section of the stratum consists of imprudent people's inability to appreciate the extent values forge their lives and to realize that intolerance to values different from their own often leads to harm and violence. Finally, the imprudent person is unable to manage life's contingencies in a healthful and constructive manner such that life becomes overwhelming and intolerable. The imprudent person vis-à-vis the Berlin notion of wisdom as expert knowledge for engaging life to its fullest is unwise and lives a truncated life filled with unrealized potential and dreams.

Ardelt's notion of the wise person and the three characteristics defining that person—cognitive, reflective, and affective—can also help to round out further the notion of imprudence or the imprudent person (Ardelt, 2004a). The imprudent person not only lacks these three characteristics but also, in some sense, is indifferent or even antagonistic to them. The imprudent person is not cognitive in terms of desiring to know truths or facts but is often disinterested in or unconcerned about them for one reason or another. Such a person might be too busy to bother with ensuring that the facts are indeed correct, or the facts might not be what fit with the imprudent person's agenda or preconceptions. Next, imprudent people are not reflective with respect to delving deeper into the issues that are generally behind phenomena. Rather, they are shallow or superficial and only interested in what is immediately palpable on the surface. They do not want to or cannot be inconvenienced or to take the time necessary to investigate a phenomenon sufficiently in order to understand it correctly or adequately. Lastly, imprudent people are not affective in terms of responding to others in an earnest or a heartfelt manner and thereby cannot commiserate with a person in need. Rather, they generally respond in a rude or an arrogant manner, because they often consider themselves as superior in some fashion. In sum, imprudent people are indifferent to and unreflective about truth and cannot connect with another person existentially to utilize truth to help another.

Finally, Achenbaum's notion of a wise act and the three features characterizing it—objective, rational, and transcendent—complete the analysis of imprudence (Achenbaum, 2004). Imprudent acts exhibit none of Achenbaum's characteristics of a wise act. First, an imprudent act is not simply unobjective, in terms of being unverifiable, but it is also subjective, with respect to mirroring the imprudent person's personal biases and prejudices. It does not represent the best course of action to take but generally the most expedient or convenient. Next, the imprudent act is irrational not simply in terms of being illogical but also with respect to failing to take into consideration potential limitations in the act itself. Imprudent people believe they can act without any restrictions or constraints, or even consequences. As such, imprudent acts often do not achieve the potential possible in a given situation. Rather, the acts simply are at best barren or at worse harmful. Lastly, imprudent acts are not transcendent in terms of being universally applicable to a great many situations but rather they are parochial and of limited application. They are partial in perspective, yielding a truncated vision of the world. Instead of inspiring greater acts, particularly with respect to anticipating and meeting the needs of others, imprudent acts implode under the weight of the provincial. Imprudent acts, then, are unobjective and subjective, irrational, and narrow in perspective.

Given the above analysis of imprudence, how is the unvirtuous physician imprudent? In terms of imprudence vis-à-vis the Berlin notion of wisdom (Baltes and Smith, 2008), the unvirtuous physician is imprudent with respect to the first stratum because he or she is not well informed about the current state of medical knowledge. Moreover, such a physician may not maintain or even attain a skill level dictated by the minimal standards of contemporary medical practice. Besides not knowing current factual knowledge, the imprudent physician in unable to apply successfully what medical knowledge he or she does have to the individual patient. As for the second stratum, the unvirtuous physician is careless about or indifferent to the patient's lifespan context and does not obtain important information about the illness story that would have a positive impact on the clinical outcome. Next, the imprudent physician fails to learn adequately or to appreciate appropriately the patient's value system, which can result in an uncooperative patient or even in a hostile patient if the physician shows disregard for what the patient values. Finally, the imprudent physician is unable to negotiate the uncertainties and contingencies in the patient's treatment because he or she is incapable of marshaling life or professional experience effectively. The imprudent physician, in terms of the Berlin notion of wisdom, simply does not have the expert knowledge needed to help the patient and often ends up harming the patient.

The unvirtuous physician is imprudent because as a person that physician does not have the characteristics—cognitive, reflective, and affective—Ardelt (2004a) deems qualifies a person as prudent. The imprudent physician is not only deficient in these characteristics but also indifferent or even hostile to them. First, the imprudent physician lacks the cognitive ability to obtain or understand the clinical facts or truths about the patient's illness, either its physical or emotional manifestations. Next, the imprudent physician is unable to reflect with any depth on the larger issues facing the patient's experience of the illness. Generally, the surface issues are sufficient for the imprudent physician, who often is too busy to take the time to explore what appear to be irrelevant, nonclinical facts about the patient's illness story. Finally, in terms of the affective characteristic of the prudent person, the imprudent physician is unable to exhibit the appropriate sympathy for the patient's pain. Unfortunately, whether intentional or not, the imprudent physician comes across to the patient as uncaring and not interested to the patient's existential needs and concerns. The unvirtuous physician *qua* imprudent person, then, is indifferent to and superficial about the patient's clinical facts and fails to connect existentially with the patient to utilize these facts in helping the patient.

The unvirtuous physician is imprudent also because the very actions of such a physician fail to display the features—objective, rational, and transcendent—Achenbaum (2004) considers necessary for acts to be prudent. First, the imprudent physician's actions are not objective in terms of following community standards for medical practice but rather subjective in that the imprudent physician imposes his or her agenda onto the patient, often forcing the patient to conform to what the physician feels is best for the patient. Next, the imprudent physician's acts, especially clinical judgments, do not conform to minimal profession's intellectual or rational standards. These acts or judgments are not simply unsound or irrational but they are also indefinite in terms of their applicability to the patient's welfare. In other words, the imprudent physician's clinical judgments fail to obtain the best outcome for patients and may even be harmful to them. Lastly, the imprudent physician's acts are generally not transcendent in terms of being commonly applicable to other patients but rather they are limited in their application to only a single patient or to just a few patients. These acts are the result of a narrow-minded worldview, which fails to inspire greater acts of caring to meet patients' physical and emotional needs.

Finally, just as prudent wisdom, according to Szawarski (2004), has healing power associated with it, so too, imprudence can prohibit patient healing and lead to patient harm, injury, or possibly death. Adapting Szawarski's three features of wisdom's healing power, imprudence's injurious potential includes deficient or faulty medical knowledge, poor clinical judgment, and lack of self-trust. First, imprudence's injurious potential generally reflects medical knowledge that is either out-dated or just simply wrong. Even if the knowledge is current, the imprudent physician is incapable of utilizing it effectively or applying it skillfully, safely, and efficiently. The next feature is poor clinical judgment in which the unvirtuous physician is unable to master the tacit dimension of medicine's art, which results in unsound diagnostic, prognostic, and therapeutic decisions. The final feature of imprudence's injurious potential is a lack of self-trust in which the imprudent physician second-guesses a clinical decision. Such second-guessing can have deleterious effects on the patient-physician relationship in that the patient may come to distrust the physician and become non-compliant. Another manifestation of this last feature is unwarranted or over confidence on the part of the imprudent physician in his or her clinical abilities. In this regard, the physician fails to appreciate personal limits and proceeds to treat a patient without regard for disciplinary boundaries and expertise. By failing to appreciate these limits, the imprudent physician exacerbates the patient's pain and suffering rather than alleviating it.

5.2.2 Lovelessness

Besides imprudence, the other major vice that defines the unvirtuous physician is lovelessness, which represents transformation of the ontic virtue of carelessness. Carelessness, both in its motivational (carelessness$_1$) and behavioral (carelessness$_2$) dimensions, is one of the major roots supporting the delivery of today's poor quality healthcare; however, as mentioned earlier, it is insufficient to support that delivery completely. Lovelessness, which reflects the loss of human emotion and deeply felt connection with the patient as another human being who is vulnerable and in need, is also necessary.[25] In terms of this vice, the patient is a commodity or consumer of healthcare goods and services—simply another cog within the machinery of the healthcare industry. This industrial or business metaphor predominates within contemporary healthcare and, unfortunately, represents often the elimination of what makes humans human—love. This love-lessness includes an inability to pull alongside compassionately the ailing patient in order to understand and possibly to experience empathically the patient's world of pain and suffering and maybe to benefit that patient altruistically at some expense to the healthcare provider. In this section, I explicate lovelessness in terms of incompassion, apathy, and selfishness. Although these manifestations of lovelessness are critical for understanding it, they

[25] As noted in Chapter 3, Hurka (2001) identifies three versions of lovelessness: avoidance of good, love of evil, and indifference to either. In this section, the focus of the discussion is not on any one of these versions but on the nature of lovelessness in terms of incompassion, apathy, and selfishness, as well as with respect to the notion of impersonal prosaic lovelessness.

do not completely capture its nature. To that end, I propose a notion of impersonal prosaic lovelessness that is the antithesis of Toner's notion of personal radical love.

As discussed in Chapter 3, incompassion and apathy are similar cognate vices of lovelessness that form a particular set. Both reflect insensitivity or aloofness towards the suffering of another person. However, each vice is different from the other. Incompassion, for example, is an inability to comprehend another person's suffering or to realize conceptually the extent of the suffering. The uncompassionate person is incapable of pulling alongside another who is suffering or in pain. Utilizing Underwood's empirical model of compassion, an important component of incompassion is the environmental context of the uncompassionate person and the physical and social factors comprising that context. This context may prohibit the uncompassionate person from appreciating or understanding another person's pain and suffering. The next component is a lack of motivation on the part of the uncompassionate person to express compassionate love for another. Employing Post's notion of compassionate love, the uncompassionate person lacks any natural desire to help another person who is suffering. Even if that person appears to express such love, the motives are mixed in that the uncompassionate person expresses compassionate love for selfish reasons. Moreover, uncompassionate people fail to discern appropriately when and how to express such love because they lack the necessary capacity to evaluate the requirements for truly loving other people. The last component is the attitude of the uncompassionate person, which reflects an indifference to another person's pain and suffering. Finally, this component feeds back onto the first component and exacerbates the uncompassionate person's incompassion towards others.

Apathy, on the other hand, is an inability to sense or experience affectively what another person is suffering or what that suffering might be like. In other words, the apathetic person has little, if any, connection emotionally with the suffering person. To use Halpern's notion of empathy to gain insight into the notion of apathy, the apathetic person is out of tune or register with the suffering of another. In other words, that person is unable to imagine or to conceive, especially at a visceral level, another person's world of pain and suffering. Consequently, the apathetic person is unable not only to understand what another person is suffering but is also incapable of appreciating what the suffering person feels or experiences. That person might be able to pull alongside another person in pain but is unable to enter into that person's sphere of pain. A comprehensive notion of apathy includes several components. The first is motivational in which the apathetic person has no desire to relate to another person's world of suffering, while the next component is cognitive in which that person makes little, if any, effort to understand that world. The third is emotional in which the apathetic person fails to connect affectively with the suffering person, while the final component is behavioral in which that person neglects to act on behalf of another person. Utilizing Spiro's definition for apathy in terms of projection, where empathy mirrors "I *am* you," apathy mirrors "I am *not* you."

Like empathy, apathy also exhibits a cyclic structure in its expression. The first phase represents a failure of the apathetic person to resonate with another's pain and suffering, i.e. an inability to comprehend both cognitively and existentially that another person is suffering or in pain. The fact that a person is suffering may register with an apathetic person but the reality of that suffering remains elusive emotionally and obscure cognitively. The next phase involves communication in which the apathetic person is unable to communicate effectively and, more importantly, meaningfully with the suffering other. By not connecting with the pain another person feels, the apathetic person is incapable of articulating the significance of the other's experience thereby resulting in frustration on the part of the other—leading to the final phase of the apathetic cycle. The last phase pertains to the reception or perception of the apathy the person in pain feels or experiences. This reception discourages the development of any meaningful relationship between the person in pain and the apathetic person. Finally, the failure to forge any meaningful relationship feeds back on to the first phase, making resonance with another person's suffering too painful to attempt for the apathetic person.

As noted earlier, Comte introduces altruism for his positive philosophy to counteract selfishness, which is the final dimension of lovelessness discussed before turning to the specific notion of impersonal prosaic lovelessness. Selfishness, self-centeredness, or egoism is fundamentally the antithesis of altruism. Whereas the altruistic person lives a life for others, especially in service to others, the selfish person lives only for the self. Comte's motto, "Live for Others," becomes for the egoist, "Live for Self." According to Kurt Baier, "egoism involves putting one's own good, interest, and concern above that of others" (1993, p. 197) Moreover, the self-centered or egoistic person may benefit himself or herself even at the expense of others. If that person does benefit others, the objective may not be intentional in that the benefit was simply secondary to the primary effort to benefit one's own welfare. Or, if the egoist does intend to benefit others, the beneficial act is predicated upon a reward for the egoist. Philosophers demarcate between three types of egoism. The first is ethical egoism, which involves the moral foundation of self-centered acts, i.e. it is right to pursue one's good and wrong not to do so (Österberg, 1988). The next type is rational egoism, which refers to the cognitive basis for self-centered acts, i.e. it is rational to aim for one's good and irrational not to do so (Shaver, 1999). The final type is psychological egoism, which pertains to the impulse for self-centered acts, i.e. we are motivated to do what is best for the self (Sober, 2000).

While incompassion, apathy, and selfishness and egoism provide an ample basis for the notion of lovelessness, still they fail to denote its nature sufficiently. To that end, I use Toner's notion of personal radical love to develop a notion of impersonal prosaic lovelessness. While Toner's notion of love is radical in the sense that the lover affectively affirms the ultimate ends of the beloved, the notion of lovelessness is prosaic in the sense that the loveless agent does not affirm adequately the ultimate ends of another person. Rather, that agent often is indifferent to those ends and substitutes others for them, especially those of the loveless agent. Such lovelessness is prosaic because it does not inspire or promote the best but simply the banal or inconsequential from people. While the radical lover provides affective affirmation for the beloved, the prosaic loveless agent is indifferent or neglectful of the other person *qua* human being. The result is anything but the union attributable to radical love, with its self-giving and affective identification, but rather it is separation and at times hatred between the loveless agent and another person, with the agent's self-taking and indifference. Finally, given its prosaic nature, lovelessness is also impersonal. While radical love is personal because it attends to the uniqueness of the beloved, prosaic lovelessness is impersonal because the loveless agent fails to recognize the other's uniqueness and treats the other indifferently or worse yet as a means towards the loveless agent's ends. Consequently, rather than the communion of lover and beloved as a composite agent, the result of impersonal prosaic lovelessness is the separation of individual agents who are unable to connect in a meaningful and loving way.

Given the above analysis of lovelessness, how is the unvirtuous physician loveless? First, such a physician is loveless in terms of being uncompassionate. As uncompassionate, the loveless physician is unable or unwilling to comprehend the patient, especially the patient's illness story and the pain and suffering associated with that story. Moreover, the uncompassionate physician is incapable of pulling alongside of the patient's world of illness and of engaging its existential anxiety. Next, the unvirtuous physician is loveless with respect to apathy for the patient's plight. In other words, apathetic physicians are indifferent to the pain and suffering associated with a patient's illness experience. They might be able to pull alongside the patient's world but for different reasons are unable or unwilling to enter that world. Apathetic physicians cannot attune themselves to the patient on a personal level and to the patient's physical and especially emotional needs. Lastly, the unvirtuous physician as loveless is self-centered and not patient-centered. Egoistic physicians think of themselves and the benefits, such as financial, social status, or job security, they may receive from practicing medicine. Self-centered physicians may not be overly conscience of these benefits; however, these benefits drive, from an existential level, such physicians to practice medicine. In sum, loveless physicians, employing Dougherty and Purtilo's notion of responsible healthcare, fail to provide the due care that is a physician's fiduciary duty.

Finally, the unvirtuous physician is loveless in terms of exhibiting impersonal prosaic lovelessness towards his or her patients. Such a physician's lovelessness is prosaic with respect to the loveless physician's failure to affirm sufficiently the ultimate ends of the patient, which may be more than simply a physical or bodily cure. Although loveless physicians might strive to achieve a cure, the motivation for their actions is not ultimately to cure the patient. Instead, personal gain or some other benefit might motivate them. In actuality, the loveless physician is indifferent to the patient *qua* person. Consequently, the physician's lovelessness is prosaic since it arouses or advances not what is the best for the patient vis-à-vis the patient's personal narrative but simply what is expedient or convenient, particularly for the physician. The physician's lovelessness is also impersonal since the loveless physician fails or refuses to acknowledge the uniqueness of patients and their illness stories and experiences. Rather, the loveless physician treats patients as simply another cog within the healthcare machinery, with little if any regard for their individual needs. In Toner's words, the loveless physician does not provide adequate affective affirmation for patients and the pain and suffering they are experiencing due to illness. The outcome is a dysfunctional patient-physician or therapeutic relationship in which the physician is unable to connect with the patient at a fundamental level, in order to gain the patient's trust that the physician can help the patient. Moreover, given the relationship's dysfunctional nature, the patient often becomes uncooperative if not hostile towards the physician, which in turn exacerbates the dysfunctional nature of the patient-physician relationship.

5.2.3 Compound Vice of Imprudent Lovelessness

Like the compound virtue of prudent love for the virtuous physician, the compound vice of imprudent lovelessness defining the unvirtuous physician also represents a combination—a combination of unwise imprudence and impersonal prosaic lovelessness. In this combination, on a superficial level, imprudence and lovelessness complement one another as the imprudently loveless physician either neglects the patient's medical needs or at times even harms the patient. As an imprudent healthcare provider, the unvirtuous physician provides care that does not reflect the demands of current professional standards or, if such care is provided, it is done with little, if any, regard for the patient's illness story. By being an imprudent person, the imprudent physician engages in imprudent acts. As a loveless healthcare provider, the unvirtuous physician delivers—whether incompetently or competently—healthcare in an uncaring and impolite manner, often with little regard for the patient's emotional or existential needs. The loveless physician is unable or unwilling to draw alongside the patient compassionately and to enter empathically the patient's world of illness, let alone to sacrifice altruistically for the patient's benefit. Rather, at worse, the loveless physician—whether intentionally or not—is dispassionate and indifferent towards the patient's pain and suffering and is anxious only for his or her own advantage. At best, the loveless physician attends to the patient's needs but in an impersonal prosaic manner—treating the patient as simply another cog in the industrial healthcare system. Consequently, the vices of imprudence and lovelessness complement one another simultaneously as the unvirtuous physician provides poor quality and unprofessional healthcare.

However, at a fundamental level, the compound vice of imprudent lovelessness is more than just a combination or even complementation of unwise imprudence and impersonal prosaic lovelessness; rather, like prudent love, it represents a species that surpasses or transcends the properties of the individual vices that compose it. The relationship, then, is synergistic in which unwise imprudence exacerbates the impersonal prosaic nature of lovelessness; and lovelessness, in turn, worsens the unwise nature of imprudence. The synergy between the two vices makes imprudence loveless and lovelessness imprudent. For example, the loveless physician in caring less for patients and their medical needs generally fails to maintain even minimal professional standards of competence and is then incapable of making wise prudent clinical decisions as to the best course of action. Consequently, the unwise imprudent physician makes decisions that may be the worst possible decisions for the patient in terms of professional standards and as a loveless agent acts on those decisions in a way that robs the patient of personal dignity. Given the synergistic relationship between the two vices, the unvirtuous physician provides the poorest quality healthcare imaginable and often runs the risk of causing more pain and suffering for the patient than the disease itself.

Imprudent lovelessness synergy often impedes the creation and development of robust patient-physician relationships that are therapeutic, particularly with respect to the formation of appropriate professional boundaries. Or, if such boundaries are formed imprudent loveless physicians often transgress them. Sexual misconduct between patients and physicians, especially in psychiatry, is the classical example of such boundary transgressions that lead to dysfunctional therapeutic relationships and patient harm (Hall, 2001; Perlman, 2009; Simon, 1999). Imprudent physicians who cross this boundary generally rationalize their behavior in order to justify it. The process of such boundary transgression is incremental, likened to a slippery slope (Galletly, 2004). Besides rationalizing their behavior imprudently, these physicians are loveless in the sense that the boundary violations also represent abuse of the power distribution between a patient and an unviruous physician, who is simply satisfying personal desires at the patient's expense (Kluft, 1993). Thus, the synergistic relationship between imprudence and lovelessness functions to deceive the unvirtuous physician into transgressing the therapeutic boundary and to engage in sexual misconduct.

Just as the cyclic interaction defines the synergistic relationship between carelessness and incompetence, so too, such an interaction defines the synergistic relationship between imprudence and lovelessness. The lovelessness-imprudence cycle (Fig. 5.3) is similar in structure to the carelessness-incompetence cycle (Fig. 4.2). Beginning with lovelessness$_1$, particularly an impersonal prosaic lovelessness$_1$, the unvirtuous physician is unmotivated in terms of connecting with the patient in order to meet the patient's comprehensive medical needs. Such a physician might meet the patient's bodily needs, if competent to do so, but he or she is uninspired to meet the patient's existential needs—if such needs even register with the loveless$_1$ physician. Being loveless$_1$, the unvirtuous physician often acts imprudently, as depicted in the first arm of the cycle, lovelessness$_1$ $\rightarrow$ imprudence, when making decisions either over technical matters or over ethical issues that arise in care of the patient. Such imprudence, in turn, leads to lovelessness$_2$ on the part of the unvirtuous physician, as depicted in the next arm of the cycle, imprudence $\rightarrow$ lovelessness$_2$, who

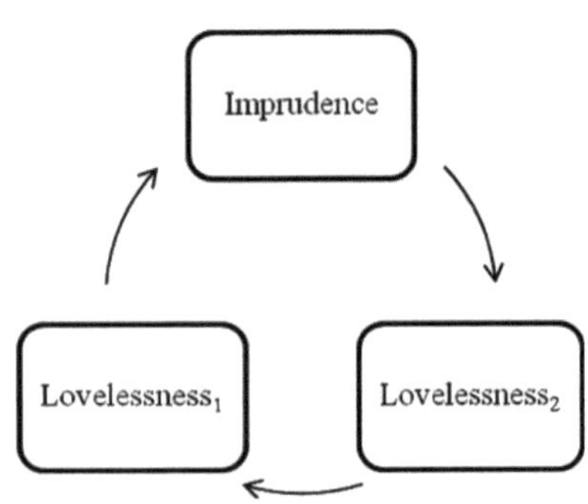

Fig. 5.3 Relationship between lovelessness and imprudence

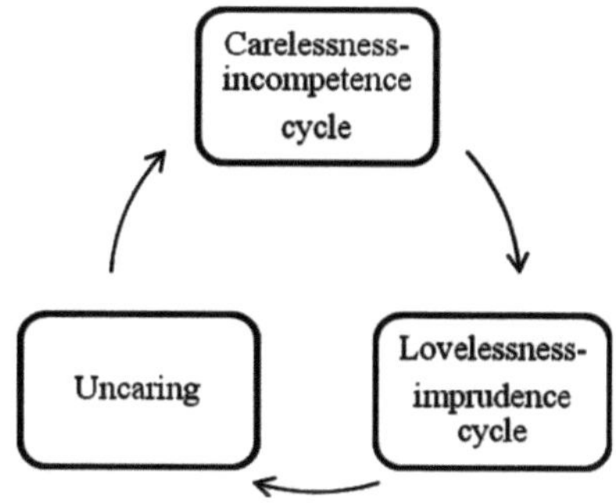

Fig. 5.4 Relationship between uncaring and the carelessness-incompetence and lovelessness-imprudence cycles

is incapable of loving the patient adequately or appropriately in order to meet the patient's comprehensive medical needs. Rather, such a physician only meets part of those needs or meets them incompetently or imprudently, or neglects them entirely. The final arm of the cycle, $\text{lovelessness}_1 \rightarrow \text{lovelessness}_2$, depicts the feedback from lovelessness_2, which exacerbates the behavior of the loveless_1 physician to love a particular patient less—if at all—as well as other patients. The overall outcome of the synergism associated with this cycle is the poorest possible healthcare in terms of quality and professionalism.

Finally, the carelessness-incompetence and the lovelessness-imprudence cycles are part of a larger structure that includes uncaring (Fig. 5.4). In this larger structure, uncaring—as the onotological vice—makes possible the carelessness-incompetence cycle, which results in an unvirtuous physician's and other healthcare providers' inability to provide quality healthcare for the patient—both physically or existentially. The patient experiences poor quality healthcare provided by an imprudent physician as callous uncaring and lack of concern or love. The unvirtuous physician's inability to care for the patient in a genuine manner has a negative impact on the patient through the feedback loop of loveless imprudence → uncaring, which, in turn, discourages the unvirtuous physician to care for other patients. This discouragement of caring leads synergistically to further careless behavior and incompetence, which continues to escalate the poor quality healthcare provided by the unvirtuous physician and other healthcare professionals. Lastly, besides the loveless imprudent physician other species of unvirtuous physicians exist—the loveless prudent or loving imprudent physician. The former engages in practical and sensible healthcare with little or no passion or conviction, while the latter in reckless and irresponsible healthcare with enthusiastic care. The former species represents Curzer's HCP who provides benevolent healthcare *as if* the provider cares but really does not (or such care *qua* liking is not required).[26]

[26] Cruzer in an attempt to avoid imprudent love or inappropriate caring with the benevolence thesis falls prey to the loveless prudent species of unvirtuous HCP.

5.3 Summary

In the previous chapter, I developed the notion of virtuous physician in terms of care and competence, while in this chapter I completed its development with the transformation of care into personal radical love and competence into prudent wisdom. The caring and competent physician *qua* lovingly prudent physician provides the best possible quality healthcare in a professional manner. The unvirtuous physician, on the other hand, is careless and incompetent. And, with the transformation of carelessness into impersonal prosaic lovelessness and incompetence into imprudence, the loveless imprudent physician provides the poorest possible quality healthcare in an unprofessional manner. Now that I have finished the explication of the notions of virtuous and unvirtuous physician, in the next chapter I illustrate them with case studies form the medical literature. In a final chapter, I compare my notion of virtuous physician with other notions in the literature and argue that my notion better serves to address the crises of quality-of-care and professionalism that are plaguing modern medicine.

References

Achenbaum, W.A. 2004. Wisdom's vision of relations. *Human Development* 47:300–303.

American Board of Internal Medicine. 1998. *Project professionalism*. Philadelphia, PA: American Board of Internal Medicine.

Anonymous. 1927. Altruism in medicine. *Journal of the National Medical Association* 19:22–23.

Anonymous. 2000. Looking back on the millennium in medicine. *New England Journal of Medicine* 342:42–49.

Ardelt, M. 2004a. Wisdom as expert knowledge system: a critical review of a contemporary operationalization of an ancient concept. *Human Development* 47:257–285.

Ardelt, M. 2004b. Where can wisdom be found? A reply to the commentaries by Baltes and Kunzmann, Sternberg, and Achenbaum. *Human Development* 47:304–307.

Aristotle. 1998. *The Nicomachean ethics,* (trans: Ross, W.D.). New York, NY: Oxford University Press.

Baier, K. 1993. Egoism. In *A companion to ethics*, ed. P. Singer, 197–204. Oxford, UK: Blackwell.

Baltes, P.B., and U. Kunzmann 2004. The two faces of wisdom: wisdom as a general theory of knowledge and judgment about excellence in mind and virtue vs. wisdom as everyday realization in people and products. *Human Development* 47:290–299.

Baltes, P.B., and J. Smith 2008. The fascination of wisdom: its nature, ontogeny, and function. *Perspectives on Psychological Science* 3:56–64.

Barrett-Lennard, G.T. 1981. The empathy cycle: refinement of a nuclear concept. *Journal of Counseling Psychology* 28:91–100.

Batson, C.D. 1991. *The altruism question: toward a social-psychological answer*. Hillsdale, NY: Lawrence Erlbaum Associates.

Batson, C.D. 2002. Addressing the altruism question experimentally. In *Altruism and altruistic love: science, philosophy, and religion in dialogue*, eds. S.G. Post, L.G. Underwood, J.P. Schloss, and W.B. Hurlbut, 89–105. New York, NY: Oxford University Press.

Bishop, J.P., and C.E. Rees 2007. Hero or has-been: Is there a future for altruism in medical education?. *Advances in Health Sciences Education: Theory and Practice* 12:391–399.

Bryan, C.S. 2007b. The seven basic virtues in medicine. *Journal of the South Carolina Medical Association* 103:135–137.

Childress, J.F., and Liverman, C.T. (eds.) 2006. *Organ donation: Opportunities for action*. Washington, DC: National Academies Press.

Clayton, V.P., and J.E. Birren 1980. The development of wisdom across the life span: a reexamination of an ancient topic. In *Life-span development and behavior*, eds. P.B. Baltes and O.G. Brim Jr., 103–135. New York, NY: Academic.

Coulter, I.D., M. Wilkes, and C. Der-Martirosian 2007. Altruism revisited: a comparison of medical, law and business students' altruistic attitudes. *Medical Education* 41:341–345.

Dougherty, C.J., and R. Purtilo 1995. Physician's duty of compassion. *Cambridge Quarterly of Healthcare Ethics* 4:426–433.

Downie, R.S. 2002. Supererogation and altruism: a comment. *Journal of Medical Ethics* 28:75–76.

Edmondson, R., and J. Pearce 2007. The practice of health care: wisdom as a model. *Medicine, Health Care and Philosophy* 10:233–244.

Edwards, K.A. 2001. Critiquing empathy: insights from professional education. *Second Opinion* 4:35–47.

Egonsson, D. 2007. Hypothetical approval in prudence and medicine. *Medicine, Health Care and Philosophy* 10:245–252.

Epstein, R.S. 1994. *Keeping boundaries: maintaining safety and integrity in the psychotherapeutic process*. Washington, DC: American Psychiatric Press.

Farber, N.J., D.H. Novack, and M.K. O'Brien 1997. Love, boundaries, and the patient-physician relationship. *Archives of Internal Medicine* 157:2291–2294.

Fehr, B., S. Sprecher, and L.G. Underwood eds. 2008. *The science of compassionate love: theory, research, and applications*. Malden, MA: Wiley-Blackwell.

Flyvbjerg, B. 2001. *Making social science matter: why social inquiry fails and how it can succeed again*. New York, NY: Cambridge University Press.

Friedman, E. 1990. The perils of detachment. *Healthcare Forum Journal* 33:9–10.

Gabbard, G.O., and C. Nadelson 1995. Professional boundaries in the physician-patient relationship. *Journal of American Medical Association* 273:1445–1449.

Galletly, C.A. 2004. Crossing professional boundaries in medicine: the slippery slope to patient sexual exploitation. *Medical Journal of Australia* 181:380–383.

Ghosh, A.K. 2004. Understanding medical uncertainty: a primer for physicians. *Journal of the Association of Physicians of India* 52:739–742.

Glannon, W., and L.F. Ross 2002. Are doctors altruistic? *Journal of Medical Ethics* 28:68–69.

Glick, S.M. 1992. The empathic physician: nature and nurture. In *Empathy and the practice of medicine: beyond pills and the scalpel*, eds. H. Spiro, M.G. McCrea Curnen, E. Peschel, and D. St. James, 85–102. New Haven, CT: Yale University Press.

Graber, D.R., and M.D. Mitcham 2004. Compassionate clinicians take patient care beyond the ordinary. *Holistic Nursing Practice* 18:87–94.

Gutheil, T.G., and G.O. Gabbard 1993. The concept of boundaries in clinical practice: theoretical and risk-management dimensions. *American Journal of Psychiatry* 150:188–196.

Hall, K.H. 2001. Sexualization of the doctor-patient relationship: is it ever ethically permissible? *Family Practice* 18:511–515.

Halpern, J. 2003. What is clinical empathy? *Journal of Internal Medicine* 18:670–674.

Hemmerdinger, J.M., S.D.R. Stoddart, and R.J. Lilford 2007. A systematic review of tests of empathy in medicine. *BMC Medical Education* 7:24.

Henry, C.W. 1993. The place of prudence in medical decision making. *Journal of Religion and Health* 32:27–37.

Hirsch, E.M. 2007. The role of empathy in medicine: a medical student's perspective. *Virtual Mentor* 9:423–427.

Hojat, M. 2007. *Empathy in patient care: antecedents, development, measurement, and outcomes*. New York, NY: Springer.

Hojat, M., J.S. Gonnella, T.J. Nasca, S. Mangione, M. Vergare, and M. Magee 2002. Physician empathy: definition, components, measurement, and relationship to gender and specialty. *American Journal of Psychiatry* 159:1563–1569.

Humphrey, N. 1997. Varieties of altruism—and the common ground between them. *Social Research* 64:199–209.

Hurka, T. 2001. *Virtue, vice, and value*. New York, NY: Oxford University Press.

Jones, R. 2002. Declining altruism in medicine. *British Medical Journal* 324:624–625.

Kalisch, B.J. 1973. What is empathy? *American Journal of Nursing* 73:1548–1552.

Kitcher, P. 2010. Varieties of altruism. *Economics and Philosophy* 26:121–148.

Kluft, R.P. 1993. The physician as perpetrator of abuse. *Primary Care* 20:459–480.

Krebs, D.L. 1970. Altruism—an examination of the concept and a review of the literature. *Psychological Bulletin* 73:258–302.

Krebs, D.L., and F. Van Hesteren 1992. The development of altruistic personality. In *Embracing the other: philosophical, psychological, and historical perspectives on altruism*, eds. P.M. Oliner, S.P. Oliner, L. Baron, L.A. Blum, D.L. Krebs, and M.Z. Smolenska, 142–169. New York, NY: New York University Press.

Landau, R.L. 1993. . . .And the least of these is empathy. In *Empathy and the practice of medicine: beyond pills and the scalpel*, eds. H. Spiro, M.G. McCrea Curnen, E. Peschel, and D. St. James, 103–109. New Haven, CT: Yale University Press.

Larson, E.B., and X. Yao 2005. Clinical empathy as emotional labor in the patient-physician relationship. *Journal of American Medical Association* 293:1100–1106.

Marcum, J.A. 2008. *An introductory philosophy of medicine: humanizing modern medicine*. New York, NY: Springer.

McGaghie, W.C., J.J. Mytko, W.N. Brown, and J.R. Cameron 2002. Altruism and compassion in the health professions: a search for clarity and precision. *Medical Teacher* 24:374–378.

McKay, A.C. 2002. Supererogation and the profession of medicine. *Journal of Medical Ethics* 28:70–73.

McLean, S.A.M. 2002. Commentary on Glannon and Ross, and McKay. *Journal of Medical Ethics* 28:74.

Mercer, S.W., and W.J. Reynolds 2002. Empathy and quality of care. *British Journal of General Practice* 52:S9–S13.

Monroe, K.R. 1996. *The heart of altruism: perceptions of a common humanity*. Princeton, NJ: Princeton University Press.

Moore, L.A. 2010. Being empathetic: benefits and challenges for the clinician and client. *Topics in Stroke Rehabilitation* 17:20–29.

More, E.S., and Milligan, M.A. (eds.) 1994. *The empathic practitioner: empathy, gender, and medicine*. New Brunswick, NJ: Rutgers University Press.

Morse, J.M., G. Anderson, J.L. Bottorff, O. Yonge, B. O'Brien, S.M. Solberg, and K.H. McIlveen 1992. Exploring empathy: a conceptual fit for nursing practice? *Image: Journal of Nursing Scholarship* 24:273–280.

Morse, J.M., J. Bottorff, G. Anderson, B. O'Brien, and S. Solberg 2006. Beyond empathy: expanding expressions of caring. *Journal of Advanced Nursing* 53:75–87.

Nadelson, C., and M.T. Notman 2002. Boundaries in the doctor-patient relationship. *Theoretical Medicine and Bioethics* 23:191–201.

Neuman, M., J. Bensing, S. Mercer, N. Ernstmann, O. Ommen, and H. Pfaff 2009. Analyzing the "nature" and "specific effectiveness" of clinical empathy: a theoretical overview and contribution towards a theory-based research agenda. *Patient Education and Counseling* 74:339–346.

Norfolk, T., K. Birdi, and D. Walsh 2007. The role of empathy in establishing rapport in the consultation: a new model. *Medical Education* 41:690–697.

Norfolk, T., K. Birdi, and F. Patterson 2009. Developing therapeutic rapport: a training validation study. *Quality in Primary Care* 17:99–106.

Österberg, J. 1988. *Self and others: a study of ethical egoism*. Dordrecht: Kluwer.

Parker, M. 2002. Whither our art? Clinical wisdom and evidence-based medicine. *Medicine, Health Care and Philosophy* 5:273–280.

Pedersen, R. 2008. Empathy: a wolf in sheep's clothing? *Medicine, Health Care and Philosophy* 11:325–335.

Pedersen, R. 2009. Empirical research on empathy in medicine—a critical review. *Patient Education and Counseling* 76:307–322.

Perlman, S.D. 2009. Falling into sexuality: sexual boundary violations in psychotherapy. *Psychoanalytic Review* 96:917–941.

Piliavin, J.A., and H.-W. Charng 1990. Altruism: a review of recent theory and research. *Annual Review of Sociology* 16:27–65.

Pilowsky, I. 1977. Altruism and the practice of medicine. *British Journal of Medicine Psychology* 50:305–311.

Post, S.G. 2003. *Unlimited love: altruism, compassion, and service*. Philadelphia, PA: Templeton Foundation Press.

Post, S.G. 2005. Altruism, happiness, and health: it's good to be good. *International Journal of Behavioral Medicine* 12:66–77.

Post, S.G. (ed.) 2007. *Altruism and health: perspectives from empirical research.* New York, NY: Oxford University Press.

Post, S.G. 2010. Humanism, posthumanism, and compassionate love. *Technology in Society* 32:35–39.

Post, S.G., and L.G. Underwood 2002. Concluding summary: future research needs on altruism and altruistic love. In *Altruism and altruistic love: science, philosophy, and religion in dialogue*, eds. S.G. Post, L.G. Underwood, J.P. Schloss, and W.B. Hurlbut, 379–386. New York, NY: Oxford University Press.

Post, S.G., Underwood, L.G., Schloss, J.P., and Hurlbut, W.B. (eds.) 2002. *Altruism and altruistic love: science, philosophy, and religion in dialogue*. New York, NY: Oxford University Press.

Post, S.G., B. Johnson, M.E. McCullough, and J.P. Schloss 2003. *Research on altruism and love: an annotated bibliography of major studies in psychology, sociology, evolutionary biology, and theology*. Radnor, PA: Templeton Foundation Press.

Reich, W.T. 1989. Speaking of suffering: a moral account of compassion. *Soundings* 72:83–108.

Reynolds, W.J., and B. Scott 1999. Empathy: a crucial component of the helping relationship. *Journal of Psychiatric and Mental Health Nursing* 6:363–370.

Schulz, A.W. 2009. *Sober & Wilson's evolutionary arguments for psychological altruism: a reassessment. Biology and Philosophy* 26: 251–260.

Scott, N., and J. Seglow. 2007. *Altruism*. New York, NY: Open University Press.

Sesardic, N. 1999. Altruism. *British Journal for the Philosophy of Science* 50:457–466.

Shaver, R. 1999. *Rational egoism: a selective and critical history*. New York, NY: Cambridge University Press.

Simon, R.I. 1999. Therapist-patient sex: from boundary violations to sexual misconduct. *Psychiatric Clinics of North America* 22:31–47.

Sober, E. 2000. Psychological egoism. In *The Blackwell guide to ethical theory*, ed. H. LaFollette, 129–148. Oxford, UK: Blackwell.

Sober, E. 2002. The ABCs of altruism. In *Altruism and altruistic love: science, philosophy, and religion in dialogue*, eds. S.G. Post, L.G. Underwood, J.P. Schloss, and W.B. Hurlbut, 17–28. New York, NY: Oxford University Press.

Sober, E., and D.S. Wilson 1998. *Unto others: the evolution and psychology of unselfish behavior*. Cambridge, MA: Harvard University Press.

Spiro, H. 1992. What is empathy and can it be taught? *Annals of Internal Medicine* 116:843–846.

Spiro, H., McCrea Curnen, M.G., Peschel, E., and St. James, J.D. (eds.) 1993. *Empathy and the practice of medicine: beyond pills and the scalpel*. New Haven, CT: Yale University Press.

Sprecher, S., and B. Fehr. 2006. Enhancement of mood and self-esteem as a result of giving and receiving compassionate love. *Current Research in Social Psychology* 11:227–242.

Spruiell, V. 1983. The rules and frames of the psychoanalytic situation. *Psychoanalytic Quarterly* 52:1–33.

Squier, R.W. 1990. A model of empathic understanding and adherence to treatment regimens in practitioner-patient relationships. *Social Science and Medicine* 30:325–339.
Stich, S. 2007. Evolution, altruism and cognitive architecture: a critique of Sober and Wilson's argument for psychological altruism. *Biology and Philosophy* 22:267–281.
Szawarski, Z. 2004. Wisdom and the art of healing. *Medicine, Health Care and Philosophy* 7:185–193.
Taylor, M.B. 1997. Compassion: its neglect and importance. *British Journal of Medical Practice* 47:521–523.
Tishler, P.V. 1992. "The care of the patient": a living testimony to Francis Weld Peabody. *Pharos* 55:32–36.
Toner, J.J. 2003. *Love and friendship*. Milwaukee, WI: Marquette University Press.
Underwood, L.G. 2004. Compassionate love. In *Encyclopedia of Bioethics,* 3rd edition, volume 1, ed. S.G. Post (ed.), 483–488. New York, NY: Thomson Gale.
Underwood, L.G. 2008. Compassionate love: a framework for research. In *The science of compassionate love: theory, research, and applications*, eds. B. Fehr, S. Sprecher, and L.G. Underwood, 3–25. Malden, MA: Wiley-Blackwell.
Zinn, W. 1993. The empathic physician. *Archives of Internal Medicine* 153:306–312.

Chapter 6
Medical Stories

The healthcare literature is replete with stories and narratives exemplifying both virtuous and unvirtuous physicians and healthcare providers. I reconstruct two clinical case stories from that literature and analyze them with respect to the notions of virtuous and unvirtuous physician and healthcare provider. The stories represent the voices of a clinician and of a patient. The first story is from the series, "On being a doctor," published in the *Annals of Internal Medicine*, and narrated in the physician's voice. It represents a powerful example of healing by what I previously called the epistemically virtuous clinician (Marcum, 2009). In the next story, a patient narrates her experience with a chronic skin disease. The story depicts the harm, especially emotional and psychological harm, which unvirtuous and uncaring healthcare providers often inflict upon patients, especially children, with chronic diseases (Marcum, 2011). Collectively these case stories reveal medicine both at its best and at its worse. I use the stories to illustrate not only the virtues of good and sound medical practitioners but also the vices of poor and unsound practitioners.[1] In terms of the chapter's mechanics or logistics, I first reconstruct the story and then provide an analysis of it.

6.1 "Communion"[2]

Richard B. Weinberg, a gastroenterologist at Wake Forest University Baptist Medical Center in Winston-Salem, North Carolina, begins by informing the reader that he is an unimposing figure (Weinberg, 1995). However, as he narrates the first encounter with his last scheduled patient of the day, the patient is huddled in a corner of the examination room. "She was in her midtwenties," observes Weinberg, "and she clutched a sheaf of medical records against her chest like a shield" (1995,

---

[1] I must caution the reader that one cannot easily categorize all stories strictly or exclusively in these terms. Often, a story reveals both the virtuous and the unvirtuous. I endeavor to keep that caution in mind during my analysis of the stories, although these two stories do provide opposite poles of the spectrum in terms of virtuous and unvirtuous healthcare.

[2] The essay is part of a collection of papers on the health problems associated with violence against women (Flitcraft, 1995).

J.A. Marcum, *The Virtuous Physician*, Philosophy and Medicine 114,
DOI 10.1007/978-94-007-2706-9_6, © Springer Science+Business Media B.V. 2012

p. 804). The chief complaint is "chronic abdominal pain," with onset in her mid-teens. Her records indicate that she has been to almost every gastroenterologist in town, who ran the right tests and prescribed almost every available drug—but to no avail. His immediate response is puzzlement over why she sought him out and he feels impotent to help her.

Weinberg begins the first consultation by taking the patient's personal medical history. Unfortunately, her description of the abdominal pain is so vague that no apparent diagnosis is immediately evident. As he questions her, however, she fascinates him and Weinberg describes her appearance in detail:

> She was anxious and withdrawn, but nonetheless she projected a desperate courage, like a cornered animal making a defiant last stand. She kept her gaze directed downward, but every now and then I caught her staring at me intensely, as if searching for something. She wore a drab, bulky sweater and oversized bluejeans, and her unkempt hair fell over her eyes. It struck me that she deliberately had done everything possible to obscure the fact that she was a very attractive young woman (1995, p. 804).

Since the patient is so uncomfortable with talking about herself, he moves on to the family history—often the next step in the overall medical history. Her parents are Italian immigrants, although her mother died when she was young, whereupon she assumed the domestic chores even though she has an older sister. The patient is a faithful Roman Catholic but does not take communion.[3] She works with her father in the family's bakery business.

Weinberg then informs the reader that although cooking is a hobby for him, he has never mastered the art of baking. In addition he is always looking for good bakeries, since he is fond of French pastries—especially Napoleons. He asks if the patient knows of a well-known bakery near the medical center, only to have the patient reprimand him that its Napoleons are inferior and that she would not feed them even to her cat. To Weinberg's surprise, he elicited from the patient a passionate response for the first time. But, it fades as quickly as it came. Unfortunately, the physical exam is equivocal. He diagnoses the problem as irritable bowel syndrome and prescribes an antispasmodic drug that she has yet to take, along with a bland diet. The patient listens but is unresponsive. Weinberg is not optimistic about the therapy he prescribes or the patient's prognosis; nevertheless, he requests that she return within a month, although he does not expect to see her again.

But, to Weinberg's surprise, the next week the patient returns. Again, she is terse in answering questions about the abdominal pain. Frustrated, he engages her in a topic that he knows interests her—Italian pastries, of which she exhibits detailed knowledge. He makes no further mention of abdominal pain during the rest of the consultation. Again, he requests that she return in a month.

After a week, however, the patient returns for a third consultation. This time Weinberg notices that she seems relax and comfortable; but, he also observes dark

[3] Not taking communion is important not so much for its facticity, although it does indicate something is amiss in the patient's life, as for the fact that Weinberg could elicit it from a patient who was reticent to talk about herself. In other words, Weinberg must have allowed the patient ample time and freedom to disclose such facts not immediately pertinent to the medical history.

rings under her eyes. He inquires whether she is getting sufficient sleep, to which the patient answers, "no." He asks why, and she answers because she has a recurring nightmare. Weinberg then asks if she could recount it for him.

> She was silent for some time, and then took a deep breath, as if she had made a decision. Then, in a barely audible monotone, she described her dream: She is running, because she must get to confession before the priest leaves. But when she enters the church it is empty, dark, cold. She calls out, but there is no answer. Suddenly, unseen acolytes seize her and drag her to the altar. Her head is pulled back and holy water is forced down her throat to drown her screams. She struggles to raise her head and sees a procession of hooded priests holding long candles headed up the aisle toward her (1995, p. 804).

Realizing the obvious implications of this horrid dream, Weinberg asks if she was ever sexually assaulted. She answers, "yes." Weinberg hesitates for a moment as to whether he should continue, but her eyes implore him to continue. She then recounts the details of the assault, about a decade earlier, by her older sister's boyfriend.

The patient has never told anyone about the sexual assault, for fear of what its exposure would do to her family. Weinberg is the first person to hear the story, which is the reason for her physical ailment. After he consoles her as best he could and after she stops crying, Weinberg advises her to see a psychiatrist or rape counselor. He tries to explain to her that as a gastroenterologist her case is well beyond his medical expertise. But, she refuses to see anyone else since—as she explains—she simply does not trust anyone else. "I then understood that having unearthed her dark secret," confesses Weinberg, "I had become responsible for her care" (1995, p. 805).

Weinberg and the patient then meet on a weekly basis for several months. He mostly listens to her story and to how she tried to exculpate the shame and guilt she felt as outcomes of the sexual assault. Besides not taking communion, she could not eat initially and then fell into a pattern of bingeing and purging herself at the bakery late at night, until her stomach ached and she was exhausted. But, nothing seemed to relieve the anguish and humiliation associated with the assault. The consultations with the patient prompt Weinberg to read the clinical literature about the relationship between rape and eating disorders but he finds little to help him with treating the patient, who is teaching him more about the relationship than the literature. Weinberg also consults a colleague in the psychiatry department, who assures him that he is doing as good a job as any psychiatrist.

Eventually the patient begins to improve. Weinberg notices that a smile, something he had not seen previously, now often replaces her anxious looks. She gains a little weight and changes her hairstyle. She returns to school and completes a high school diploma program. She also informs him that she is receiving communion. The consultations become less frequent as her improvement continues. Finally, after a three-month interval, she appears one day in the clinic. At first, Weinberg does not recognize the patient, "such was the extent of her transformation. She was vibrant, alive. And she looked beautiful—elegantly attired as if for a night on the town. I realized," he continues, "she had dressed up for me. I also sensed that something was completed, that this was a leave-taking" (Weinberg, 1995, p. 805).

The patient tells Weinberg that she is quitting the bakery and traveling to Italy for the summer, after which she intends to matriculate to college. She brought him a gift of six Napoleons that she made, and she thanks him for believing in her. Weinberg, in turn, thanks her for believing in him. After kissing his cheek, she warns him, while leaving, not to eat the pastries all at once since it would not be healthful. He then informs the reader that this case reminds him of an aphorism taught him during his clinical training: clinicians do not choose their patients, rather patients choose their clinicians. Weinberg concludes the essay noting that after the day's evening dinner, "I opened my present and partook of the communion from the baker's daughter" (1995, p. 805).

Obviously, Weinberg is a caring and virtuous physician. Because he is deeply caring—in terms of caring as an ontological virtue—he is present to the patient in a genuine way, which makes it possible for him to connect with her at an existential level through the ontic virtues of care and competence, through the compound virtue of prudent love, and through other traditional intellectual, ethical, and theological or transcendent virtues. In Mayeroff's terms, Weinberg exhibits the structure of caring with respect to "being with" and "being for" the patient. With respect to the first structural element, he enters the patient's world (interestingly through baking, even though the patient's chief complaint is the initial cause for her seeking medical assistance) by simply being there with her, which allows him to forge eventually a bond of care with the patient. With respect to Mayeroff's second structural element, Weinberg is solicitous for the patient's well-being (physical and mental health) and as such is able to meet her care needs. His concern is to help the patient grow beyond the restraints illness imposed not only on her gastrointestinal tract but also and, more importantly, on her life in order to realize her potential as a person, especially in terms of reaching beyond the sexual assault and even beyond the confines of the family's bakery business. In Heidegger's terms, Weinberg is solicitous not only about the patient's present condition (abdominal pain) but also her future condition (a robust active life) and is able to "leap ahead" of the patient in a caring fashion to anticipate her physical and existential needs and to meet them.

Besides the caring structure, Weinberg's clinical practice also exhibits Mayeroff's caring pattern. First, he forges a caring bond with the patient through their shared interest in baking and pastries. Next, the patient recognizes Weinberg's solicitous caring for her and affirms that caring by returning weekly—even though the appointments are initially scheduled monthly. Finally, the patient responds in a positive fashion to Weinberg's caring, which he provides in a committed and non-controlling fashion. Weinberg also exhibits several of Mayeroff's features of caring, including knowing in terms of understanding what the genuine needs of the patient are after recognizing the cause of her chief complaint, courage with respect to moving ahead when he realizes that the basis of the patient's chief complaint is psychological, humility in seeking assistance from a psychiatric colleague, and honesty with the patient by admitting he is not a rape counselor. In sum, Weinberg's caring illustrates well McKinnon's notion of caring with respect to its rational, emotion, and ethical dimensions for meeting another's needs.[4]

Weinberg's caring, as an ontological virtue, can be instantiated in terms of both care and competence as derived ontic virtues. In terms of care as an ontic virtue, Weinberg certainly exhibits what Pettersen calls mature care, i.e. a "co-feeling" with the patient in terms of her pain and suffering from the sexual assault. And, he moves from Pettersen's notion of a thin mature care in which Weinberg provides care in a general fashion, as he would for any patient—such as prescribing an antispasmodic drug and bland diet for the patient's complaint of abdominal pain and then requesting a return visit after a month to assess her condition—to a thick mature care in which he meets with the patient on a weekly basis and basically listens to her narrate the illness story. Besides the development of Weinberg's care from thin to thick, he also develops in terms of Pettersen's symmetric and asymmetric notions of care. Initially, the relationship between Weinberg and the patient is asymmetric, with Weinberg providing care for the patient and receiving only the gratification of helping her. However, as the therapeutic relationship develops Weinberg's sense of himself as a caregiver expands from just a gastroenterologist to that of a healer. In other words, the patient begins to teach him, as he freely acknowledges, especially about sexual assault cases. In the end, the patient exhibits the final act of care with a kiss and gift of six Napoleons.

Weinberg's ontic care also illustrates the difference between $care_1$ and $care_2$. In terms of $care_1$, he is genuinely interested in the patient and motivated to help her, as evident by his desire to connect with the patient and by his ability to recognize that she responds positively to questions about baking. Moreover, Weinberg exhibits a natural incentive to connect with patients as evident from his frustration to elicit an ardent response from the patient, during the initial medical consultation. In other words, Weinberg cares about the patient first as a person, exhibiting what Peabody calls general interest in humanity. This caring about people ($care_1$) and particularly about patients allows him then to take care of ($care_2$) the patient in terms of her medical needs, i.e. addressing the chief complaint of abdominal pain, and her psychological and existential needs, i.e. attending to the medical fallout from the sexual

[4] Weinberg's clinical behavior as caring is also analyzable in terms of Hamington's embodied care and Roach's five C attributes of caring. With respect to Hamilton's embodied care, Weinberg exhibits caring knowledge, which he obtains through observing and assessing tacit cues the patient exhibits in terms of her behavior, e.g. unkempt yet attractive; caring habits, such as Weinberg's shifting from questions about the patient's chief complaint to those about baking, during the second interview; and caring imagination, especially when the patient recounts her sexual assault dream. With respect to Roach's five C attributes of caring, Weinberg exhibits compassion in terms of genuine sensitivity to the pain and suffering the patient is experiencing, especially after he learns the real cause of her ailment; competence with respect to not only his specialty but also to areas outside his specialty, particularly for victims of sexual assault; confidence that he inspires in the patient in his professional abilities, even though he is insecure at times in them (such as when he consults a colleague in psychiatry); conscience in terms of his moral obligation to the patient, once he learns of her sexual assault; and, finally, commitment to help and not abandon the patient because her needs are outside his medical specialty.

assault. His care then is not only motivational ($care_1$), it is also behavioral ($care_2$). In terms of $care_2$, Weinberg is able to help the patient with the trauma she experienced from the sexual assault. He provides this care by obtaining a comprehensive clinical picture, what Peabody calls an impressionistic painting, of the patient through listening to her illness story.

To provide such comprehensive or holistic $care_2$, he is both technically and ethically competent at the science and the art of his profession. In terms of technical competence, he is able to address her medical needs with respect to her chief complaint of abdominal pain. But, he is also competent medically to realize that the etiological reason or cause for that complaint is something other than an organic lesion. He is aware the patient has seen a number of competent gastroenterologists in the Winston-Salem area and none could help her, which arouses his suspicion that more is at play medically for this patient than simply an organic lesion. In other words, Weinberg's is technically competent with respect to the limitations of his specialty, which permits him to entertain possible causes for the patient's distress outside the specialty. Even when he realizes that the etiology of the patient's disease is outside his specialty, he seeks expert advice to ensure that his treatment of the patient is within professional standards of care. Weinberg is also ethically competent in terms of accepting the moral responsibility of the patient's care, not only physically, but also emotionally, psychologically, existentially, and even spiritually. For example, emotionally he braces himself during the third consultation to proceed even though he realizes at that point that the cause of the patient's chief complaint is something other than a physical or mechanistic gastroenteral malfunction. To his credit, Weinberg assumes the responsibility of caring for the patient as competently as he could and does not abandon her. To abandon the patient, he realizes, would have been both ethically and morally reprehensible. Finally, Weinberg exhibits many of ACGME's and Epstein and Hundret's competencies, especially interpersonal and communication skills, necessary for competent medical practice.

Certainly, the instantiated ontological virtue of caring in terms of the derived ontic virtues of care and competence account for Weinberg *qua* virtuous physician; but, a fuller account requires transformation of the ontic virtues into the compound or composite virtue of prudent love. With respect to the first virtue of the compound virtue, Weinberg is an excellent example of an epistemically virtuous physician in terms of prudence, as well as the other intellectual virtues. Prudence allows Weinberg to take the correct (epistemically) and right (ethically) course of action in caring for the patient. With respect to Pellegrino and Thomasma's notion of prudence in medical practice, prudence guides Weinberg in times of uncertainty to move forward with respect to providing effective care by balancing the benefits and risks of treatment—a treatment that is foreign to him, i.e. a counselor for sexual assault victims. Upon learning of the patient's sexual assault, Weinberg reads the literature on the subject and consults a colleague in psychiatry to ensure how he is treating the patient meets professional standards of care. Moreover, Weinberg's treatment of the patient reflects the Aristotelian-Thomistic three steps associated with the prudent agent: (1) "taking counsel" or discovering the significant clinical evidence to ascertain the facts, (2) "forming judgment" in terms of a diagnosis based on the evidence and facts, and (3) "commanding action" with respect to devising a treatment plan to meet the patient's physical and psychological needs.

Importantly, Weinberg also exhibits the virtue of prudent wisdom in terms of the Berlin notion of wisdom, Ardelt's notion of wise person, and Achenbaum's notion of wise action. With respect to the Berlin notion of wisdom as expert knowledge, Weinberg is prudently wise *vis-à-vis* the first stratum of factual and strategic knowledge. He knows how to apply current medical knowledge and practices strategically to assist the patient in healing from the sexual assault. Weinberg is also prudently wise in terms of the second stratum by tapping into the patient's lifespan context, especially the nightmare, to reveal the underlying cause of her illness. In terms of Ardelt's notion of wise person, Weinberg exhibits the three characteristic of the wise person: cognitive, in prudently understanding the factual content of her nightmare; reflective, in stepping back from the emotional trauma of the sexual assault to guide the patient's recovery; and affective, in expressing appropriate concern over the patient's years of suffering from the assault. Finally, with respect to Achenbaum's notion of wise action, Weinberg's actions on behalf of the patient are objective, in the sense that he does not allow personal bias to affect his treatment; rational, in conforming to community standards in treating the patient; and transcendent, in grasping the larger context of the relationship between sexual assault and gastrointestinal disorders. In sum, Weinberg is a prudent physician, in terms of his expert knowledge, personhood, and actions, who mobilizes the necessary clinical resources available to him to treat the patient caringly and competently.

In addition, Weinberg exhibits a number of the cognate intellectual virtues associated with prudence, especially curiosity, during clinical consultations with the patient. With respect to curiosity, he is inquisitive as to why the patient, who is rather attractive, dresses in an unkempt and unattractive fashion. His curiosity prompts him to ask additional questions of the patient in order to understand her better, not simply as a patient *qua* clinical object but as a patient *qua* person. Hence, he refrains from the patient's personal medical history in the first clinical consultation and moves on to the family medical history to obtain information about the patient from a larger social context, information that eventually pays off in later consultations in terms of making an accurate diagnosis. In addition, he is curious about the dark rings under the patient's eyes, which prompts him to inquire about them. Once assured they represent an important clue to something critical in the patient's life, i.e. sexual assault, he then exhibits further curiosity about the sexual event. Weinberg also exhibits intellectual curiosity in trying to understand better the relationship between sexual assault and gastrointestinal problems. His curiosity motivates him to read the literature on the relationship, in order to treat the patient. Besides curiosity, Weinberg also exhibits the cognate intellectual virtues of insightfulness in grasping the intelligibility of the clinical evidence *vis-à-vis* sexual assault and of truthfulness in recognizing the correspondence of the clinical facts to the reality of the patient's sexual assault.

Besides the above cognate intellectual virtues, Weinberg, as an epistemically virtuous clinician, particularly relies on the cognate intellectual virtues of theoretical and practical wisdom, to help the patient clinically. First, he chooses wisely in terms of the theory or worldview in which he practices clinical medicine. Although he operates in terms of the current biomedical model rather than a complementary or an alternative medical model, he does not restrict himself to that model in his professional practice. Rather, he is open to humanizing the biomedical model and thereby stepping outside his professional comfort zone of gastroenterology to engage a theoretically challenging model of humanized biomedicine (Marcum, 2008). Such wise clinical practice pays big theoretical dividends with respect to the patient's prognosis and allows Weinberg to act wisely in a practical manner, especially in terms of grasping the significance or meaning of the clinical evidence *vis-à-vis* the cause of the patient's abdominal pain. For example, his initial diagnosis and therapy recommendation are well within the clinical standard of gastroenterology, given the patient's early medical history. But, he does not limit himself to that standard, which has important clinical ramifications. Rather than ignore certain clinical signs—like the dark rings under the eyes—he pursues them, which provide a more accurate medical history.[5] Also, operating outside the conventional model for gastroenterology allows Weinberg to listen to the patient narrate her illness story—again, such clinical practice for a gastroenterologist is uncommon given the biomedical model.

Besides analyzing Weinberg as an epistemically virtuous physician in terms of the traditional intellectual virtues, Weinberg can also be analyzed as an epistemically virtuous physician in terms of both reliabilist and responsibilist virtue epistemology. As a reliabilist epistemic agent, his sensory faculties, especially sight or vision, are properly functioning during consultations with the patient and serve as a mechanism of belief-formation under appropriate conditions (i.e. his medical training and years of clinical experience) to assist in an accurate diagnosis of the patient's disease etiology.[6] He is keenly observant of the patient's overall appearance during the initial consultation and this serves to stimulate other intellectual virtues, especially the intellectual virtues associated with the conceptual faculties and intellectual curiosity. For example, he initially observes that the patient is unkempt even though she is attractive. This observation does not cause Weinberg to dismiss the patient but rather rouses his intellectual curiosity about the patient and serves to stimulate further questions that lead to revealing the origin of the patient's chronic abdominal pain.

Of course, probably the most important observation is the clinical sign of dark rings under the patient's eyes. Observation of this sign allows Weinberg access to what is responsible for the patient's physical symptom of abdominal pain. It also serves to engage the intellectual virtues associated with his conceptual powers, such as insightfulness, especially the inference he makes in terms of interpreting

---

[5] The patient was seen by other gastroenterologists, who may or may not have observed the dark rings and if they had observed them chose for one reason or another not to enquire about the patient's sleeping pattern. Dark rings under the eyes may not be a clinically significant sign for gastroenterologists to enquire about given the clinical boundaries of biomedical model for gastroenterology. Only someone, like Weinberg, operating outside those boundaries and with a humanized version of the model might inquire.

[6] These conditions can also include environmental factors like adequate lighting or appropriate observation distance (not too far or not too close).

the patient's nightmare as sexual assault.[7] It is interesting to note that Weinberg's observation of the dark rings under the patient's eyes does not occur until the third consultation. Possibly the rings are present at earlier consultations, since the patient's nightmare is recurring, and Weinberg misses them because of diminished powers of observation, *vis-à-vis* this clinical sign, due to the limitations of his medical specialization of gastroenterology. In other words, as a gastroenterologist he would not think to look for clinical signs other than those associated with his specialty. However, once he steps back from that specialty, he is able to observe this sign.

Besides properly functioning sensory faculties of a model reliabilist epistemic agent, Weinberg also demonstrates many of the reliabilist epistemic virtues associated with properly functioning conceptual faculties or powers. For example, his accurate memory of the first clinical consultation with the patient serves him well during the second consultation. By remembering that the patient is responsive to a discussion on baking, he is able to engage the patient in a "successful" consultation. The consultation is successful because the patient begins to trust Weinberg as a person and more importantly as a clinician. Such trust is critical since it pays off epistemically in the third consultation, when the patient decides to recount her nightmare to Weinberg. Of course, as noted above, Weinberg displays insightful analysis of her recurring nightmare by inferring that the patient has been sexually assaulted. He also exhibits keen intuition into the natural end of the clinical relationship. The patient's healing is complete, as evident from her transformed physical appearance after months of consultations, and there is no need to prolong the relationship.

Besides properly functioning sensory and conceptual faculties of the reliabilist epistemic agent, Weinberg also displays responsibilist epistemic virtues; however, these virtues represent not just the traditional intellectual virtues as already discussed but also traditional ethical virtues functioning as intellectual virtues. For example, the traditional ethical virtues of courage and caution serve Weinberg as intellectual virtues, during clinical consultations with the patient. He demonstrates authentic intellectual courage when he discovers that the patient was sexually assaulted. Although he hesitates for a moment, certainly because he is aware of the implications of additional inquiry, he realizes that the courageous act—particularly in terms of making an accurate clinical diagnosis—would be to inquire about the assault and thereby to actualize its epistemic potential.[8] As Weinberg acknowledges later, he recognized that he was responsible for the care of the patient once he uncovered the reason behind her chronic abdominal pain. Only an authentic intellectual courageous act could genuinely help the patient. At the same time, he is cautious about continuing with the case, once he learns of the sexual assault. Intellectual caution, the flipside of intellectual courage, keeps Weinberg from acting recklessly

---

[7] Weinberg also exhibits intellectual insightfulness when he observes that the patient's physical appearance is transforming before his eyes.

[8] It is well known among physicians that patients who are victims of abuse can be difficult and frustrating to treat (Bligh, 2000).

clinically. He tells the patient that he is a gastroenterologist, not a psychiatrist or rape counselor, and that her case is outside his clinical expertise. Realizing his clinical limitations, he is cautious not to overstep specialty boundaries, and even seeks the opinion of a colleague in the psychiatry department and reads the clinical literature on the relationship between rape and eating disorders.

In addition, Weinberg relies on the cognate virtue of honesty to the ethical virtue of justice, as an intellectual virtue on a number of occasions to deliver an accurate clinical diagnosis and effective therapy. First, he is honest with himself in terms of the initial consultation in that he is not optimistic about her return for further consultations. Weinberg is also honest with the patient (as well as with himself)—once she tells him of the sexual assault—in that he does not deceive her (or himself) into believing that he knows how to provide an effective therapy for a victim of sexual assault. His intellectual honestly, in terms of his limitations concerning the relationship between sexual assault and gastrointestinal problems, also serves to motivate him to consult the professional literature and a psychiatrist colleague. Throughout the clinical consultations with the patient, he is forthright about his abilities and inabilities, as well as his knowledge and ignorance, "when it counts." This last qualifier is important because Weinberg does not tell the patient everything he believes about her case, especially when his beliefs are speculative and might hinder her recovery; rather, he shares with the patient what he is thinking when he has good reason to believe sharing it would help in the patient's healing. Importantly, Weinberg does not deceive himself about the significance of the relationship between the patient's sexual assault and her presenting symptom. Because he exhibits intellectual honesty, he takes special care to ensure that he can help the patient therapeutically.

In terms of the second component virtue of the compound virtue, prudent love, Weinberg certainly displays a genuine love and concern for the patient in that he is not simply discharging a duty to care for the patient but he is authentically committed, at an emotional level, to her personally and to her wellbeing. In Swanton's terms, he draws close to the patient emotionally with respect to viewing her in a caring manner and appreciating her as a person of value and worth. As Swanton notes and Weinberg's case study illustrates, the source of such love stems from a robust self-love that animates a universal love for others, a love that is unrestricted or expansive and yet at the same time particular or individual. Such love allows Weinberg to meet Frankena's injunction, "Be loving!", not as a command or Kantian duty or obligation but as a fiduciary charge or trust to care deeply for the patient. Moreover, Weinberg's love for the patient is a liberating act, which permits him to treat the patient even beyond his clinical expertise. Such freedom grounded in love satisfies Augustine's directive, "Love, and do what you will." Lastly, Weinberg's love for the patient also liberates the patient to respond positively to treatment.

Although the above discussion certainly describes Weinberg's love for the patient adequately, Toner's notion of personal radical love captures more sufficiently Weinberg's caring and love for the patient. Specifically, his notion of radical love—including one's whole being—aptly portrays Weinberg's loving or caring solicitude for the patient. Weinberg does not restrict himself simply to the clinical situation when initially taking the patient's medical history but also incorporates his interest in the culinary arts. Through his and the patient's interest in these arts, he gains the patient's trust as someone who cares about (care$_1$) her; and, in turn, she allows him to take care of (care$_2$) her. In Toner's words, Weinberg exhibits one of the key features of radical love—response of the lover to the beloved's whole being—in that he responds not just to her diseased body part but to her as a person. He also exhibits its other two key features. First, Weinberg connects with the patient at a fundamental level to form a union between himself and the patient. Out of this union emerges an effective therapeutic relationship, which gives rise to radical love's final feature—affective affirmation. Weinberg's affirmation of the patient's pain and suffering from the sexual assault provides the patient with the strength needed to overcome the assault's medical fallout.

Besides the three key features of personal radical love, Toner's five essential components of this love facilitate analysis of Weinberg's love for the patient. First, Weinberg's radical love for the patient includes a cognitive component involving his perceptions, memories, conceptions, and especially clinical judgments. For example, not until the third consult does Weinberg recognize the dark rings under the patient's eyes. Since the nightmare is reoccurring, the rings are most likely present at the other two consults. He recognizes them during the third consult because the cognitive component comes in to play as Weinberg develops a deep sense of love and care for the patient. He now sees things about her that he was unable to see before because his love for her allows him to see her differently. Next is the affective component in which Weinberg's feelings for the patient *qua* person—and not simply as a clinical specimen—leads him to explore her life world context, as exemplified during the second consultation when he and the patient discuss mainly the culinary arts. This component sets up the third one, the "act of affectivity," in which Weinberg, now realizing what precipitated the patient's illness, chooses to respond positively to the patient's medical needs. This component relies on the next component, freedom, in which Weinberg decides voluntarily to treat the patient—even though such treatment exceeded his expertise. The final component involves the physiological dimensions of love, as when the patient first smiles during a later consult.

Finally, the culmination of Weinberg's care and love for the patient *vis-à-vis* Toner's notion of radical love is personal communion of the lover and beloved. This personal communion, as Toner articulates it, involves mutual belief in each other and exchange of each other's affections. Weinberg's medical case study bears this personal dimension out exquisitely. Specifically, the essay is entitled "Communion" and plays off that title in several ways *vis-à-vis* Toner's notion of personal communion. First, the Napoleons represent the patient's gift or exchange in appreciation for Weinberg's talent as a healer. They are a gift affirming Weinberg as an authentic person, who genuinely cares about her as a person and not simply as a clinical case in need of resolution. Also, the patient dresses up for Weinberg at their final meeting, indicating the level of care and respect she feels for him—in contrast to the first time she met Weinberg when she dressed shabbily. Importantly, as acknowledged at the end of the essay, they both thank each other for believing in one another. This belief or trust in each other is critical for forging the union that composes the therapeutic/loving relationship in which both patient and physician function in unison—Toner's "one composite act of one composite agent." Finally is the religious dimension, which is critical for the patient as a devote Roman Catholic of which Weinberg is keenly aware. An important indicator of the patient's healing is her taking communion again. The significance of this communion resides in her union with God, the ultimate source of love.

Besides employing love as a theological or transcendent virtue, especially in its transformative sense, or even as an ethical virtue, Weinberg also adapts the virtue for epistemic purposes. For example, he has love for clinical knowledge, in that he is passionate about determining what is wrong with the patient. This passion allows him to mobilize the intellectual virtues—whether traditional, reliabilist, or responsibilist—for making an accurate diagnosis and for providing an effective therapy. He is not satisfied with his lack of knowledge concerning the relationship between rape and eating disorders, for instance, and seeks out authoritative sources—such as a colleague in psychiatry and the professional clinical literature on the relationship—to ensure that he is assisting the patient in the healing process. This love of knowledge allows Weinberg to marshal and connect the various intellectual virtues to achieve genuine insight into the patient's illness. For example, his passion for helping the patient allows him to overcome an earlier inaccurate diagnosis of irritable bowel syndrome and an ineffective therapy of an antispasmodic drug and a bland diet. His keen and well-trained observational skills enhance his inferential powers, along with the intellectual courage to forge ahead when the intellectual landscape looks threatening or hopeless in terms of his professional comfort zone as a gastroenterologist. Finally, he is certainly not indifferent to the clinical challenge that the patient presents but meets it head on to help the patient heal.

Moreover, Weinberg exhibits the cognate virtue of humility, as an epistemic virtue, to the theological virtue of love. This cognate virtue definitely contributes to his ability to treat the patient effectively. First, he is humble epistemically *vis-à-vis* the patient's seeking him out even though she has seen almost every other gastroenterologist in town. What more could he possibly know to help treat the patient than what the other gastroenterologists knew? Weinberg is intellectually humble in admitting that he is unsure how helpful he could be for her, especially in terms of relieving her chronic abdominal pain, and, once knowing about the sexual assault, that he is not an expert in terms of psychiatry or rape counseling. Most importantly, however, he is intellectually humble often by simply listening to the patient, during their many consultation sessions, and by having the patient teach him. He does not dismiss or abandon her because she is unkempt but attributes to her a worth that elicited his best clinical practice. He also exhibits epistemic humility in revising the initial diagnosis of irritable bowel syndrome. Moreover, Weinberg does not exhibit the intellectual vice of arrogance in which he thought or acted that he knows the best course of therapeutic action; rather, he seeks the advice of a colleague in the psychiatry department and gathers information from the clinical literature. Weinberg's epistemic humility certainly makes possible clinical success in terms not only of finding the cause of the patient's illness but also of providing efficacious treatment.

Weinberg also exhibits several of the cognate virtues of love. Primary is empathy in which he is able to project himself into the patient's situation and "leap ahead" of her in a Heideggerian sense of solicitous caring for her emotional and psychological needs, especially as the patient often simply narrates her illness story. In addition, Weinberg's empathic care of the patient demonstrates Halpern's notion of empathic resonating through associative reasoning, by incorporating her illness narrative into treatment of the patient. Moreover, he is a good example of Zinn's empathic physician in that Weinberg enters into the patient's world of pain and suffering without that world absorbing him and thereby robbing him of his personal integrity or ability to treat the patient. Weinberg is obviously sympathetic towards the patient's efforts to exculpate her shame and guilt over the sexual assault incident by being with the patient, in Mayeroff's terms, as she struggles to articulate the impact of the incident on her life in terms of binging and purging herself late at night in the bakery. In being with or alongside the patient, Weinberg fulfills his duty, as Dougherty and Purtilo (1995) define it, to identify the source of the patient's illness and to act efficaciously to alleviate it. Importantly, Weinberg's empathy and compassion complement each other so that he could provide the patient with quality healthcare in a professional manner.

Weinberg also exhibits the cognate virtue of loyalty to the patient, by devoting himself to her care and by striving to promote her welfare through making good clinical judgments and decisions about how best to treat her. He does not abandon her, either because of social stigma or personal bias attached to the sexual assault, or because of a felt perception that the case is outside his clinical expertise. Rather, Weinberg exhibits a genuine level of altruism in treating the patient by broadening and stretching his clinical comfort zone and by finding time in a busy practice to meet once a week with the patient to listen to her narrate the events of the sexual assault and its effects on her life. Moreover, his treatment of the patient is altruistic in that Weinberg exposes himself in a genuine and vulnerable fashion to the patient's pain and suffering. He not only draws alongside the patient's world of illness compassionately and enters that world empathetically but he also, and more importantly, occupies and lives in it altruistically. In other words, Weinberg exhibits an agape or altruistic love through his unselfish and generous bestowal of time and presence for a person who is suffering intensely and in enormous need of care.

The compound or composite virtue of prudent love, especially its synergy, is responsible for forging a robust therapeutic relationship between Weinberg and the patient. Specifically, Weinberg's prudent love allows him to make not only the wisest but also the most loving decisions for what is best in treating the patient. Moreover, it provides him the strength and resources to help the patient in the first place, especially when he is unsure of himself in terms of his ability to take care of the patient effectively. A critical part of that relationship is the patient-physician boundary that Weinberg establishes in taking care of the patient. Prudent love empowers him to help the patient recover from the devastating effects of the sexual assault by generating a therapeutic boundary between himself and the patient—a boundary that serves to demarcate the patient's world of illness from his own. For, without such a palpable boundary, Weinberg could inadvertently allow other concerns to jeopardize the relationship. Through that boundary, Weinberg could identify and connect with the patient in order to treat the patient's medical needs rather than meet his own needs that might be detrimental to the patient's welfare. Importantly, prudent love synergy prevents Weinberg from taking advantage of the power differential between himself and the patient. Rather, it serves to harness his power as a healer, particularly as a listener, to assist the patient in recovering from the sexual assault and regaining her integrity and life.

Besides the compound virtue of prudent love, Weinberg exhibits other cardinal and theological virtues in treating the patient. As an ethically virtuous physician, Weinberg utilizes the traditional ethical virtues of courage, temperance, and justice, as well as the cognate virtues associated with them. With respect to courage, as noted already for the virtue's epistemic use, the best illustration is his decision to pursue examination of the patient's sexual assault experience, when she tells him about the nightmare. Weinberg is well aware of the ethical and moral demands uncovering such information places upon him as a clinician. To turn a jaundice eye towards this story would have devastating effects for the patient's recovery. The patient visited a number of physicians prior to Weinberg, without entrusting herself to them. In other words, she did not trust them with her story about the assault. Her decision to tell the story to Weinberg for the first time is a true act of courage, one in which she risks much not only personally and emotionally but also in terms of her family relations. As an ethically courageous person, Weinberg is aware of how courageous the patient is to tell her story and responds in a morally appropriate fashion, i.e. courageously. To fail to respond in such fashion would have been pusillanimous and disclosed moral weakness on Weinberg's part to cave in to the demands and possible risks of pursing the patient's story. Finally, he displays valor even to the extent of being heroic. Why?, because the patient's illness is far afield for him as a gastroenterologist. He places himself at great risk in terms of professional liability, especially if anything detrimental would happen to the patient while he is treating her.

Weinberg also exhibits the ethical virtue of temperance when treating the patient. He is able to manage his emotions appropriately upon learning of the patient's sexual assault and does not lose control of himself. He illustrates well Osler's notion of *aequanimitas* with respect to imperturbability and equanimity. As to imperturbability, Weinberg remains levelheaded as he learns about the assault and treats the patient, while for equanimity he maintains composure in the face of the patient's struggle to exculpate herself of the shame and guilt associated with the assault in order to regain her health and future. Weinberg's temperance also bears out Carr's defense of Osler in that the virtue allows Weinberg to achieve "a sympathetic attunement toward the patient" and a genuine connection to her. The cognate virtues of temperance, especially modesty, also come into play as Weinberg treats the patient. Weinberg is well aware that the best resource for treating the patient is either a psychiatrist or a rape counselor. He exhibits no delusions that he has the training or know-how to treat her. He also displays the virtue of discipline, in terms of seeking out professional help either from a colleague or from the literature. Finally, he is very patient by not over reacting when the patient informs him that she has chosen him as her healer. Rather, he remains civil and deferent to the patient as a person of worth and dignity—and treats her as such. Weinberg's behavior with respect to these latter cognate virtues of temperance is probably the main reason the patient decides to inform Weinberg of her sexual assault—she feels safe with him in that he would not lose control and thereby exacerbate her physical and emotional condition.

For the final traditional virtue associated with ethics, justice, Weinberg utilizes it effectively to treat the patient. Again, upon learning of her sexual assault Weinberg is just in his conduct towards her. He is just in the sense, especially from the notion of distributive justice, in that he mobilizes valuable medical resources to meet the patient's needs rather than limit their availability to her. Probably the most important resource is his time, which he gives generously while he often simply listens to the patient narrate her sexual assault story and its aftermath on her health. He is also just with respect to the notion of restorative justice, in that he reinstates not only the patient's health *vis-à-vis* elimination of the abdominal pain but he also helps her regain *herself* as a person of worth. In other words, he rectifies the injustice perpetrated on her through the sexual assault. Most importantly, however, especially from the patient's perspective, Weinberg displays the virtue of trustworthiness. As the patient confides in him at one point, she trusts only him with her care. She obviously realizes that Weinberg is a person whom she could entrust her secret without moral recrimination. Finally, Weinberg is an honest person, who tells her the truth about himself and his capabilities as a clinician, and a person of integrity, who tells the truth because he embodies it. He does not deceive the patient in terms of the truth or distort it, and more importantly, the patient realizes this about him. In sum, Weinberg's virtue of justice is the foundation of his relationship with the patient without which she would not have chosen him to help her or he could help her.

Besides the ethical cardinal virtues, Weinberg also relies on the theological or transcendent virtues of faith and hope to play a role in forging a therapeutic relationship with the patient. First, Weinberg and the patient have faith in each other, as both admit freely at the end of their relationship. This faith consists of Polanyian tacitly held commitments in which both Weinberg and the patient presuppose that the truth concerning the patient's medical condition is palpable or knowable apart from each other. In other words, both physician and patient assume that the other is an authentic source of knowledge and that each would divulge that knowledge truthfully and responsibly. This faith also has a transcendental quality to it in that neither Weinberg nor the patient could justify rationally or empirically their faith in each other. For example, when the patient first discloses the account of her sexual assault, Weinberg has faith in her account as accurate and reliable of the transpired events. Such faith allows Weinberg to act and is necessary or critical for a successful clinical outcome. The cognate virtues of faith are also operative in the relationship. Both Weinberg and the patient believe in, trust, and are committed to each other and to the truth that each has important contributions to make to the therapeutic relationship and to the patient's overall welfare. In addition, attitudinally both are confident, assured, and convicted that each is a responsible member of the therapeutic relationship. Again, neither Weinberg nor the patient could justify these cognate virtues either rationally or empirically but they could only hold them tacitly or implicitly.

Hope is also an important operative theological or transcendent virtue in the therapeutic relationship between Weinberg and the patient, although at first it is not present. In fact, Weinberg exhibits hopelessness when faced initially with the patient's chief complaint of chronic abdominal pain. Her description of the complaint, as he acknowledges, is so ambiguous and diffuse that he could not make an accurate diagnosis, settling initially for irritable bowel syndrome. However, once the patient divulges the sexual assault episode hopelessness turns to hope, although restrained at first, as Weinberg and the patient realize that a resolution to the patient's chief complaint is possible. In other words, the object of expectation, i.e. curing the patient's abdominal pain, is now probable given that the causative or etiological factor is evident. In terms of Downie's locutions of hope, Weinberg goes from "I hope that I can help the patient with respect to her chief complaint" to "I hope to do something about helping the patient recover from the aftermath of the sexual assault." A large part of this transformation includes Weinberg's caring for the patient, since, as Quinn notes, an intimate connection between hoping and curing exists, i.e. caring includes some hope for a better world. Weinberg also displays several of the cognate virtues of hope while treating the patient, including expectation, as he looks forward to the patient's reinstatement of health as she narrates her sexual assault story; aspiration, as he seeks advice from a colleague and the literature; and, gratefulness, for a better future for the patient as smiles eventually replace her anxious looks. Finally, Weinberg does not offer too much hope to instill a sense of false expectation that her recovery is inevitable or too little hope to cause the patient to despair about recovery.

The clinical case also illustrates the connection between epistemic, ethical, and theological or transcendental virtues, especially through the fundamental virtue of caring, as expressed in the compound virtue of prudent love, for delivering the clinical goods.[9] Thus, the virtuous agent cares deeply enough to deliver those goods for the patient in a rationally, morally, and transcendentally responsible fashion. In this way, caring functions ontologically to make possible the epistemic, ethical, and transcendental virtues, especially the compound virtue of prudent love. Weinberg certainly exhibits this virtue of caring by not only caring about the clinical goods but also by taking care to see that the clinical goods are therapeutically efficacious. For example, he realizes at one point that having discovered that the patient has been sexually assaulted he is bound by an epistemic, an ethical, and a transcendental obligation to provide the appropriate and necessary clinical goods to help the patient heal. To that end, for example, he requires rational courage as a prudent agent to proceed not only with questioning the patient concerning the assault but also with treatment when there is little help from the medical literature. At the same time, he also exhibits moral courage to take on the case and to see it to its conclusion, especially given the legal risk he exposes himself to in that he is neither a psychiatrist nor rape counselor. In addition, he utilizes transcendental courage as a radical hope believing that through his solicitous caring and prudent love he

[9] Vrinda Dalmiya (2002) also proposes a "care-based epistemology" in which care is the fundamental virtue that connects all other virtues, whether epistemic or ethical.

could benefit the patient clinically. Finally, he employs epistemic and moral caution *qua* prudently loving physician, not to promise more than he could possibly deliver.[10]

Lastly, the Weinberg case study illustrates the cyclic relationships between care and competence and between love and prudence, as well as the larger prudent love-caring relationship that embeds both the care-competence and the love-prudence cycles. In terms of the care-competence cycle, Weinberg's caring is certainly instantiated with a genuine desire about or concern for ($care_1$) the patient and her physical and emotional welfare. In other words, he cares about the patient because of his general interest in humanity. This is evident from efforts to connect with the patient during the first clinical consultation and subsequent consultations, through a common interest in baking. Because of such $care_1$, Weinberg insures he is professionally competent both technically and ethically to take $care_2$ of the patient as best he could, given the circumstances. Technically, he consults a colleague in psychiatry and the literature on rape and eating disorders to ensure he is meeting professional standards. Ethically, he does not abandon the patient but embraces her personally and her care needs *qua* patient. Finally, the success of this case in terms of taking $care_2$ of the patient reinforces Weinberg's desire to $care_1$ about other patients and their medical needs.

In terms of the love-prudence cycle, Weinberg's $care_1$ becomes a personal radical $love_1$ as he learns more about the patient, especially the horrid reason for her physical symptom of chronic abdominal pain. This love emerges from his compassionate and empathic response to the patient's pain and suffering. Weinberg also responds to the patient prudently by acknowledging his clinical limitations and, once realizing that she has chosen him to treat her, by seeking assistance to ensure he treats her competently according to accepted professional standards. Such prudence guides Weinberg's $love_2$ for the patient to actualize her healthcare needs. Importantly, prudent-love synergy empowers Weinberg to make wise and loving clinical decisions that aid in forging a robust therapeutic relationship to assist the patient heal. Lastly, to come full circle, Weinberg's caring as an ontological virtue undergirding the care-competence cycle is the foundation of his clinical practice as a prudent-loving physician, which in turn reinforces his virtuous caring as well as high-quality and professional healthcare. In sum, his narration of the case study's conclusion, by partaking comm*union* with the patient through her gift of Napoleons,

[10] Caution must also be exercised in taking on the patient in the first place, given the potentially large number of patients who may demand more intense healthcare attention. In response to a letter to the editor of the journal where Weinberg's essay was originally published and which questioned whether physicians have enough time to spend on every patient needing such attention (Rixey, 1996), Weinberg writes: "The intense relationship described in my article does not imply that a physician must serve as a personal counselor for every patient. . .Occasionally, however, the needs of a patient call us to commit ourselves beyond screening questions, beyond referrals, beyond the convenient or the comfortable. The main point of my article," he goes on to stress, "is that such intervention is not to be feared. What I really learned was that to '. . .own the problem, fix it, be responsible. . .' can provide one of the most exquisite joys of our profession" (1996, p. 427).

belies a deep sense of satisfaction with the successful outcome of the clinical case and could not but help kindle caring in Weinberg for other patients and the medical profession.

6.2 "Lifelong Effects of Chronic Atopic Eczema"[11]

In a powerful essay, "Lifelong effects of chronic atopic eczema," Shelley Diamond (1996), then in her late thirties, reconstructs a narrative about her lifetime illness experience. She has never known a day without suffering from full-body atopic eczema. As a child, she constantly scratched her skin, often with destructive consequences. At age eight or nine, she vividly remembers going to the hospital for an operation to remove a tumor from her left clavicle. Because a sling restrained her left arm and an IV her right, she was unable to use her hands for almost a week. The impact was devastating for her, since just before undergoing the operation she was finally learning to control her hands, *vis-à-vis* her eczema, through a variety of non-injurious actions, such as pinching or stroking the skin instead of scratching it with her nails. Another non-destructive mechanism for coping with her chronic eczema was reading, an activity denied her because of the restraints—she simply could not hold a book or turn its pages. Although the intent of the health profession was to address the risks related to the tumor, the outcome of this experience for Diamond was to increase the pain and suffering associated with her chronic eczema. "Put bluntly," to quote Diamond, "it was torture for me to be unable to use my hands just when I was learning to use them in positive ways rather than destructively...I was in agony" (1996, p. 1).

The young Diamond's hospital stay had a certain rhythm to it. During the day, besides meals, the only distraction for Diamond was her parent's daily visits. However, even her parents were not terribly understanding or supportive of their daughter's suffering from chronic eczema. For example, because of the arm restraints the only parts of her body Diamond could move were her head and legs. At times, the eczema symptoms were so maddening that she agitatedly shook her head and legs, while moaning and crying. In an effort to console her, Diamond's father would say, "C'mon, stop that crying. Crying doesn't do any good. What doesn't kill you makes you stronger—you know that" (1996, p. 2). Sadly, Diamond received even less support from her mother, who often apologized with platitudes for the failure of the healthcare profession to address the pain and suffering associated with the chronic disease. During the night, sleep—generally under the best conditions was difficult and fitful—was next to impossible because of her restraints. Diamond would remain awake struggling to cope with the situation through mental gymnastics, such as bargaining with God, reciting spells from a book on Merlin, or reenacting in her mind great escapes from prisons and concentration camps she had

[11] I thank Kay Toombs for drawing my attention to this story.

witnessed in movies. All was to no avail, unfortunately, as she frantically searched for relief from sleeplessness as a rat seeks escape from a maze.

Desperate for relief from the constant itching caused by the eczema, Diamond mustarded courage one day to confide in a physician attending her in the hospital that she found the itching uncontrollable and did not know what to do. The physician assured her that she would receive the best possible treatment and that an antihistamine he was prescribing for her should relieve the symptoms. When Diamond mentioned that in the past the drug was ineffectual for alleviating her symptoms, the physician's response was mortifying:

> "Well, you're in a different place now, and not being able to use your hands is the best thing for you. A lot of that itching is all in your mind. I think you've just gotten into a bad habit, and that's what you've got to work on. OK? [said brightly, as if that solved everything]. You do want to get better, don't you?" I meekly nodded. "I'm sure that this medication is what you need, but you've got to give it time to work, kiddo. I'll see if I can get one of the nurses to read you a story or something" (1996, p. 2).

And, just like that, the physician stifled Diamond's young voice, belittled her illness experience, and negated her possible contribution to her treatment. Again, her parents were unsupportive when Diamond voiced her disappoint in the physician's behavior. Her mother defending the medical professional simply stated in characteristic fashion, "There are a lot of people here much sicker than you are," she said. "You're not going to die, honey. First they have to take care of the people who're dying. You understand that, don't you? Now you've got this nice private room, you lucky girl. Look out the window and enjoy the view" (1996, p. 2).

Diamond also voiced to the hospital nursing staff concerns over the restraint of her hands, but the staff simply ignored her. She began to realize that even the nursing staff was disinterested in her pain and suffering or unable to connect with her emotionally. During her weeklong convalescence after the operation, for example, she confided in a nurse, whom Diamond initially considered sympathetic, that she needed to talk with someone about her eczema and the suffering it caused her. The nurse's reply was disappointing, at best. The nurse responded, "That's not my job, honey." And, she went on to explain,

> I'm sorry but I have to make my rounds with medications and do my paperwork. What if all the patients wanted me to sit and talk? I wouldn't have time for all the important work I have to do. If I had nothing else to do, that would be fine, but I'm busy, child. Maybe the morning nurse will have a few minutes to read you a story or something. I'll make a note on your chart about that, OK? (Diamond, 1996, p. 3).

To add injury to insult, the nurse later refused to answer Diamond's buzzer in order to assist the child to the toilet. The result was that Diamond urinated in bed—but rather than feeling shame or humiliation, she felt an overwhelming sense of gratitude from the release of expressing herself spontaneously.

The use of restraints during Diamond's hospitalization did improve her skin's condition, as she admits. However, her parents then used constraints at home, including handcuffs, rope, straightjackets, and other devices, to keep their daughter from scratching herself. Try as hard as they might, Diamond was able to foil their attempts to restrain her. Like Houdini, to whom she compared herself, she found a way to escape their restraints. Once liberated, she would savagely tear at her flesh in fits of rage. "The frenzied scratching that ensued was," confesses Diamond, "orgasmic" (1996, p. 3). Besides tearing her flesh apart, the disease tore Diamond herself apart. "My developing ego," recounts Diamond, "had to separate my self from my body in order to survive" (1996, p. 3). Once separated from the body, she wanted to destroy it. The eczema not only split Diamond in two but it also split her from her parents. Her parents fretfully sought treatment for their daughter through traditional and nontraditional means, only to retreat from the ineffectiveness of these treatments. Eventually the mother retreated to telling Diamond to "Think of something else." Diamond later realized that her mother was not consoling her so much as she was consoling herself. Her mother distracted herself with household chores, hoping for a miraculous cure. Her father also retreated domestically but to the basement, where he found solace in his electrical workshop—vowing "on not letting anything 'get to' him."

The overall effect of such poor quality healthcare from both medical professionals and parents was devastating on the life of the maturing Diamond. She slowly began to realize and accept that the healthcare profession was incapable of addressing or unwilling to attend not only to her physical or bodily needs but also to her psychological or emotional ones. "I held on to the point of emotional exhaustion," claims Diamond, "telling myself that if I just asked the right person at the right moment in the right way so that they understood, then somebody would just listen to me about what was going on. But now I realized it was hopeless" (1996, p. 3). Diamond's response to the above realization and acquiescence was to withdraw and to seek solace for her anger and outrage at an uncaring healthcare profession. In addition, as noted earlier, she began separating herself from her body. Her body became a foreign object, something to which she had difficulty connecting. "I felt my mind drift out of my body," recalls Diamond during her hospitalization, "and it seemed as if I was on the ceiling looking down, knowing what 'she' was feeling, yet at the same time separated from 'her'" (1996, p. 3). The result of this early trauma and harm foisted on her by an uncaring healthcare profession led to drug abuse, social isolation, and suicide attempts, among other posttraumatic, stress-related behavior. At the age of twenty-one, she underwent voluntary sterilization.

Diamond's story aptly illustrates not only unvirtuous healthcare providers but also an unvirtuous healthcare system, whose overall impact was detrimental not only on the patient but also on the family. The root cause of the unvirtuous behavior, for both the system and its providers, is the ontological vice of uncaring or an inability to feel concern or compassion for another person and, consequently, a failure to extend beyond the secure boundaries that ensure a respectable distance between the patient and provider, especially with respect to emotions. Neither the system nor its providers involved in the Diamond case are intentionally vicious, i.e. in seeking to harm or to inflict injury on the patient,[12] but neither are they

[12] The nurse's refusal to assist Diamond to the bathroom, however, borders on the vicious.

adequately motivated to help the patient—except within minimal standards of professional behavior. As evident from Diamond's illness narrative, such behavior is exemplified by trivial platitudes, such as "A lot of that itching is in your mind" or "We'll talk more about this next time," on the part of the healthcare providers. Rather than exemplifying Mayeroff's notion of caring and the "being with and being for" structure of caring, the healthcare providers show little interest in being with or for the young Diamond. For example, the nurse flat out tells Diamond that she is much too busy to attend to her need to talk about the agony from eczema. Indeed, both the nurse and attending physician shrug off the duty to be with or to talk to her to another healthcare provider, e.g. they write in her chart that a provider should read her a book. Unfortunately, these healthcare providers completely miss Diamond's request. Diamond did not simply want someone to be with her in such a trivial capacity as reading a book. She needed someone to be with her at a substantive existential level to discuss with her the pain and suffering associated with her eczema. Such inability to be with the patient prohibits these healthcare providers from attending to or meeting the patient's needs or, in Mayeroff's words, "being for" the patient. Ultimately, Diamond's healthcare providers fail to provide satisfactory healthcare in order to assist the patient towards healing, again in Mayeroff's words, in "actualizing" her potential *qua* person. Rather, such inhumane care had devastating effects upon the patient in later life.

Diamond's case study also exemplifies Halldórsdóttir's wall model of uncaring. Rather than building a bridge to connect themselves to Diamond, her healthcare providers construct a wall around Diamond to separate them from her. Part of the defense for building the wall around her was Diamond's awareness that her healthcare providers perceived her as a troublesome patient. "I heard many heave a sigh of relief," narrates Diamond of the healthcare professionals attending her, "outside my room. Clearly everyone felt sorry for me, but no one was willing to engage me in a conversation about my predicament. All I heard," she laments, "was the murmur of predictable platitudes followed by hush sessions of unintelligible whispering in the hallways" (1996, p. 2). Moreover, the actions of these healthcare providers exhibit the various stages associated with development of the wall model of uncaring. For example, the healthcare providers initially show indifference or little interest in Diamond's existential and emotional needs—none of the providers really want to talk with her about the pain and suffering she experienced from her chronic eczema, their interest is helping her recover postoperative from the removal of the tumor. This indifference leads to insensitivity to eczema's impact on Diamond's physical and emotional wellbeing and to coldness or diffidence in their treatment of her. Finally, the results of constructing a wall of uncaring are inhumane acts, such as when the nurse refuses to respond to Diamond's call for assistance to go to the toilet with the outcome of Diamond urinating in bed. As Halldórsdóttir notes, the outcome of an uncaring attitude on healthcare providers is the development of instrumental behavior—behavior Diamond's providers clearly exhibit.

Unfortunately, healthcare professionals generally do not hear or believe the voices of children in pain or suffering, which often leads to a wall of uncaring as illustrated in the Diamond case study. Gary Walco and colleagues identify three reasons why such professionals disbelieve children's accounts of pain (Walco et al., 1994). First, healthcare professionals generally provide care based on an "appropriate" child standard. The standard is subject to various myths, such as infants are incapable of experiencing pain or children do not remember pain and so no need to worry about long lasting adverse effects. Or, children may overestimate the magnitude of their pain and suffering. These myths often hinder healthcare professionals from responding appropriately to a child's pain or suffering. The next reason is that healthcare providers are anxious about the consequences of relieving pain, especially with addictive drugs. The final reason is that pain in children may be useful in treating them, since it may produce character in them. All three reasons, according to Walco and associates, are inadequate not only ethically but also technically.

Although the above reasons certainly operate in the healthcare profession to stifle the voices of suffering children, the vice of uncaring is operative as well. Diamond's story of chronic eczema supports the role of this vice. For example, the uncaring physician attending Diamond holds to the first myth in telling Diamond that her perception of itching is really in her mind. Hence, the healthcare provider does not believe that the pain and suffering Diamond is experiencing is real or even possible, or of any serious consequence. In other words, he simply does not care about her existential state *vis-à-vis* the pain and suffering associated with chronic eczema, since he could not break out of his world of myth to join the patient in her world of reality.[13]

In Diamond's case study, moreover, the ontic vices of carelessness and incompetence clearly instantiate the uncaring actions of healthcare providers. The carelessness these providers exhibit is both motivational (carelessness$_1$) and behavioral (carelessness$_2$). As for carelessness$_1$, for example, the nurse's response to Diamond's request to talk about her eczema demonstrates that the nurse is incapable of caring about Diamond's physical and psychological or emotional needs. Simply put, the nurse is just not motivated to care about Diamond's needs or Diamond herself but only about the demands placed upon her with respect to her employment as a nurse. In other words, the nurse could care less about Diamond and the pain and suffering the child was experiencing from chronic eczema. Why is the nurse careless$_1$? She may have been careless$_1$ for a number of reasons. The first is that she may not have cared for Diamond personally. Diamond may have struck the nurse as a difficult patient at the outset and therefore one to avoid at all costs. Another reason is that the nurse may have had little, if any, compassion for people in general. In other words, she simply cared less about the suffering of others, and Diamond is just one instance in the nurse's careless$_1$ attitude towards others.

The physician presents an interesting twist to the carelessness Diamond experiences during her hospitalization. He gives the impression of caring about Diamond but he really does not. In fact, Diamond acknowledges that she knew intuitively that

[13] Even Diamond's father illustrates the third myth with his comment, "What doesn't kill you makes you stronger."

the physician, as well as the rest of the hospital staff, is $careless_1$. "On a superficial level," as Diamond narrates, "they appeared to be concerned, but I could tell they were just being polite, doing a job; they didn't have the time *to care about* [$carelessness_1$] the frustrated soul trapped in my little body" (1996, p. 2, emphasis added). Why does Diamond believe that the physician and hospital staff are $careless_1$? Because, as she explains, they simply do not listen to her. If the physician genuinely cared about her, then he would listen to Diamond and would take into consideration her comments about how ineffectual antihistamines are in alleviating her painful symptoms. Instead, the physician placates Diamond and marginalizes her illness experience. The reason for such $carelessness_1$ may be attributable to the reasons Walco and colleagues offer to account for why healthcare professionals do not listen to children in pain. For example, the attending physician may have been approaching Diamond's healthcare with a preconceived notion of the "appropriate" child standard. Finally, Diamond's recognition of the healthcare providers' insincerely in treating her undermines Curzer's notion that caring is a vice and that these providers need to act only *as if* they are caring while simply being benevolent. As a child, Diamond obviously could distinguish between the providers' sincerity and insincerity and such insincerity has devastating consequences for her relationship with these providers, as well as for the quality of her healthcare she receives from them.

The hospital staff's $carelessness_1$ is, in part, the basis for its technical and ethical incompetence and hence its inability to take care of ($carelessness_2$) Diamond. For example, the physician is technically incompetent because he fails to use the best possible therapeutic protocols to treat Diamond's eczema.[14] The physician's obligation is to ensure that he is providing the most appropriate treatment, whether antihistamine drugs or something else, and if he does not know the appropriate treatment then he should consult an expert in the field if he is not one. But, because of his $carelessness_1$, he is incapable of providing quality healthcare. The physician is also ethically incompetent for failing to discharge his duty to provide such care for Diamond by listening to her illness narrative. Rather than doing the right thing by listening intently and sympathetically, the physician does the wrong thing by marginalizing or stifling Diamond's voice. The net result of the physician's actions is an inability to take care of Diamond, thereby exhibiting the vice of $carelessness_2$. Such $carelessness_2$ is the result of the physician's previous habits of $carelessness_1$ and incompetence.

The nurse is also incompetent, both technically and ethically, to take care of Diamond. Technically, she may not have been trained either in nursing school or on the job to take care of children with eczema and does not want to extend herself into an unknown area of healthcare to become technically competent to do so. Ethically, she may feel no obligation to take care of Diamond for a number of reasons, such

[14] Although the attending physician and nurse fail with respect to treating Diamond's eczema, they are technically competent in helping her to recuperate from her surgery. The point is that both the physician and nurse have an opportunity to provide quality care for Diamond's eczema, an opportunity they miss because they simply do not care to realize it.

as the excuse she offers Diamond as being too busy to have the time to talk with her. The net result of the nurse's incompetent and careless$_2$ actions is an inability to provide quality care for this young child, who was suffering.

The failure of the hospital staff attending her, according to Diamond, is not so much the lack of cure or even management of—to take care of—her physical illness but the inability to care about (carelessness$_1$) her at a psychological or an emotional level, particularly with the pain and suffering she had endured since childhood. As Diamond pleads with the healthcare profession in general, especially in terms of treating patients with chronic illnesses, "the chronically ill require an emotional response. I'm talking about sincere human empathy for physical conditions that must be endured for the rest of one's life" (1996, p. 4). Such psychological and emotional care is necessary for any genuine hope on the patient's part for authentic healing. By failing to care about (carelessness$_1$) her, the hospital staff is unable to take care of (carelessness$_2$) her, because it is technically and ethically incompetent. Technically, the hospital staff is obviously not trained properly to treat the pain and suffering of children. Ethically, the staff is unable to realize that its actions caused Diamond greater harm than the disease.

More generally, according to the adult Diamond, the healthcare profession incurs a responsibility to assist chronically ill patients, especially children, to heal from the ravages of such devastating and debilitating illnesses, not to add to the pain and suffering already associated with chronic diseases. "I urge the medical community," counsels Diamond, "to take responsibility for assisting chronically ill patients in finding emotional support among peers and professionals. Doctors and patients actively working together can diminish the human tragedy of chronic illness" (1996, p. 4). To that end, healthcare providers must avoid many of the vices, whether intellectual, ethical, or even transcendental or theological, as the Diamond case study illustrates—but before discussing those vices I must examine next the role of the compound vice of imprudent lovelessness.

The compound vice, imprudent lovelessness, certainly plays an essential and pivotal role in the healthcare staff's inability to provide Diamond with quality healthcare in a professional manner. The healthcare providers attending Diamond during her convalescence from surgery for removing the clavicle tumor clearly exhibit the traditional intellectual vice of imprudence. As imprudent, the attending physician, for example, is rash and incautious in forming the clinical judgment that an antihistamine would relieve the symptoms associated with Diamond's eczema, especially since she tells him that this type of drug was ineffectual in the past for relieving them. A prudent physician would take into account what the patient is saying and look further into the type(s) of antihistamine(s) used previously to treat Diamond and to investigate whether a better drug is currently available. If not, then he would explore possible non-pharmaceutical options for addressing and relieving the patient's pain and suffering. Importantly, the physician's and hospital staff's imprudent treatment of Diamond represents the antithesis of the Berlin notion of wisdom, Ardelt's notion of the wise person, and Achenbaum's notion of wise action.

In terms of the Berlin notion of wisdom (Baltes and Smith, 2008), the hospital staff fails to demonstrate either theoretical or practical wisdom associated with the first stratum of wisdom. Again, the attending physician, probably relying on an "appropriate" child standard in terms of pain, shows little theoretical aptitude for the therapeutic limitations of antihistamines in treating the symptoms of Diamond's eczema. Rather, he relies on standards that most likely meet community standards with respect to treatment but none-the-less these standards are certainly inappropriate with respect to Diamond's healthcare needs. Practically, he should have listened to Diamond's past-history about the lack of relief from taking antihistamines. In all likelihood, a low-tech option of talking to her, which is apparently what she was requesting of the hospital staff, might have had tremendous therapeutic benefits. Moreover, the hospital staff fails to exhibit the second stratum of the Berlin notion of wisdom in their unwillingness to discover and appropriate Diamond's lifespan context in their treatment of her. For example, the nurse is imprudent in rejecting Diamond's request to talk about the pain and suffering associated with the eczema. The staff, then, lacks the expert knowledge or wisdom, in terms of both current humane practice and the patient's illness story, to provide quality, professional healthcare.

The hospital staff is also imprudent with respect to Ardelt's notion of the wise person (Ardelt, 2004). First, the staff fails to demonstrate the cognitive dimension of a wise person, i.e. it lacks the intellectual ability to obtain the necessary information required to treat the symptoms associated with Diamond's eczema. Even when told about the ineffectiveness of antihistamines, the physician, for instance, fails to understand its significance for relieving Diamond's symptoms. Next, the staff is unable to step back from the daily contingencies of treating Diamond and to reflect on the bigger picture, especially the challenges facing Diamond when her hands are immobilized just as she is learning to use them constructively to deal with the annoying eczema symptoms. A wise healthcare provider would realize the implications such immobilization would have on her. Lastly, the hospital staff is imprudent in failing to convey to the patient that it cares about her as a person. In Ardelt's terms, it lacks the affective characteristic of a wise person. Because it lacks this characteristic, the staff causes serious patient harm as when the nurse fails to answer Diamond's call for assistance to the toilet. The staff, then, is deficient in the characteristics that define a wise person.

Finally, the hospital staff is also imprudent in terms of Achenbaum's notion of wise action (Achenbaum, 2004). First, the hospital staff physician imposes a subjective standard of treatment that he thinks is best for Diamond. Even though the treatment appears to meet community standards for treating the painful symptoms of eczema, the patient tells the physician that antihistamines are ineffectual. The physician fails to listen to the patient and to incorporate objectively this information into treatment; rather, his action is imprudent and is simply the most expedient means for dealing with the patient's complaint about the eczema symptoms. Next, the physician's clinical judgments and decisions concerning the use of antihistamines are not appropriate for treating the patient, since previous experience indicated the drugs are ineffective. Instead of benefiting the patient, his actions are imprudent and result in significant patient harm, especially emotionally and psychologically. Lastly, the hospital staff's acts are not transcendent in that, for example, the nurse's failure to answer the patient's call for assistance to the toilet is certainly not a standard of care applicable to the treatment of all patients, let alone a child suffering from a chronic disease like eczema. The hospital staff's actions, then, are imprudent in not providing the patient with a standard of care needed to help her heal; instead, their imprudent actions cause the patient irreparable harm.

Besides being imprudent, the healthcare providers treating Diamond also exhibit intellectual vices cognate to imprudence. For example, the incidence surrounding prescribing antihistamine illustrates the intellectual vices of close-mindedness and indifference. The physician allows the vice of close-mindedness to limit the questions he asks to understand the impact the disease has on Diamond's life, through a possible prejudice akin to one of Walco and colleague's myths about children's pain and suffering. Moreover, he is indifferent with respect to engaging the young Diamond in any meaningful discussion of her chronic eczema, other than offering platitudes to pacify and dismiss the child. Rather than tapping into the healing power of prudent wisdom, as Szawarski (2004) advocates, the staff mine the injurious force of imprudence to exacerbate the patient's pain and suffering. Particularly, it is incapable of utilizing the tacit dimension of the clinical arts to appropriate sound clinical judgment in treating Diamond. Instead, the staff simply trusts in platitudes and the expedient to treat her.

Importantly, the healthcare providers are unwise in failing to grasp eczema's meaning and significance for Diamond's daily life. Rather than identify the true meaning and significance, they substitute false meaning and significance or the meaningless and insignificant based on, e.g. Walco and colleague's myths for dismissing or diminishing children's pain and suffering. The nurse, for instance, thought she is responding in a wise fashion by informing Diamond of the important nursing duties that require her immediate and expert attention. But, rather than being wise, the nurse is unwise by responding in such a manner. If she is a thoughtful—instead of a thoughtless—person, the nurse could step back from the situation and place herself in Diamond's circumstances. What is Diamond asking of the nurse? Simply to take a moment to talk about how to cope with the pain and suffering associated with the chronic eczema. Would the time investment on the part of the nurse had been that great to make her negligent of her other nursing duties? Most likely not, the amount of time would probably have been negligible. The wise and caring action would have been to talk with Diamond and to explore what the child needed to help relieve her symptoms. In addition, if the nurse is wise she would realize that she alone could not address Diamond's every healthcare need in terms of eczema and would require help from additional members of the healthcare team.[15] The nurse is injudicious and careless$_2$ in belittling Diamond's request to talk about the pain and suffering associated with her eczema.

Besides the traditional intellectual vices, the epistemic unvirtuous behavior of the healthcare providers in Diamond's case study can also be analyzed in terms

[15] Diamond does not mention that during her hospitalization a dermatologist visited her. The hospital staff should have consulted a dermatologist, given the severity of Diamond's eczema.

of virtue epistemology, whether reliabilist or responsibilist. With respect to reliabilist virtue epistemology, the hospital staff apparently has properly functioning or dependable sensory and cognitive faculties for delivering the epistemic goods in treating Diamond. However, closer inspection reveals that the staff's inability to care for Diamond *qua* person diminishes its epistemic faculties. In other words, the staff's prejudices and biases towards the extent of Diamond's pain and suffering from eczema dull, or even distort, its sensory information concerning Diamond's physical and emotional condition and decouple that information from cognitive faculties thereby prohibiting the staff from properly evaluating it intellectually. The physician, for instance, may have seen Diamond's skin but his inability to enter into her world of suffering keeps him from appreciating the extent of the child's pain and thereby responding clinically in an efficacious manner. In terms of responsibilist virtue epistemology, the hospital staff displays a number of epistemic vices in its treatment of Diamond. For example, the nurse is an epistemic coward by shrinking with fear from talking with Diamond because she believes that she would not have time for her other nursing duties. Fundamentally, the staff really does not want to know the truth about Diamond and the impact eczema has on her life. In a way, it has distain—not desire—of knowledge.

With respect to the second vice of the compound vice, imprudent lovelessness, the hospital staff certainly exhibits lovelessness, while treating Diamond. In Swantonian terms, the staff is incapable of drawing close to Diamond, either receptively or appreciatively, to meet her physical or emotional needs. In contrast to being receptive, the hospital staff members avoid Diamond, scurrying about outside her room and muttering in hushed tones so that Diamond could not hear them. Also, the staff members heave an enormous sigh of relief when leaving her room after attending to her. In contrast to being appreciative, the staff fails to value Diamond as a person of genuine worth in her own right. Rather, it judges her worth *qua* person in terms of her compliance with its ineffectual treatment or management of her eczema symptoms. In short, the hospital staff is unable to pull alongside Diamond in order to gain an understanding or appreciation of the suffering she is experiencing from being unable to use her hands positively to manage the symptoms associated with her eczema.

In particular, the hospital staff exhibits an impersonal prosaic lovelessness towards Diamond. Its lovelessness is prosaic in the sense that the staff does not ultimately affirm the patient's unique healthcare needs—Diamond's coping in an effective manner with the excruciating pain associated with her eczema symptoms. Rather, it treats her in a banal manner. For example, the staff nurse prioritizes her need to meet job obligations and duties of paperwork and other patients over taking a few minutes to discuss with Diamond the suffering associated with her eczema and to explore possible strategies to deal with it while Diamond's hands are immobilized after the surgery. Of course, the nurse's work obligations are important, but she is care-less about Diamond's needs in her response to the child, a care-lessness that comes across as love-lessness to the patient. Thus, the nurse is indifferent to the child's healthcare needs and substitutes banal nursing duties for the chance to love Diamond radically by attending to her ultimate healthcare needs and not the immediate needs of recuperating from surgery. The staff's lovelessness is also impersonal, since it fails to acknowledge or recognize Diamond's uniqueness as a person of value or worth. The staff treats her impersonally in terms of being remote and distant to the child's suffering. In contrast to union of an effective therapeutic relationship, the hospital staff's impersonal prosaic lovelessness results in a dysfunctional relationship that harms rather than helps the patient.

The hospital staff also displays several of the cognate vices associated with lovelessness. For example, the staff is unsympathetic or uncompassionate towards the pain and suffering associated with Diamond's eczema in that it is unable or unwilling to draw alongside her through the patient narrating her illness experience, in order to understand that world of pain and suffering. Moreover, the staff is apathetic in that it is incapable of any feeling for Diamond's pain and suffering, let alone capable of co-feeling with her. In other words, it is unable to enter her illness world, if even vicariously, to gain perspective on what impact eczema has on her life. Moreover, the staff is self-centered and not patient-centered in that it attends to its duties and obligations, forging a wall between itself and Diamond's ultimate healthcare needs. Lastly, the staff exhibits the vice of pride in that it is incredibly arrogant in presuming that Diamond has little to offer in terms of suggestions concerning her treatment. To dismiss her input that antihistamines were previously ineffectual in treating or managing her eczema symptoms, as the hospital physician does, is the acme of hubristic clinical behavior. In sum, the staff fails to provide the quality care Dougherty and Purtilo (1995) claim is a fiduciary responsibility of professional healthcare providers.

The compound vice, imprudent lovelessness, especially its synergy, prohibits the hospital staff from developing an effective therapeutic relationship with Diamond in order to meet her healthcare needs. Specifically, the imprudent loveless staff is incapable of developing distinct therapeutic relationship boundaries in which the staff could identify its professional obligations towards the patient. Rather, it develops indistinct and amorphous boundaries in which the staff takes advantage of or abuses the power differential between itself and the patient. The nurse's refusal to answer Diamond's call for assistance to the toilet is an apt example of this abuse, because it represents a blatant transgression of the boundary between the patient and the nurse. The nurse simply could not distinguish between herself as healthcare provider and Diamond as someone in need of healthcare. Instead, she blurs the boundary between herself *qua* nurse and Diamond *qua* patient thereby prohibiting any chance of connecting with the patient and meeting her call or need to go to the toilet. This boundary is critical for the professional healthcare provider to pull alongside the patient because it demarcates the region for therapeutic interaction. Without the boundary, the healthcare provider cannot distinguish between what helps or harms the patient, as the nurse's behavior illustrates.

The healthcare providers also exhibit the traditional ethical vices of cowardice, intemperance, and injustice, as well as their cognate vices. With respect to cowardice, the hospital staff succumbs to its supposed fears about the danger Diamond presents as a patient. The staff perceives her as a troublesome and demanding patient who could disrupt not only the professional lives of the hospital staff but also the efficient and effective running of the hospital itself. Consequently, based on this fear the staff avoids Diamond and her need to talk about the suffering she experiences from her eczema. Not only does the staff avoid her; but, it demeans her by speaking in hushed tones outside her room and trivializing her request by shunting the responsibility for care to another healthcare provider, especially in terms of reading her a story. The staff is ignoble and fainthearted in dealing with the child. Rather than exhibiting the moral fortitude of engaging Diamond in substantive discussion over the pain and suffering of eczema and helping her to heal, the staff caves into pusillanimous behavior that exacerbates the pain and suffering she was already experiencing. Because of its moral frailty and weakness, the staff fails its ethical calling to treat the patient caringly and competently.

The hospital staff also exhibits the ethical vice of intemperance in their treatment of Diamond. In other words, it responds to Diamond more or less fanatically. For example, the physician when discussing with Diamond antihistamines for treating her eczema intimidates her with demanding whether she wants to get better. Such extreme behavior towards a child is not only unprofessional and uncaring but also immoral. Would this physician treat his own child in a similar manner, or if he were a child would he like to be treated in such a manner? The staff fails to act in a way Osler would find justifiable for clinical behavior, i.e. with *aequanimitas*. Diamond perturbs the staff with her need to talk about her eczema to such a point that the nurse refuses the common decency to assist Diamond to the toilet. The staff is simply unable to compose itself around her, shuttling about and whispering in inaudible tones that upset the young Diamond. The result is a discordant relationship between the staff and the patient rather than Carrian/Oslerian "sympathetic attunement." The staff is just unable to manage its response to Diamond in a professional and ethical manner. At times, it is both impatient and uncivil towards her. For example, the nurse, in upbraiding Diamond for wanting to talk, simply yields to an emotional desire to force Diamond to enter her demanding world of nursing. A temperate nurse, however, would enter the patient's world of suffering and respond in a caring and competent fashion.

The hospital staff also displays the vice of injustice, particularly distributive and restorative injustice. In terms of distributive injustice, the staff fails to supply healthcare goods for Diamond in an equitable way. Rather, it withholds such goods to her detriment. For example, the physician certainly neglects his ethical obligation as a just provider of healthcare goods by not exploring pharmacological or therapeutic options other than antihistamines to treat Diamond's eczema. A just provider would investigate other options to treat Diamond and not limit options for treating Diamond to antihistamines alone. Another important healthcare resource that the staff limit in an unjust manner, with respect to distributive injustice, is time spent treating or attending to Diamond. Rather than spend a few extra minutes discussing the impact of eczema on Diamond's overall quality of health and life, the staff unjustly withholds its time and distributes it inequitably to other patients and tasks. As for restorative injustice, the hospital staff fails its ethical duty to provide Diamond a quality of life for which she is searching. The staff misses an opportunity to impart a balance to the patient's life. As Diamond notes, it is not so much that the staff fails to cure her as it fails to care about her as a person. The injustice of the staff's actions in terms of restorative injustice is that the staff misses an opportunity to restore her dignity as a person. The staff also exhibits several of the cognate vices associated with injustice in its treatment of Diamond. Besides being woefully unfair to Diamond in allowing external factors to influence its behavior towards the patient, the hospital staff is prejudiced towards her because it perceives her as a difficult patient. And, the staff is untrustworthy in the sense that it is not loyal to Diamond and her efforts to address the pain and suffering associated with eczema; rather, it abandons her.

Besides the compound vice of imprudent lovelessness and the traditional intellectual and ethical vices, the hospital staff exhibits the transcendental or theological vices of faithlessness and hopelessness, as well as their associated cognate vices. In terms of faithlessness, the staff does not hold to a set of Polanyian tacit commitments that the truth concerning Diamond's eczema is knowable apart from either itself or the patient. In fact, the staff has little, if any, faith in Diamond as an authentic source of knowledge or truth concerning her medical condition. Rather, it completely dismisses any contribution from her as to treatment of the chronic disease. For example, the attending hospital physician stifles Diamond's protestation that previous attempts to use an antihistamine to manage the symptoms of her eczema failed. Rather, the physician informs Diamond that what he is prescribing in terms of an antihistamine is what she needs. As such, he also exhibits several of the cognate vices associated with faithlessness. For instance, the episode with the antihistamines also belies the vices of doubt and skepticism on the part of the physician. In terms of doubt, the physician is certainly apprehensive of Diamond as an authority of genuine knowledge concerning the treatment of her disease, while with respect to skepticism he is cynical that Diamond could contribute any useful information concerning the effects of the drug upon her condition. Rather, the physician impresses upon Diamond that only he is an authentic source of information for using antihistamines to relieve the symptoms associated with eczema.

The hospital staff also exhibits the transcendental vice of hopelessness, during its treatment of Diamond. The type of hopelessness is not that the staff could not help the patient to recover from or to manage the symptoms of eczema but rather a false hope that it could help. And, if Diamond would only comply with its therapeutic protocol, i.e. the antihistamines, then she would get better and not need to talk about the pain and suffering she experiences with the chronic disease. This sense of false hope that the staff foists on Diamond has a detrimental clinical impact on her. First, she is well aware that antihistamines are ineffectual for managing the eczema symptoms. In fact, she wants to discuss with the staff alternatives for possibly managing the symptoms—alternatives she is discovering on her own, such as stroking or gently pinching the skin. Rather, the staff's response is the use of restraints to keep her from scratching the skin. A second fallout from the staff's false hope is a sense of hopelessness on the Diamond's part when she begins to realize that the staff does not care to discuss alternatives with her, or that it really cares less for her and the pain and suffering she is experiencing from the ravages of the chronic disease. The staff also exhibits a number of the cognate vices *vis-à-vis* its false hope in managing Diamond's eczema. For example, it exhibits the vice of anxiety because it is pessimistic about its success to treat Diamond, other than with antihistamines, since it perceives her as a noncompliant or troublesome patient. However, if the staff had engaged her in meaningful dialogue about the treatment of the chronic disease instead of perpetuating a false hope then such anxiety would have been unnecessary.

As the Weinberg clinical case illustrates the connections among the virtues and the power of prudent love to deliver the clinical goods for treating the patient effectively and safely, so the Diamond case exemplifies the connections among the corresponding vices and the power of imprudent lovelessness in failing to deliver clinical goods—which resulted in further patient harm and injury. The hospital staff attending Diamond during her hospitalization is fundamentally uncaring in its treatment of her because the staff is unable to enter into the patient's world of chronic eczema and to experience or co-feel her pain and suffering—which in turn prohibits it from prudently loving her. Since the staff is uncaring, it cares less about Diamond and whether she effectively manages the painful symptoms associated with eczema—which in turn keeps the staff from taking care of her. For example, the nurse refuses to discuss Diamond's struggle to cope with eczema and the pain and suffering associated with the disease, since she fears that taking time to talk with Diamond would hinder her from performing her obligatory nursing duties. The nurse is an ethical coward in her blunt and crude dismissal of the child. She caves into her fears, to the detriment of the patient's wellbeing. Of course caution is necessary, but the nurse's behavior is not only unprofessional but also demeaning and to some extent cruel, especially when she refuses to answer Diamond's call for assistance to the toilet. The nurse's cowardice also has an epistemic dimension in that she imprudently fails to explore possible alternatives Diamond is discovering for safely using her hands to manage her eczema. Finally, the nurse's cowardice is transcendental in nature since she hides behind a false hope that the hospital staff is capable of managing the symptoms of Diamond's eczema. This false hope ends in hopelessness for Diamond, who eventually loses trust completely in the medical profession.

Lastly, the Diamond case study illustrates the cyclic relationships between carelessness and incompetence and between lovelessness and imprudence. With respect to the carelessness-incompetence cycle, the hospital staff's uncaring attitude is instantiated through its care less approach towards Diamond and her pain and suffering from eczema. The staff—perceiving the child as a difficult and demanding patient—avoids her, giving Diamond the distinct impression that it does not care or could care less about her (carelessness_1). Because of its careless approach towards her, the staff is incompetent to meet not only her physical needs, needs as simple as assisting her to the toilet, but also her emotional and psychological needs, such as the separation she begins to experience between her body and mind. As a result, the staff is incapable of taking care of Diamond (carelessness_2) because of its careless approach towards her. In other words, it is careless not only in attending to Diamond in a professional manner, such as answering her call for assistance to the toilet, but also it could care less in doing so. Finally, its carelessness_2 feeds back onto its carelessness$_1$ with the outcome of making the staff even more incompetent to treat her and to meet her healthcare needs.

With respect to the lovelessness-imprudence cycle, the hospital staff's carelessness$_1$ quickly devolves into lovelessness$_1$ as it cares less about (carelessness$_1$) Diamond and the impact eczema has on her life. The devolution is represented by the staff's loveless or apathetic attitude towards Diamond, as it treats her with indifference and, at times, even contempt or cruelty (lovelessness$_1$). Because of its lovelessness$_1$, the staff is not only incompetent, either technically or ethically, to take care of her (carelessness$_2$), but it is also imprudent in its clinical judgments and decisions in taking care of her. Such imprudence supports a loveless performance in which the staff seeks simply to survive the task of attending to Diamond and her postoperative recovery (lovelessness$_2$). Moreover, the imprudent lovelessness synergy prevents any chance of the staff's ability to forge a robust therapeutic relationship with Diamond in order to meet her healthcare needs. And, because of this synergy, the staff's lovelessness$_2$ feeds back onto its lovelessness$_1$ making it less likely to meet the healthcare needs of future patients. Finally, to come full circle, the hospital staff's uncaring—as an ontological vice grounding the carelessness-incompetent cycle—is the foundation for its healthcare practice as loveless-imprudent providers, which, in turn, reinforces its uncaring and its poor quality and unprofessional healthcare.

6.3 Summary

The two clinical stories illustrate well the difference between virtuous and unvirtuous physicians and healthcare providers. As for the virtuous provider, Weinberg depicts a physician who is not only caring at a fundamental level motivationally but who follows through to ensure that he employs his clinical skills for the benefit of the patient. For example, Weinberg pursues a connection to the patient at a visceral level, a connection involving the patient's employment—baking. Through this connection, he gains access into the reason for the patient's chief complaint from which he is then able to help the patient recover from the torment of her sexual assault. In helping the patient heal, Weinberg utilizes the various traditional virtues—often in novel ways. He exhibits courage, for instance, not only customarily as an ethical virtue but also as an intellectual virtue. Importantly, Weinberg employs the compound virtue of prudent love that exemplifies the virtuous physician. He is prudent in making clinical decisions under uncertain clinical conditions because of his love for the patient and concern for her welfare and recovery. This prudent love grounds and improves his overall clinical caring of the patient through its synergy, which, in turn, feeds back to enhance his caring for other patients.

Diamond's healthcare providers, on the other hand, reflect providers who are not only uncaring motivationally but who are also unable to employ their clinical skills to benefit the patient but rather to harm her. For example, the physician belittles and ignores Diamond's input on the past ineffectiveness of antihistamines to manage the painful symptoms associated with her chronic eczema. Instead, he intimidates the child through bludgeoning her with the realization that he is in control and that what he prescribes for her (another antihistamine) would take care of her symptoms. Rather than connect with the patient through building a bridge of caring, he builds a wall of uncaring that separates him from Diamond. That wall keeps him from entering the patient's world of pain and suffering, and it prevents the formation of an effective therapeutic relationship in which the physician could employ his clinical skills to help the patient. In other words, as an unvirtuous physician, fueled by the various traditional vices, especially the compound vice of imprudent lovelessness, he is impotent to provide quality healthcare professionally. Again, as noted earlier, Diamond's healthcare providers are not vicious in treating her but, given their unvirtuous behavior, they are not only unable to help but actually cause the patient irreparable harm and add to the pain and suffering associated with her chronic disease.

Finally, given the difference between Weinberg's approach to healthcare and Diamond's healthcare providers' approach, an interesting question arises as to how Weinberg would treat Diamond.[16] In commentary on Diamond's illness story, Kay Toombs notes that rather than placating Diamond the physician attending Diamond in the hospital during her recuperation from the surgery has an opportunity to explore the pain Diamond is suffering from eczema, in a genuine effort to heal her. "If one responds by asking further questions—'Tell me more about the itching. When is it the worst? Is there anything we can do to make it tolerable?—the patient is encouraged to give more detailed information which," according to Toombs, "may ultimately suggest a different therapeutic approach" (1996, p. 6). Weinberg, as a virtuous physician, would certainly follow Toombs' lead and take time to pursue such questions about the pain and suffering Diamond experienced from her eczema. He would not succumb to the myths, identified by Walco and colleagues, surrounding children's pain and suffering. Thus, the notion of virtuous physician addresses the inhuman treatment Diamond experiences at the hands of healthcare providers. In the final chapter, I look more closely at this notion and issues surrounding its implementation in modern medicine, especially in terms of medical pedagogy.

References

Achenbaum, W.A. 2004. Wisdom's vision of relations. *Human Development* 47:300–303.

Ardelt, M. 2004. Wisdom as expert knowledge system: A critical review of a contemporary operationalization of an ancient concept. *Human Development* 47:257–285.

Baltes, P.B., and J. Smith. 2008. The fascination of wisdom: Its nature, ontogeny, and function. *Perspectives on Psychological Science* 3:56–64.

Bligh, J. 2000. Whole doctor education. *Medical Education* 34:496–497.

Dalmiya, V. 2002. Why should a knower care? *Hypatia* 17:34–52.

[16] An equally interesting question that arises is how Diamond's healthcare providers would treat Weinberg's patient. Obviously, they would not take the time to pursue questions surrounding the sexual assault. In fact, I doubt that Weinberg's patient would feel secure enough around them to divulge her secret.

Diamond, S.F. 1996. Lifelong effects of chronic atopic eczema. *Making the Rounds in Health, Faith, & Ethics* 2:1–4.
Dougherty, C.J., and R. Purtilo. 1995. Physician's duty of compassion. *Cambridge Quarterly of Healthcare Ethics* 4:426–433.
Flitcraft, A. 1995. From public health to personal health: Violence against women across the life span. *Annals of Internal Medicine* 123:800–802.
Marcum, J.A. 2008. *An introductory philosophy of medicine: Humanizing modern medicine*. New York, NY: Springer.
Marcum, J.A. 2009. The epistemically virtuous physician. *Theoretical Medicine and Bioethics* 30:249–265.
Marcum, J.A. 2011. "First, do no harm": Uncaring and incompetent healthcare. *International Journal of Health, Wellness and Society* 1:1–11.
Rixey, S. 1996. Domestic violence and physician intervention. *Annals of Internal Medicine* 125:427.
Szawarski, Z. 2004. Wisdom and the art of healing. *Medicine, Health Care and Philosophy* 7: 185–193.
Toombs, S.K. 1996. Commentary: Heeding the call of suffering. *Making the Rounds in Health, Faith, & Ethics* 2:5–6.
Walco, G.A., R.C. Cassidy, and N.L. Schechter. 1994. Pain, hurt, and harm: The ethics of pain control in infants and children. *New England Journal of Medicine* 331:541–544.
Weinberg, R.B. 1995. Communion. *Annals of Internal Medicine* 123:804–805.

Chapter 7
The Virtuous Physician and Medicine's Crises

I employ the notion of virtuous physician to address the quality-of-care and professionalism crises introduced in the first chapter. To that end, I initially summarize the notion of virtuous physician and compare it to other notions within the literature. Importantly, the notion of virtuous physician I develop serves as a foundation for integrating EBM and PCM into what I call virtuous holistic medicine, which I discuss in a subsequent section. The virtuous physician embodies the spirit of both EBM and PCM by applying the latest medical knowledge and technology to meet prudently and lovingly the patient's individual needs. EBM and PCM are not separate and antagonistic approaches to the practice of medicine but two sides of the same medical coinage. Indeed, a virtuous physician cannot practice effective medicine without engaging both approaches. In contrast, the unvirtuous physician practices an unvirtuous fragmented medicine in which EBM and PCM diverge. In a succeeding section, I utilize the notions of virtuous physician and virtuous holistic medicine to resolve the quality-of-care and professionalism crises plaguing modern medicine. The virtuous physician as a genuine professional can deliver both the ethical and epistemic healthcare goods, particular in terms of virtuous holistic medicine, in a wise or prudent and compassionate or loving fashion.

In a final section of the chapter, I explore the implications of the notion of virtuous physician for medical education and the medical humanities movement. As William Stempsey so aptly argues, medical education has been instrumental in changing the public's perception of the physician, especially in the United States, "from the kindly and caring individual to the unfeeling technocrat or, even worse, the greedy entrepreneur" (1999, p. 3). Although Stempsey offers a useful prescription for this "illness," I propose the notion of virtuous physician to reorient that perception and to include the humanities in medical curricula to balance their almost exclusive focus on the scientific and technical. One of the major questions facing these curricula is the pedagogy of virtues, i.e. whether virtues can be taught or learned. My answer to that question is two-fold. Educators cannot teach virtues and students cannot learn them as easily or completely as scientific facts. However, the medical faculty can introduce virtues in a factual manner initially, sensitive to a student's developmental stage; but equally, if not more importantly, virtues can and must be modeled. Only through introducing and modeling virtues can students be taught and learn virtues. Modeling virtues in the clinic may help

J.A. Marcum, *The Virtuous Physician*, Philosophy and Medicine 114,
DOI 10.1007/978-94-007-2706-9_7,

to address the detrimental effects of the hidden curriculum medical students face in the last two years of medical school and residents and interns in their post-graduate programs.

7.1 Virtuous Physician

The notion of virtuous physician or the role of virtue in the practice of medicine is certainly not new. For example, Drane (1995) proposed a notion of the "good doctor" based on the function of virtues in medical practice. "To be a good doctor," claims Drane, "modern physicians. . .need personal qualities which enable them to be fully engaged with their patients" (1995, p. 30). Specifically, he identifies virtues like benevolence, truthfulness, respect, and justice to operate as requisite qualities for physicians to practice medicine virtuously. Damian Clarke offers another example of the virtuous physician. For him, the virtues distinguish a medical professional's practice from simply engaging biological or clinical facts about the patient. "Medicine has always prided itself," according to Clarke, "on producing men and woman who have a duty to something higher than self interest. As doctors," he argues, "we make a commitment to the patient, an assurance that we will never abandon the patient, that we will never willfully harm the patient, that we will strive to do everything in our power in the interests of the patient. Medicine, which deals so intimately with humanity, has a very pragmatic approach within its framework of the virtuous doctor" (2009, p. 57).[1] In this section, I examine briefly the notion of virtuous physician developed by Pellegrino and Thomasma and by Peter Toon and compare their notions to mine.

Pellegrino (1985) developed over a quarter of a century ago the notion of virtuous physician. For him, a virtuous physician is someone who advances the goal or end of medicine, which is to seek a patient's good—whether that good is curing or managing the patient's disease or simply helping the patient to cope with the disease or possibly end of life issues.[2] For the most virtuous physician, a patient's good may trump even the physician's good in that the physician may act altruistically. To achieve the patient's good, Pellegrino's virtuous physician employs the traditional virtues towards that goal of medicine. These virtues, then, shape the physician *qua* person, on whom a patient must rely. That reliance or trust, which depends on the virtues, is at the core of the therapeutic relationship. "What we expect of the virtuous physician," argues Pellegrino, "is that he [virtuous physician] will exhibit them [virtues] when they are required and that he will be so habitually disposed to do so

[1] Jack Coulehan (2005) also yokes the notion of virtuous physician to medical professionalism in an attempt to address the dehumanization associated with current healthcare practices. Indeed, Elaine Adamson and colleagues found in an empirical study of orthopaedists that patients are less likely to sue for malpractice those clinicians exhibiting virtuous practice (Adamson et al. 2000).

[2] According to Pellegrino, the patient's good includes: "(1) clinical or biomedical good; (2) the good as perceived by the patient; (3) the good of the patient as a human person; and (4) the Good, or ultimate good" (1985, p. 244).

that we can depend upon it" (1985, p. 246). Importantly, he identifies no particular virtue or set of virtues that define the virtuous physician. Rather, as long as a physician achieves the goal of medicine through employing virtues the physician behaves in a virtuous fashion.

Pellegrino, in collaboration with Thomasma, further develops the notion of virtuous physician, especially with respect to employing both the head and heart in medical practice, but only in terms of applying the virtues generally with no specific emphasis on any particular virtue or set of virtues. Specifically, in *The virtues of medical practice*, Pellegrino and Thomasma (1993) employ the cardinal virtues of *phronesis*, justice, fortitude, and temperance, to explicate the notion of virtuous physician *vis-à-vis* medicine's goal of the patient's good. Although they identify no one virtue as critical for demarcating the virtuous physician, pride of place goes to *phronesis*, which they claim is "indispensible" for the practice of medicine. What makes the virtue indispensible is that *phronesis* as prudence or practical wisdom provides the virtuous physician with the "moral insight" needed to determine when he or she achieves the patient's good. Besides the cardinal virtues, Pellegrino and Thomasma also discuss the role of other virtues, particularly fidelity, compassion, integrity, and self-effacement, in the practice of virtuous medicine. Overall, their notion of virtuous physician pertains almost exclusively to the ethical or moral nature of medicine so that the virtuous physician is equivalent or restricted to the ethically virtuous physician. Importantly, they do recognize the significance of a virtuous medical community to support and aid the practice of virtuous physicians.

In a subsequent book, *The Christian virtues in medical practice*, Pellegrino and Thomasma (1996) employ the theological virtues of faith, hope, and charity, to transform their notion of virtuous physician into what they call the "Christian personalist physician." To that end, they adopt John Paul II's "personalist principle" that people are truly human when they give themselves to others through loving and caring for others. In other words, people have intrinsic worth and value because God created them. The principle, insist Pellegrino and Thomasma, has direct application for medicine *vis-à-vis* the patient's good and shaping the physician's professionalism. In addition, it has an impact on the moral relationship of the patient-physician dyad and ethical decisions made during the practice of medicine. "When the Christian concept underlies the physician-patient relationship," write Pellegrino and Thomasma, "that relationship becomes a relationship of love, not in a sentimental or physical sense but in a sense of giving oneself and one's knowledge for the benefit of others, as Christ would have done" (1996, p. 147). Importantly, for Pellegrino and Thomasma, the Christian personalist physician not only engages in the practice of virtuous medicine but also in his or her salvation, since the end sought by such a physician is not simply good medical practice but the beatific vision as well.

Toon also develops a notion of virtuous physician, particularly in terms of general healthcare providers and their clinical practice. In *What is good general practice?*, Toon (1994) explores the biomedical, humanistic, and virtuous approaches to the practice of quality healthcare by general practitioners, with no particular recommendation made. However, in *Towards a philosophy of general practice: a study of the virtuous practitioner*, Toon (1999) opts for virtues in the general practice of good medicine. He recognizes that modern technology drives the practice of contemporary medicine and its practitioners often to the detriment of the patient. He proposes that virtues would play an important role in reversing the harm associated with this technological dependence. Although he identifies no specific set of virtues particular to the medical profession, he acknowledges that some virtues are more relevant to medical practice than others. For example, in discussing the notion of the wounded healer, Toon spells out the advantages of humility on the part of the physician for assisting patients left frail and vulnerable by their disease. Physicians, who overcome such frailties and vulnerabilities themselves, can better serve patients, "since an important aspect of the therapeutic relationship is empathy and compassion, frailties which place the doctor alongside the patient struggling with a problem, rather than in the position of a superior being" (Toon, 1999, p. 40). Moreover, he recognizes that virtues in the practice of medicine may depend on the medical specialty and that even within a particular specialty physicians may stress one virtue over another; however, he acknowledges that virtue is critical to successful and fulfilling medical practice. "Doctors cultivate their own virtues through the practice of medicine," concludes Toon, "and must do so if they are to practise satisfactorily" (1999, p. 41).

Importantly for Toon (2007), the virtuous physician is one who can negotiate the boundaries of clinical practice to provide quality medical care under a variety of conditions. He presents a number of medical case stories in which the virtues operate to assist the general practitioner in meeting patients' needs and discusses the significance of these stories *vis-à-vis* virtuous ethics and practice. For example, he reconstructs a clinical story in which a young couple, who are themselves medical professionals, seek advice about the birth of their first child who has severe brain damage. Toon concludes from this story that the virtuous practitioner acts as a "wise friend in authority," who draws upon virtues like practical wisdom, courage, and benevolence, to provide sage clinical advice concerning the most appropriate action. In another case story, he recounts the events of a practitioner who responds to the call of an elderly women suffering from heart failure. The physician assists the woman to the hospital and cares for her dog overnight until he finds housing for it the next day. Toon concludes from this story that the general practitioner acts as a "good neighbor," who draws upon virtues, such as justice and temperance, associated with the notion of Good Samaritan. From these stories, Toon construes, "The concept of doctor playing different roles that require different virtues and a different sort of response is valuable not only in helping us to characterize the virtuous practitioner but also in helping the doctor decide what to do in specific situations" (2007, p. 97).

The virtues for Toon, especially in terms of virtue ethics, provide the general practitioner with the moral resources necessary not only to address the problematic ethical conundrums that arise on occasion in practice but also the common or mundane ethical issues that are part of an everyday medical practice. The virtuous physician is a species of the virtuous person. "Being a doctor," asserts Toon, "does not mean that you are not also human—something that traditional approaches to medical ethics tend sometimes to overlook" (2007, p. 96). The virtues allow physicians to meet the needs of patients, even though the needs are not part of their professional role. Just because a person is a physician does not mean that he or she cannot function as a wise friend or a good neighbor in an appropriate fashion. Toon concludes concerning the nature of the virtuous physician, especially as a professional, "The virtuous doctor will be flexible and compassionate, not limiting themselves to a formally professional role, and will act in these other roles with courage, temperance and justice as appropriate" (2007, p. 96). According to Toon's notion of virtuous physician, then, the virtues play an important function in assisting physicians to determine not only their professional role but also their role as a human beings responding appropriately to other human beings in need.

Although I dispute neither Pellegrino and Thomasma's notion of virtuous physician *qua* ethically virtuous physician or even *qua* Christian personalist physician nor Toon's notions of the virtuous general practitioner and good general practice, I do contend that they simply do not go far enough in identifying and explicating specifically the virtues responsible for the virtuous practice of medicine and healthcare. To that end, I propose that the ontological virtue of caring and the ontic virtues of care and competence, including their transformation into the compound virtue, prudent love, are critical for defining the virtuous physician and for describing the virtuous practice of a holistic medicine in terms of EBM and PCM. The virtuous physician is definable then in terms of a professional who delivers quality healthcare, because he or she is caring in terms of loving (transformed $\text{care}_{1\&2}$) the patient and of prudently attending to (transformed technical and ethical competence) the patient's healthcare needs, either bodily or psychological or both—depending on the patient's needs at the time. A strong desire motivates the virtuous physician to help the patient, who is in a vulnerable position, and animates the virtuous physician both ethically and epistemically.[3]

In contrast, the notion of unvirtuous physician involves an agent who is unprofessional and delivers poor quality healthcare because he or she is uncaring in terms of being loveless (transformed $\text{carelessness}_{1\&2}$) towards the patient and imprudently attending to (transformed technical and ethical incompetence) the patient's healthcare needs, either bodily or psychologically. What defines such a physician then is the compound vice of imprudent lovelessness, and what defines his or her practice is a fragmented medicine that stresses either EBM or PCM to the exclusion of the other. What motivates the unvirtuous physician is generally not the vulnerable patient who needs healthcare but often the physician's own agenda or needs, whether financial, social status, or egoistic. Unfortunately, after years of delivering poor quality healthcare, fear too often animates the unvirtuous physician, who then practices medicine from a defensive posture, both ethically and epistemically.

---

[3] My integration of virtue ethics and epistemology with respect to the compound virtue of prudent love differs from Zarkovich and Upshur's integration based on the virtues of conscientious and judicious (Zarkovich and Upshur, 2002). See Pellegrino (2002b), for additional discussion of Zarkovich and Upshur's thesis.

In summary, my notion of virtuous physician (and of unvirtuous physician) is more specific and yet comprehensive for defining what a virtuous (and an unvirtuous) physician is, than either Pellegrino and Thomasma's or Toon's notions, by including both the ethical virtues (and vices) and the epistemic or intellectual virtues (and vices), in the formulation of a compound virtue, prudent love (and a compound vice, imprudent lovelessness). In other words, their notions serve to address the ethical challenges of medical practice, while mine incorporates the intellectual dimensions of medical practice thereby covering more of what physicians do in terms of their professional practice on a daily basis. I cannot stress strongly enough that my notion provides a wider perspective of how the virtues (and vices) operate in providing good (or bad) quality care in a professional (or unprofessional) manner. Importantly, my notion thereby opens up an avenue for addressing the quality-of-care and professionalism crises plaguing modern medicine—to which I now turn.

7.2 Virtuous Holistic Medicine: Integrating EBM and PCM

Before examining the resolution of the quality-of-care and professionalism crises *vis-à-vis* the notion of virtuous physician, I demonstrate how the notion predicated upon the ontological virtue of caring and the two ontic virtues of care and competence, along with their transformation into the compound virtue of prudent love, facilitates the integration of both EBM and PCM into a holistic practice of medicine. Importantly, as argued so far, virtues are not just something that physicians or healthcare providers add to enhance their practice but rather they are its foundation. In other words, good and caring medicine is the product of good and competent practices by good and virtuous physicians. In contrast, the notion of unvirtuous physician provides a means to examine why those who maintain separation or even conflict between EBM and PCM often exacerbate the problems associated with the crises. Unvirtuous physicians provide bad or poor quality medicine as the product of bad or incompetent practices. In this section, I discuss how prudent wisdom is necessary not only for the practice of EBM but also for PCM and personal radical love for PCM as well as for EBM. I call the integration of EBM and PCM through prudent love, virtuous holistic medicine (VHM), in contrast to unvirtuous fragmented medicine (UFM) in which EBM and PCM remain divergent or incompatible.

As noted in Chapter 1, several proposals for integrating EBM and PCM are available in the scholarly literature. For example, Lacy and Backer (2008) propose evidence-based patient-centered care (EBPCC), which represents an overlapping or intersecting region between EBM and PCM. They employ Hegel's notion of dialectical tension to discuss the relationship between EBM and PCM and to motivate EBPCC conceptually. However, they acknowledge that a philosophical approach to warrant that integration is not available presently. Lacy and Backer also identify a number of barriers or challenges to the implementation of their model in clinical practice. These barriers include three major challenges: (1) system-related factors such as restrictions particularly on time, since many healthcare professionals believe that engaging both evidence-based and patient-centered healthcare would require too much time, (2) relational challenges between patients and physicians such as new and difficult or demanding patients, and (3) problems associated with discovering mutual frameworks between patients and physicians especially for developing effective treatment plans.[4] As I proposed in Chapter 1, the notion of virtuous physician represents a good means for integrating both EBM and PCM and for addressing these barriers or challenges to their integration. I now turn to the mechanics of that integration.

Caring, as discussed in Chapter 4, is the chief ontological virtue of the medical and healthcare professions (Fig. 7.1). It makes possible the two prevalent ontic virtues of medical practice—care and competence. Moreover, the cyclic relationship between these two virtues, in which care enhances competence and competence care, is important for augmenting the overall quality of healthcare professionally. Although care and competence are critical for providing quality healthcare proficiently, they are insufficient for providing the best possible healthcare. To that end, as laid out in Chapter 5, a transformation must first take place: competence into prudent wisdom and care into personal radical love. Just as care and competence are reciprocally related, so are prudence and love yielding a compound virtue—prudent love. The relationship between prudence and love is also synergistic in that the compound virtue exhibits features transcending the features of each virtue individually. The best possible quality care delivered in the most professional fashion, then, is the

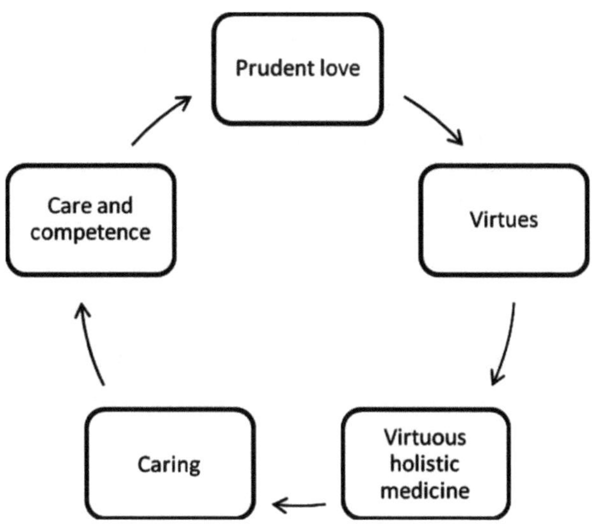

Fig. 7.1 Relationship among the components comprising VHM

[4] Other barriers include ineffective communication, EBPCC's complexity with respect to language and cultural differences or ethical principles, lack of role assumption particularly for physicians, uncertainty or inappropriateness of clinical evidence, multiple morbidities of the patient, and difficulties associated with translating population studies effectively to the individual patient.

synergistic product of prudent love that utilizes the various virtues. The mechanism by which the compound virtue achieves this quality care is through the integration of EBM and PCM to form VHM, to which I now turn.

Caring as an ontological virtue makes possible VHM through the transformation of the ontic virtues of competence and care into prudence and love, respectively. First, competence is the foundation for the cardinal virtue of prudent wisdom into which it is transformed. This transformed competence *qua* prudence provides a foundation for EBM by allowing the virtuous physician to make competent and prudent use of the best clinical evidence available for diagnosing and treating a particular patient. As such, the triad of competence/prudence/EBM makes possible the best technical care of the patient. Next, care is the basis for the virtue of personal radical love into which it is transformed. This transformed care *qua* love supplies the underpinning for PCM by letting the virtuous physician not only care about the individual patient but also take care of his or her medical needs with loving compassion and empathy (Stevenson, 2002). Moreover, these triads of care/love/PCM and competence/prudence/EBM triad feedback upon each other to produce a synergistic relationship between the various elements of each triad such that the integration of EBM and PCM yields a medicine with properties that transcend the properties of either EBM or PCM individually. The following is a schematization of the various elements comprising VHM.[5]

Caring ⇨ Competence/Prudence/EBM ⇔ Care/Love/PCM ⇨ VHM

This medicine is holistic since it combines and integrates both EBM and PCM to treat not only the physical but also the existential needs of the patient, thereby providing quality comprehensive healthcare professionally. Although VHM represents an ideal form of medicine, it is an ideal that can help stem the tide of today's poor quality healthcare delivered in an unprofessional and uncaring manner.

The integration of EBM and PCM as VHM, in terms of the notion of virtuous physician, also provides a means for addressing the barriers or challenges identified by Lacy and Backer (2008), especially the three major barriers. As for the system-related challenge of time limitations, VHM empowers virtuous physicians to prioritize clinical duties efficiently and effectively so to provide patients who need extra time the opportunity to disclose fully their illness narrative. As for the relational challenges between patients and physicians, VHM enables physicians to embrace new or even difficult or demanding patients because the virtuous physician is able to empathize with the patient's illness experience—especially the pain and trauma the illness imposes on the patient's life. Through this embrace, virtuous physicians are able to form deeply caring relationships with patients that are not only therapeutic for the patient but also gratifying for the physician. As for discovering mutual frameworks between physicians and patients, VHM facilitates the virtuous physician's capacity to incorporate the patient's preferences and values into

[5] VHM can also feedback onto caring to promote further interaction between EBM and PCM triads, as well as their integration.

a treatment plan—even if those preferences and values run counter to those held by the physician. However, if a patient's preference and values lead to unnecessary patient harm, the virtuous physician makes that potential harm known in a manner that is not belittling or humiliating to the patient holding those preferences or values.

The Weinberg clinical case study is an excellent example of VHM. First, genuine caring was the animating virtue for Weinberg's treatment of the patient. As a caring clinician, Weinberg is highly competent in his specialty; however, he is also competent in the general practice of medicine. He exhibits that competence in terms of prudently diagnosing the patient and providing an initial treatment plan that complied with the best clinical evidence available. Although Weinberg is accurate in that diagnosis, he was unaware of the causal basis for the patient's chief complaint. But, because he authentically cares about the patient, he eventually discovers its cause. Importantly, the patient recognizes the $care_1$ that Weinberg exhibits for her, which leads to the formation of an effective therapeutic relationship between physician and patient. Because of that relationship, Weinberg is able to take care of the patient in an individualized manner. Indeed, his $care_2$ stimulates an effort to take care of the patient competently through seeking counsel from a psychiatrist and the literature on sexual assault so he could treat the patient according to the best clinical evidence available. Weinberg certainly cares about and takes care of the patient in a loving manner that places the patient's need for healing in high priority within his healthcare practice. In sum, Weinberg practices VHM in which he centers on the patient's individual clinical needs (PCM) and meets those needs through the best clinical practice available to him (EBM).

Uncaring, as discussed in Chapter 4, is the ontological vice responsible for the delivery of poor quality healthcare that often leads to patient harm (Fig. 7.2). This vice makes possible the two ontic vices of the unvirtuous physician, carelessness and incompetence. Although these two ontic vices and their reciprocal relationship are

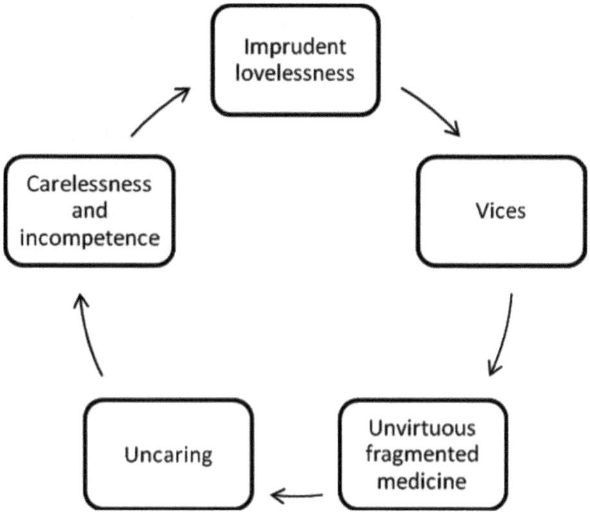

Fig. 7.2 Relationship among the components comprising UFM

adequate for providing substandard healthcare, they are insufficient for delivering the poorest quality healthcare. To that end, as detailed in Chapter 5, a transformation must first occur: incompetence into imprudence and carelessness into impersonal prosaic lovelessness. Just as carelessness and incompetence are reciprocally related, so are imprudence and lovelessness yielding the compound vice—imprudent lovelessness. The relationship between imprudence and lovelessness is synergistic in that the compound vice exhibits features transcending the features of each individual vice, especially in utilizing other vices, yielding the worse possible healthcare in the most unprofessional manner. The mechanism by which the compound vice achieves this impoverished care is through the separation or divergence of EBM and PCM to form UFM, to which I now turn.

Uncaring as an ontological vice makes possible UFM through the transformation of the ontic vices of incompetence and carelessness into imprudence and lovelessness, respectively. First, incompetence is the underpinning of the vice imprudence into which it is transformed. This transformed incompetence *qua* imprudence then provides the foundation for some unvirtuous physicians to assert that PCM is the only way to practice medicine and that EBM is unnecessary. These physicians make this assertion because of their incompetence, especially technical incompetence, to practice medicine and consequently their imprudence to make sound medical decisions in diagnosing and treating patients. In turn, the triad, incompetence/imprudence/PCM only, makes possible inauthentic care of patients through UFM. The following depicts this structural relationship among these elements.

Uncaring ⇨ Incompetence/Imprudence/PCM only ⇨ UFM

Next, carelessness is the source for the vice of impersonal prosaic lovelessness into which it is transformed. This transformed carelessness *qua* lovelessness is also a basis for constructing UFM by permitting the unvirtuous physician to care less about the patient, thereby making such a physician unable to take care of the patient in a compassionate or empathetic manner. Importantly, given this inability, some unvirtuous physicians generally stress EBM to the exclusion of PCM to compensate for this inability. The following depicts this structural relationship among these elements.

Uncaring ⇨ Carelessness/Lovelessness/EBM only ⇨ UFM

Finally, the belief that either EBM or PCM is the only way to practice medicine ultimately leads to complete divergence between EBM and PCM fragmenting the practice of unvirtuous physicians.[6] Patients then appear to have either their physical or existential needs met but not both. However, closer inspection of UFM reveals that neither a patient's physical nor existential needs are really met since the fragmentation of medical practice completely suppress either prudent wisdom and personal radical love, so that those unvirtuous physicians advocating EBM only

[6] UFM, whether the result of either triad, can feedback onto uncaring to exacerbate the level of uncaring and the poor quality of healthcare provided in an unprofessional manner.

abuse clinical technology in practicing defensive medicine and those advocating PCM only abuse the clinical narrative in practicing ineffectual medicine.

The Diamond case story aptly illustrates UFM. Ontologically, what animates the healthcare professionals providing care for the young Diamond is their uncaring, which makes possible their carelessness and incompetence. The physician, for example, certainly cares less what the young Diamond has to say about how ineffective prior use of antihistamines are in relieving her eczema symptoms than about prescribing the drug to treat her. Prescribing the drugs represents an expedient therapeutic path rather than listening compassionately to Diamond's illness story and responding to her with credulity. Given this carelessness_1, he is unable to take care of (carelessness_2) her because he is not only incompetent but also imprudent in terms of his clinical decision to prescribe only the antihistamines and to believe naively that that is what Diamond needs with respect to treatment. Moreover, the physician exhibits an impersonal prosaic lovelessness towards Diamond in that he treats her as he would treat any patient under similar circumstances, particularly in terms of the "appropriate" child standard (Walco et al. 1994). As an incompetent and imprudent physician, he may console himself with the thought that he is addressing Diamond's medical needs by chatting with her briefly about the antihistamine prescription; but, as evident from Diamond's own comments, he genuinely does not care about her and thus cannot take care of her. Likewise, as a careless and loveless physician, he may reassure himself with the thought that he is treating her with the best possible therapeutic option for her; but, unless he is an expert, he is simply following standard protocol based on pathophysiology and not on the best available clinical evidence. Consequently, the type of medicine the physician practices, even though asked he probably would claim EBM, is neither EBM nor PCM but a fragmented medicine in which neither Diamond's physical nor her emotional or psychological needs are met.

7.3 Resolving the Quality-of-Care and Professionalism Crises

In a *Lancet* editorial on the quality-of-care crisis, John Bignall proposes a possible solution to the crisis, "doctors might profitably learn how to like people more, rather than learn techniques for coping with them" (1994, p. 249). As he continues, however, he insists that "like" is not robust enough to solve the crisis and suggests that agape or service-type love would better serve to resolve it. Indeed, a number of medical pundits recognize the need for love as the basis for a good therapeutic relationship and for delivering quality healthcare. Later in his career, for example, Donabedian claims, "Ultimately, the secret of quality [healthcare] is love. You have to love your patient" (Best and Neuhauser, 2004, p. 472).[7] And again, Gregory Larkin and colleagues identify love or charity as an essential virtue for the modern professional physician. "In an era where physicians perceive threats to

[7] Interestingly, Donabedian goes on to encourage physicians to love both their profession and God.

their autonomy and to their financial status," they claim, "charity remains the pinnacle of virtue, because, at the root, it is about genuine caring and selfless giving" (Larkin et al. 2009, p. 54).[8] Finally, "The prescription, love thy patient," recommends Rosamond Rhodes, "is good medicine for the good doctor" (1995, p. 441). For, not loving patients could lead to patient harm. Commenting on the atrocities associated with Nazi doctors, for instance, Emily Friedman cautions, "We must be utterly conscious of not loving patients at all. We must guard against the creeping callousness that makes it easier for us to provide bad care" (1990, p. 10). Love, then, is essential for quality healthcare provided in a genuine professional manner.

Although love, particularly its personal radical form as discussed in Chapter 5, is certainly a vital and necessary component in providing quality healthcare, especially in a professional fashion, it is not sufficient. Physicians must inform or temper their love towards patients with prudent wisdom. Without such prudence, love is often times blind and could result in harm not only to the patient but also to the physician. The resolution of the quality-of-care and the professionalism crises, then, depends not simply on loving patients but in prudently loving them. In other words, physicians must know how best to love and serve patients by wisely deciding the most appropriate and efficacious means for treating their healthcare needs. In this way, the physician *qua* virtuous healthcare provider delivers professional quality medical care. Hence, the chief virtue for the virtuous physician is the compound virtue, prudent love. In contrast, the unvirtuous physician—who is imprudently loveless—delivers poor quality care in an unprofessional manner. Such a physician exacerbates the quality-of-care and professionalism crises, leading to the erosion of medicine's reputation among patients and society.

Prudent love addresses and resolves the quality-of-care crisis in terms of both its technical and interpersonal dimensions. In terms of its technical dimension, the compound virtue provides the virtuous physician with the ethical and epistemic resources required to perform competently, skillfully, and safely medical procedures that meet not only the profession's standards of excellence but also the patient's expectations of quality care. Moreover, such a physician is motivated for the right reasons, i.e. to meet the patient's needs and to be the best he or she can be as a provider of quality technical healthcare. The compound virtue also addresses and resolves the interpersonal dimension of the quality-of-care crisis. Because the virtuous physician cares deeply or loves patients and desires to restore them to health or wholeness, he or she takes the necessary time to forge with patients robust therapeutic relationships, by listening intently, compassionately, and thoughtfully to their illness stories. Such a physician, then, incorporates these stories into clinical reasoning and decision-making, especially in terms of formulating therapy that takes into consideration the values and preferences of patients. In sum, the compound virtue resolves the quality-of-care crisis because it equips the virtuous physician to practice

[8] Pellegrino (2002a) also intimately unites the virtue of charity and medical professionalism, although he gives preference to the virtue of prudence.

VHM, thereby providing both patients and the larger communities with the quality healthcare they expect and deserve.

In like manner, the compound virtue of prudent love addresses and resolves the practical or technical and ethical or moral dimensions of the professionalism crisis. With respect to the practical dimension, the compound virtue first allows the medical community to realize that increasing a physician's technical knowledge of clinical medicine does not necessarily define the medical provider *qua* professional. Unfortunately, the medical profession often deceives itself into thinking that no crisis exists in terms of the practical dimension of professionalism. The medical professional, however, is simply more than a repository of medical facts and technical skills. Rather, what is imperative for professional behavior is apposite application of pertinent facts and skills to a patient's individual healthcare needs. By prudently loving patients, the virtuous physician develops the required skill to appropriate the technical knowledge of medicine in a practical and judicious manner that meets patients' individual healthcare needs. For the ethical or moral dimension of the professionalism crisis, the virtuous physician, who prudently loves patients, behaves in a fashion to fulfill their healthcare needs that conforms not only to the medical community's ethical standards but also to those of society. The prudently loving physician realizes that as a professional an obligation exists to treat patients with dignity and respect, and in so doing discharges that obligation not because of any ethical imperative but because it is good to behave as such. In sum, the compound virtue resolves the professionalism crisis because it furnishes the virtuous physician with the initiative and capacity to practice VHM, thereby meeting both patient and community standards of professionalism.

In contrast, the unvirtuous physician's compound vice of imprudent lovelessness exacerbates both the quality-of-care and professionalism crises. The compound vice worsens the quality-of-care crises in that the unvirtuous physician, through unwise or imprudent decisions and uncompassionate or loveless actions, alienates patients, thereby restricting opportunities of forming viable therapeutic relationships with them. With an inability to forge such relationships, the physician cannot provide the quality healthcare necessary to treat either their bodily or emotional needs. Imprudent lovelessness also aggravates the professionalism crisis within medicine. The unvirtuous physician makes imprudent clinical decisions that fail to conform to a community's minimal professional standards for competency, which may lead to serious bodily harm for patients and to feelings of inadequacy by the physician. In addition, the unvirtuous physician treats patients in a loveless manner, with little regard for their values or preferences, thereby failing to meet professional standards for performance that instill within patients the sense that the physician respects them as persons of worth. Again, the outcome of such unprofessional behavior is generally patient harm, especially emotional or psychological harm. In sum, the unvirtuous physician's imprudent lovelessness undercuts professional quality healthcare, as the physician practices UFM.

In the remainder of this section, I use the two medical case studies reconstructed in Chapter 6 to illustrate and discuss the role of virtuous physician in resolving the quality-of-care and professionalism crises and the role of the unvirtuous physician in exacerbating them. First, the Weinberg clinical case provides an apt instance of how the virtuous physician delivers quality care in a professional manner and thereby resolves the quality-of-care and professionalism crises. As for the quality-of-care crisis, the case exemplifies how Weinberg *qua* virtuous physician in terms of prudent love provides quality care for the patient in both its technical and interpersonal dimensions. Technically, Weinberg delivers quality care to the patient because he is competent not only in his specialty but also in the practice of good Toonian general medicine; and, he ultimately makes prudent decisions that yields quality healthcare for the patient. Although he initially prescribes a drug and diet combination to treat the patient's chief complaint, chronic abdominal pain, after learning of the patient's sexual assault he ensures that he is sufficiently knowledgeable about how to provide quality technical care for the patient in order to assist her in healing from the traumatic event. Interpersonally, Weinberg spends sufficient time listening to the patient divulge her narrative of sexual assault and the detrimental impact it has upon her life. Through her narrative, he forges a therapeutic relationship based on mutual trust.

With respect to the professionalism crisis in modern medicine, the Weinberg case epitomizes the best in contemporary medical professionalism, especially in terms of its practical or technical and moral dimensions. Practically, Weinberg *qua* virtuous physician treats the patient according to professional technical standards ensuring that he not only attends the patient competently but that he also reasons and makes clinical decisions and judgments prudently. For example, he consults a psychiatrist colleague about the patient to certify that he is treating the patient according to professional standards. Morally, Weinberg exhibits behavior conforming to the ethical ideals of the medical profession by selflessly assisting on a weekly basis the patient struggle through the sordid details of the sexual assault. At no time does he compromise those ideals by withdrawing support or condemning the patient for what happened to her. Rather, Weinberg believes in the patient and provides the safe and loving refuge of the consulting room for her to disclose the suffering she experienced from the assault. Through his compassionate and empathic or radical love for the patient and desire to restore the patient to wellness, he acts professionally according to the highest ethical standards of the medical profession. In sum, the compound virtue of prudent love is the basis of Weinberg's practice of VHM, which allows him to provide quality professional healthcare exceeding current professional standards.

In contrast to the Weinberg clinical case, the Diamond clinical case reveals how the unvirtuous physician, in terms of the compound vice imprudent lovelessness, exacerbates both the quality-of-care and professionalism crises. In this case, Diamond receives from the hospital staff poor quality healthcare in an unprofessional manner. With respect to the technical dimension of quality care, Diamond's attending physician, for instance, fails to consider what the patient tells him about her previous experience with antihistamines. Such incompetence on the part of the physician belies a deep-seated inability to make prudent decisions about effective therapy in order to provide quality technical healthcare to treat the patient. With respect to the interpersonal dimension, the staff nurse exemplifies how loveless she is in rejecting Diamond's request to talk about the suffering she is experiencing from her chronic eczema. She is loveless in terms of ignoring the patient's desperate cry for help in coping with the pain and anguish associated with her illness. In all, Diamond's healthcare team delivers the poorest possible healthcare, in terms of its quality, because of their general disregard for her welfare.

The Diamond case also reflects the worse possible instance of professionalism, both in terms of its practical or technical and moral dimensions. With respect to its practical dimension, Diamond's healthcare providers fail to provide her with the technical care she needs to cope with her eczema according to professional standards; rather, they utilize what Walco and associates call the "appropriate" child standard and imprudently reason that Diamond is too young to appreciate or understand the technical aspects of her treatment. With respect to the moral dimension of the professionalism crisis, the staff nurse does not treat Diamond according to the ethical standards of her profession, let alone to the expectations of common decorum, when she refuses to answer the patient's call for assistance to the toilet. The compound vice of imprudent loveless undergirds the unprofessional behavior of these unvirtuous healthcare providers. Finally, the Diamond case raises the question of how to retrain the healthcare providers to deliver quality healthcare in a professional manner, which in turn raises the question of the role of education in initially training virtuous physicians and healthcare providers in order to resolve the quality-of-care and professionalism crises plaguing modern medicine.

7.4 Virtues and Medical Education

In an essay on virtue in medical education, Jack Coulehan and Peter Williams (2001) from Stony Brook Health Sciences Center in New York recount the story of a first year medical student who arrives on campus with a heart full of empathy to meet the needs of patients and to have an impact on society for the good. However, by the end of the student's medical education she is no longer enthusiastic about medicine and serving patients or about the good of society but turns inward in order to survive the personal hardships, if not abuse, she faces daily as a medical student. In response to a questionnaire, she confesses, "I've become numb. So much of what I do as a student is stuff that I don't fully believe it. And rather than try to change everything that I consider wrong in the hospital or the community at large," she confesses, "I just try to get through school in the hope that I will move on to bigger and better things when I have more control over my circumstances" (Coulehan and Williams, 2001, p. 599). However, one fears what her story is after residency and further exposure to the deleterious effects of the hidden curriculum (Chuang et al. 2010). Unfortunately, her story is not uncommon for many medical students. Indeed, recent studies report that the empathy of medical students declines during their training to become physicians. For example, Bruce Newton and colleagues find that vicarious empathy—the ability to respond to the perceived feelings of others—begins to decrease noticeably during the first year of medical school and falls significantly after the third year of clinical rotations (Newton et al. 2008).[9]

The question is how best to stem the tide of vanishing empathy and compassion in medical students as these students progress in their training to become professional caregivers. For as Coulehan and Williams note, the question is exacerbated since medical students are exposed to a grave conflict between two sets of commitments present in the medical curriculum. On the one hand, the medical faculty teaches "an *explicit* commitment to traditional values of doctoring—empathy, compassion, and altruism among them," while on the other hand, it models "a *tacit* commitment to behaviors grounded in an ethic of detachment, self-interest, and objectivity" (Coulehan and Williams, 2001, p. 604).[10] Unfortunately, the faculty often abandons students to resolve the conflict on their own or adds to it by exemplifying exclusively the tacit commitment. Coulehan and Williams propose the development of courses, such as Stony Brook's Medicine in Contemporary Society, which reconnect students to medicine's social contract of providing compassionate care for patients. They are less sanguine about bioethics and medical humanities courses in resolving the conflict, since "the culture of clinical training is often hostile to professional virtue" (Coulehan and Williams, 2001, p. 602). In response to their proposal, William Branch (2001) contends that small-group discussion classes and bedside teaching with humanistic emphasis can reverse the deleterious effects of the conflict on medical students.[11] He cites a number of studies to support his contention. Indeed, a recent study with third-year medical students demonstrates that medical humanities courses can provide an opportunity for students to process their clinical experience and thereby prevent decline in empathy (Rosenthal et al. 2011).

Although many medical schools make a concerted effort in teaching medical humanities courses, as Hirsch confesses about his experience as a medical student taking such courses, "a large number of students did not take the [medical humanities] curriculum seriously, seeing it as a waste of time that could have been better

---

[9] Newton and colleagues contrast vicarious empathy with role-playing or imaginative empathy, in which the former is a "gut" response while the latter is an intellectualized response.

[10] Toon also acknowledges that the medical curriculum stunts students' natural inclination to react compassionately to a patient's illness experience. "Medical education and the training of other health professionals," writes Toon, "encourages students to suppress their natural human response to suffering either overtly through exhortations not to get too involved or covertly through the modelling of detachment by teachers and senior members of the profession" (2007, p. 97).

[11] Jochanan Benbassat and Reuben Baumal (2004) provide a multistep procedure for training medical students to enhance their skills in empathizing with patients. In a review of empirical studies examining the effectiveness of teaching empathy to medical students, Kathy Stepien and Amy Baernstein (2006) report that although problems plague these studies the overall outcome is positive in terms of enhancing students' ability to empathize with patients. Pedersen (2010) also reviews the literature and recommends that medical educators integrate the humanities into the basic and clinical science courses to bridge the gap between students' technical competence and empathic competence.

spent studying" (2007, p. 425). Why do medical students see this curriculum as a waste of time? As he acknowledges, the medical school culture, including the pre-medical curriculum, stresses the objective and quantifiable to the exclusion of the subjective and qualitative. "The challenge for medical educators," concludes Hirsch, "is to present the information in a format that makes it relevant and actively engages the students. Although students may not immediately see the value of this type of education," he admits, "it is to our benefit that my generation of physicians is specifically instructed in empathy and professionalism" (2007, p. 426). Consequently, what needs to change is students' realization and appreciation of the relevance such courses provide in terms of becoming physicians who provide quality care in a professional manner. My proposal is that educating medical students to become virtuous physicians, who practice VHM, cannot only reverse the loss or attenuation in empathy that occurs during medical education but also enhance empathy by providing a medical school culture that values virtue thereby assisting medical students to appreciate the significance of virtue for good medical practice.

An important question arises as to whether the faculty can teach or students can learn virtue(s) to resolve the conflict between implicit and explicit commitments. Critics generally claim that the medical faculty cannot teach medical students virtues since the students matriculate to medical school with a well-formed moral compass and the faculty can do little, if anything, to reorient that compass. However, as Pellegrino and Thomasma respond to this criticism, students need not learn the virtues necessary to become a virtuous person but rather those to become a virtuous physician. "Medical students come to medical school," claim Pellegrino and Thomasma, "precisely for the purpose of being educated to be physicians. There is a relevance and an inevitability about this fact that make character evaluation a de facto reality. Whether the faculty wishes it or not," they insist, "they do teach virtue and vice in everything they do and say" (1993, p. 176). Wayne Shelton (1999) also believes that the medical faculty can effectively teach virtues as part of the medical curriculum. To that end, he proposes the notion of a "good" physician who exhibits the virtues needed to practice virtuous medicine.[12] "Any approach to attempting to prepare medical students to function as fully competent physicians trained to care for the total needs of the patient," according to Shelton, "will resemble a virtue approach" (1999, p. 674).[13] In other words, virtues are a necessary component of medical education, since medicine is more than simply a scientific discipline; it is

[12] Drane, also espousing the notion of a "good" physician, admits that physicians can learn virtues. "After learning that certain forms of moral conduct," notes Drane, "are indispensible to good medicine, a doctor can learn to be helpful, kind, caring, respectful, promise-keeping, friendly, and the rest" (1995, p. 159). He also recognizes that learning to be virtuous requires work, dedication, and practice.

[13] Shelton does acknowledge the numerous challenges facing the medical faculty in teaching virtues, such as how to conceptualize virtues and which or whose virtues to teach in a pluralistic western society.

also a moral enterprise, and the medical faculty must equip its students with the skills needed to address the moral and ethical issues of medical practice.[14]

The question then is not simply whether the medical faculty should or can teach virtues to medical students but rather how. The interest here is not on the precise types or contents of the courses but rather on the general strategy for incorporating virtues into the medical curriculum. Although the faculty cannot teach and students cannot learn virtues as easily as medical or scientific facts, it can set the conditions for learning and practicing virtues throughout the curriculum in innovative courses that make the virtues relevant to medical practice. The faculty needs to design courses that introduce early on in the student's educational experience the nature and types of virtues required for practicing medicine that meets the patient's physical and emotional or psychological needs. With an understanding by students of general virtue theory and of the array of virtues available, the faculty can then design courses later on in the student's formal training that instantiate the virtues and translate them from the classroom into the clinic. These courses must provide realistic conditions taken from actual clinical experience that instructs and challenges the students to incorporate virtues into medical practice. Without such practical relevance, students—as Hirsch notes—simply will not take seriously the importance of virtues for practicing VHM.

Besides teaching formal courses, the faculty must also model informally the virtues of medical practice within both the classroom and clinic. It must also serve as mentors to shepherd students in the art of VHM, especially when situations arise where no clear-cut way of proceeding is obvious. Mary O'Flaherty Horn's essay, "The other side of bed rail" (1999), aptly illustrates the failure of modeling virtue in the clinic and the devastating effect it has on the patient. Horn recounts the degrading experience in which Dr. L., administering a electromyography test to confirm an ALS diagnosis, treats Horn brusquely and rudely. As Horn concludes from the experience, "It is a lesson in healing. Although my physicians may not be able to cure my illness, their encouragement, time, patience, and trust build bridges that enable me to cope one day at a time. Encounters such as mine with Dr. L., the antithesis of caring, could become more common as medical care becomes more fragmented and long-term relationships with patients become relics. Physicians are the vital human link that can give patients the strength they require. As the pace of change in medicine quickens," she warns, "physicians who teach will bear a special responsibility to provide strong examples of empathy and professionalism to students and residents. After all, one day we may all find ourselves on the other side of the bed rail, and those young physicians will become what we model for them today" (1999, pp. 940–941). The issue is not whether medical faculty should teach virtues, for if the faculty does not teach students virtues explicitly students may learn vices implicitly.

---

[14] Clinical medicine requires skills in what Alfred Tauber (2005) calls "moral epistemology," and virtues are important for developing those skills.

Finally, besides medical education, training in the virtues for practicing quality medicine according to professional standards should begin formally at the undergraduate level. Just as undergraduate natural sciences courses can prepare premedical students for clinical courses in medical school, so undergraduate humanities courses can prepare them for medical school humanities courses. For premedical education, educators can use general humanities courses in literature, religion, fine arts, sociology, and psychology; specific medical humanities courses, such as philosophy of medicine, medical ethics, literature and medicine, history of medicine, and healthcare economics; and practical clinical courses, such placing undergraduate students in clinical settings, to introduce the necessary epistemic and ethical virtues for practicing clinical medicine. Such sustained training is necessary in order to make the virtues part of the student's eventual clinical habits. The notion of virtuous physician and its associated notion of VHM can serve as a general philosophical foundation for reformulating both premedical and medical education, for which recent appeal has been made (Kanter, 2008; Whitcomb, 2007). My proposal is that both notions will reorient medical training to provide the type of "whole doctor education" required to graduate physicians who deliver quality healthcare in a professional manner (Bligh, 2000).

7.5 Summary

I propose a notion of virtuous physician to resolve the quality-of-care and professionalism crises facing contemporary medicine, especially a medicine overly dependent on science and technology that often brackets the existential dimension of the patient's illness experience. Although the notion is not new, the notion I propose is unique in the sense that it involves not only the ethical or moral dimension of medical practice but also its epistemic and metaphysical dimensions thereby providing a comprehensive notion robust enough to address and resolve the crises. To that end, the virtue of caring is the ontological foundation of the virtuous physician, which makes possible the two ontic virtues of care and competence (Fig. 7.1). Although these two ontic virtues are necessary, they are insufficient for adequately defining the virtuous physician. Hence, I transform the ontic virtues into a compound virtue, prudent love, which serves as the basis for defining sufficiently the virtuous physician, who utilizes virtues to deliver quality professional healthcare. In contrast, the compound vice—imprudent lovelessness—serves as the foundation for defining the unvirtuous physician, who employs vices to provide poor quality healthcare in an unprofessional manner (Fig. 7.2).

However, before I demonstrated how the notion of virtuous physician resolves the quality-of-care and professionalism crises, the type of medicine practiced by the virtuous physician required explication. That medicine, which I call VHM, represents an integration of EBM and PCM in terms of the compound virtue, prudent love. The compound virtue integrates EBM and PCM through a structural relationship in which prudence and competence ground EBM and love and care PCM in a reciprocal feedback fashion, such that both approaches to medicine form an integrated whole that cannot be separated without making practice ineffective or jeopardizing the other approach. This virtuous medicine is holistic since it provides medical care that meets the patient's individual needs, whether bodily or emotional, using the best clinical evidence available. In contrast, the unvirtuous physician practices UFM in which EBM and PCM remain divergent with no possible integration or even significant overlap. The result is either incompetent and imprudent or careless and loveless healthcare, or both, in which the patient's healthcare needs go untreated—including even possible patient harm.

The notion of virtuous physician and its associated notion of VHM resolve the quality-of-care and professionalism crises specifically through the compound virtue of prudent love. For the quality-of-care crisis, the compound virtue provides the ethical or moral and epistemic or factual resources for virtuous physicians to deliver quality technical healthcare, i.e. VHM, by caring deeply for individual patients and thereby forging robust therapeutic relationships with them. For the professionalism crisis, prudent love equips virtuous physicians with the technical and interpersonal skills necessary to appropriate clinical knowledge practically and judiciously for individual patients, i.e. VHM, in a caring manner that not only satisfies professional standards but also, at times, exceeds them. In contrast, the compound vice of imprudent lovelessness leads unvirtuous physicians to exacerbate the crises through the practice of UFM. Specifically, unvirtuous physicians are unable to forge robust therapeutic relationships with patients because of incompetent performance and imprudent decisions and because of treating patients in a careless and loveless manner. The result is the delivery of poor quality healthcare that fails to meet even minimal professional standards.

Finally, the notions of virtuous physician and VHM have important implications for revising premedical and medical education. Society invests a large number of resources into training physicians, beginning early in their professional career, in order to make them competent technically. For example, entrance into medical school requires specified natural science courses. However, courses in the arts and humanities remain unspecified. The outcome of not requiring specific courses in the arts and humanities is generally a failure to equip physicians with the skills necessary to address the existential and ethical issues facing them daily in the clinic or hospital. Although medical humanities courses represent an attempt to instill and nurture humane care within prospective physicians, it is often too little and too late in the curriculum (Campo, 2005). My proposal of virtuous physician and VHM involves reorientation in the vision of medical educators by training both premedical and medical students in ethical and intellectual virtues. The success of this reorientation is going to require endorsement not only from the top-down by medical administrators but also support from the bottom-up by virtuous physicians as mentors.

In conclusion, contemporary medicine is facing a number of crises as it moves into the twenty-first century. No two crises are more looming and daunting than the quality-of-care and professionalism crises, which slash to the heart of medicine itself. Any solution to these crises is going to meet challenges, and the introduction of virtue into the medical curriculum is not immune from them. However, as Sulmasy warns, "there is really no morally acceptable alternative. We need to create environments that cultivate professional virtue in our schools, and in our practice settings" (2000, p. 515). Otherwise, one of the alternatives to teaching virtues intentionally is teaching vices unintentionally—especially through the hidden curriculum. My proposal of both virtuous physician and VHM creates a medical culture that addresses and resolves the quality-of-care and professionalism crises plaguing contemporary medicine.

References

Adamson, T.E., W.H. Bunch, D.C. Baldwin, Jr., and A. Oppenberg. 2000. The virtuous orthopaedist has fewer malpractice suits. *Clinical Orthopaedics and Related Research* 378:104–109.

Benbassat, J., and R. Baumal. 2004. What is empathy, and how can it be promoted during clinical clerkships? *Academic Medicine* 79:832–839.

Best, M., and D. Neuhauser. 2004. Avedis Donabedian: Father of quality assurance and poet. *Quality and Safety in Health Care* 13:472–473.

Bignall, J. 1994. Learning and loving. *Lancet* 343:249–250.

Bligh, J. 2000. Whole doctor education. *Medical Education* 34:496–497.

Branch, W.T. Jr. 2001. Small-group teaching emphasizing reflection can positively influence medical students' values. *Academic Medicine* 76:1171–1172.

Campo, R. 2005. The medical humanities, for lack of a better term. *Journal of American Medical Association* 294:1009–1011.

Chuang, A.W., F.S. Nuthalapaty, P.M. Casey, J.M. Kaczmarczyk, A.J. Cullimore, J.L. Dalrymple, L. Dugoff, E.L. Espey, M.M. Hammound, N.A. Hueppchen, N.T. Katz, and E.G. Peskin. 2010. To the point: Reviews of in medical education—taking control of the hidden curriculum. *American Journal of Obstetrics & Gynecology* 203:316.e1–316.e6.

Clarke, D.L. 2009. Moral principlism alone is insufficient, and traditional moral theories remain important for practical ethics. *South African Journal of Bioethics and Law* 2:54–58.

Coulehan, J. 2005. Today's professionalism: Engaging the mind but not the heart. *Academic Medicine* 80:892–898.

Coulehan, J., and P.C. Williams. 2001. Vanishing virtue: The impact of medical education. *Academic Medicine* 76:598–605.

Drane, J.F. 1995. *Becoming a good doctor: The place of virtue and character in medical ethics*, 2nd edition. Kansas City, MO: Sheed & Ward.

Friedman, E. 1990. The perils of detachment. *Healthcare Forum Journal* 33:9–10.

Hirsch, E.M. 2007. The role of empathy in medicine: A medical student's perspective. *Virtual Mentor* 9:423–427.

Horn, M. O'F. 1999. The other side of the bed rail. *Annals of Internal Medicine* 130:490–491.

Kanter, S.L. 2008. Toward a sound philosophy of premedical education. *Academic Medicine* 83:423–424.

Lacy, N.L., and E.L. Backer. 2008. Evidence-based and patient-centered care: Results from an STFM group project. *Family Medicine* 40:417–420.

Larkin, G.L., K. Iserson, Z. Kassutto, G. Freas, K. Delaney, J. Krimm, T. Schmidt, J. Simon, A. Calkins, and J. Adams. 2009. Virtue in emergency medicine. *Academic Emergency Medicine* 16:51–55.

Newton, B.W., L. Barber, J. Clardy, E. Cleveland, and P. O'Sullivan. 2008. Is there hardening of the heart during medical school? *Academic Medicine* 83:244–249.

Pedersen, R. 2010. Empathy development in medical education—a critical review. *Medical Teacher* 32:593–600.

Pellegrino, E.D. 1985. The virtuous physician, and the ethics of medicine. In *Virtue and medicine: Explorations in the character of medicine*, ed. E.A. Shelp, 237–255. Dordrecht: Reidel.

Pellegrino, E.D. 2002a. Professionalism, profession and the virtues of a good physician. *The Mount Sinai Journal of Medicine* 69:378–384.

Pellegrino, E.D. 2002b. Medical evidence and virtue ethics: A commentary on Zarkovich and Upshur. *Theoretical Medicine* 23:397–402.

Pellegrino, E.D., and D.C. Thomasma. 1993. *The virtues in medical practice*. New York, NY: Oxford University Press.

Pellegrino, E.D., and D.C. Thomasma. 1996. *The Christian virtues in medical practice*. Washington, DC: Georgetown University Press.

Rhodes, R. 1995. Love thy patient: Justice, caring, and the doctor-patient relationship. *Cambridge Quarterly in Healthcare Ethics* 4:434–437.

Rosenthal, S., B. Howard, Y.R. Schlussel, D. Herrigel, B.G. Smolarz, B. Gable, J. Vasquez, H. Grigo, and M. Kaufman. 2011. Humanism at heart: Preserving empathy in third-year medical students. *Academic Medicine* 86:350–358.

Shelton, W. 1999. Can virtue be taught? *Academic Medicine* 74:671–674.

Stempsey, W.E. 1999. The quarantine of philosophy in medical education: Why teaching the humanities may not produce humane physicians. *Medicine, Health Care, and Philosophy* 2:3–9.

Stepien, K.A., and A. Baernstein. 2006. Educating for empathy: A review. *Journal of General Internal Medicine* 21:524–530.

Stevenson, A.C. 2002. Compassion and patient centred care. *Australian Family Physician* 31:1103–1106.

Sulmasy, D.P. 2000. Should medical schools be schools for virtue? *Journal of General Internal Medicine* 15:514–516.

Tauber, A.I. 2005. *Patient autonomy and the ethics of responsibility*. Cambridge: MIT Press.

Toon, P.D. 1994. *What is good general practice? A philosophical study of the concept of high quality medical care, occasional paper 65*. Royal College of General Practitioners.

Toon, P.D. 1999. *Towards a philosophy of general practice: A study of the virtuous practitioner, occasional paper 78*. Royal College of General Practitioners.

Toon, P. 2007. Setting boundaries: A virtue approach to the clinician-patient relationship in general practice. In *Primary care ethics*, eds. D. Bowman and J. Spicer, 83–99. Abington, UK: Radcliffe Publishing.

Walco, G.A., R.C. Cassidy, and N.L. Schechter. 1994. Pain, hurt, and harm: The ethics of pain control in infants and children. *New England Journal of Medicine* 331:541–544.

Whitcomb, M.E. 2007. Medical education reform: Is it time for a modern Flexner report? *Academic Medicine* 82:1–2.

Zarkovich, E., and R.E.G. Upshur. 2002. The virtues of evidence. *Theoretical Medicine* 23: 403–412.

Index

J.A. Marcum, *The Virtuous Physician*, Philosophy and Medicine 114,
DOI 10.1007/978-94-007-2706-9, © Springer Science+Business Media B.V. 2012

D

MIX
Papier aus verantwortungsvollen Quellen
Paper from responsible sources
FSC® C105338

If you have any concerns about our products,
you can contact us on
ProductSafety@springernature.com

In case Publisher is established outside the EU,
the EU authorized representative is:
Springer Nature Customer Service Center GmbH
Europaplatz 3, 69115 Heidelberg, Germany

Printed by Libri Plureos GmbH
in Hamburg, Germany